MTP International Review of Science

Volume 2

Molecular Structure and Properties

Edited by **G. Allen**
University of Manchester

Butterworths · London
University Park Press · Baltimore

THE BUTTERWORTH GROUP

ENGLAND
Butterworth & Co (Publishers) Ltd
London: 88 Kingsway, WC2B 6AB

AUSTRALIA
Butterworths Pty Ltd
Sydney: 586 Pacific Highway, NSW 2067
Melbourne: 343 Little Collins Street, 3000
Brisbane: 240 Queen Street, 4000

NEW ZEALAND
Butterworths of New Zealand Ltd
Wellington: 26–28 Waring Taylor Street, 1

SOUTH AFRICA
Butterworth & Co (South Africa) (Pty) Ltd
Durban: 152–154 Gale Street

ISBN 0 408 70263 X

UNIVERSITY PARK PRESS

U.S.A. and CANADA
University Park Press Inc
Chamber of Commerce Building
Baltimore, Maryland, 21202

Library of Congress Cataloging in Publication Data

Allen, Geoffrey, 1928–
Molecular structure and properties.

(Physical chemistry, series one, v. 2) (MTP international review of science)
1. Molecules. 2. Chemistry, Physical and theoretical
I. Title. [DNLM: W1 PH683K v. 2 1972. XNLM: [QD 453 M718 1972]]
QD453.2.P58 Vol. 2 [QD461] 541′.3′08s [541′.22]
ISBN 0–8391–1016–2 72–5160

MTP MEDICAL AND TECHNICAL PUBLISHING CO. LTD.
Seacourt Tower
West Way
Oxford, OX2 OJW
and
BUTTERWORTH & CO. (PUBLISHERS) LTD.

Filmset by Photoprint Plates Ltd., Rayleigh, Essex
Printed in England by Redwood Press Ltd., Trowbridge, Wilts
and bound by R. J. Acford Ltd., Chichester, Sussex

MTP International Review of Science

Molecular Structure and Properties

MTP International Review of Science

Publisher's Note

The MTP International Review of Science is an important new venture in scientific publishing, which we present in association with MTP Medical and Technical Publishing Co. Ltd. and University Park Press, Baltimore. The basic concept of the Review is to provide regular authoritative reviews of entire disciplines. We are starting with chemistry because the problems of literature survey are probably more acute in this subject than in any other. As a matter of policy, the authorship of the MTP Review of Chemistry is international and distinguished; the subject coverage is extensive, systematic and critical; and most important of all, new issues of the Review will be published every two years.

In the MTP Review of Chemistry (Series One), Inorganic, Physical and Organic Chemistry are comprehensively reviewed in 33 text volumes and 3 index volumes, details of which are shown opposite. In general, the reviews cover the period 1967 to 1971. In 1974, it is planned to issue the MTP Review of Chemistry (Series Two), consisting of a similar set of volumes covering the period 1971 to 1973. Series Three is planned for 1976, and so on.

The MTP Review of Chemistry has been conceived within a carefully organised editorial framework. The over-all plan was drawn up, and the volume editors were appointed, by three consultant editors. In turn, each volume editor planned the coverage of his field and appointed authors to write on subjects which were within the area of their own research experience. No geographical restriction was imposed. Hence, the 300 or so contributions to the MTP Review of Chemistry come from many countries of the world and provide an authoritative account of progress in chemistry.

To facilitate rapid production, individual volumes do not have an index. Instead, each chapter has been prefaced with a detailed list of contents, and an index to the 13 volumes of the MTP Review of Physical Chemistry (Series One) will appear, as a separate volume, after publication of the final volume. Similar arrangements will apply to the MTP Review of Organic Chemistry (Series One) and to subsequent series.

Butterworth & Co. (Publishers) Ltd.

Physical Chemistry
Series One
Consultant Editor
A. D. Buckingham
Department of Chemistry
University of Cambridge

Volume titles and Editors

1 **THEORETICAL CHEMISTRY**
Professor W. Byers Brown, *University of Manchester*

2 **MOLECULAR STRUCTURE AND PROPERTIES**
Professor G. Allen, *University of Manchester*

3 **SPECTROSCOPY**
Dr. D. A. Ramsay, F.R.S.C., *National Research Council of Canada*

4 **MAGNETIC RESONANCE**
Professor C. A. McDowell, F.R.S.C., *University of British Columbia*

5 **MASS SPECTROMETRY**
Professor A. Maccoll, *University College, University of London*

6 **ELECTROCHEMISTRY**
Professor J. O'M Bockris, *University of Pennsylvania*

7 **SURFACE CHEMISTRY AND COLLOIDS**
Professor M. Kerker, *Clarkson College of Technology, New York*

8 **MACROMOLECULAR SCIENCE**
Professor C. E. H. Bawn, F.R.S., *University of Liverpool*

9 **CHEMICAL KINETICS**
Professor J. C. Polanyi, F.R.S., *University of Toronto*

10 **THERMOCHEMISTRY AND THERMODYNAMICS**
Dr. H. A. Skinner, *University of Manchester*

11 **CHEMICAL CRYSTALLOGRAPHY**
Professor J. Monteath Robertson, F.R.S., *University of Glasgow*

12 **ANALYTICAL CHEMISTRY—PART 1**
Professor T. S. West, *Imperial College, University of London*

13 **ANALYTICAL CHEMISTRY—PART 2**
Professor T. S. West, *Imperial College, University of London.*

INDEX VOLUME

Inorganic Chemistry
Series One
Consultant Editor
H. J. Emeléus, F.R.S.
Department of Chemistry
University of Cambridge

Volume titles and Editors

1 **MAIN GROUP ELEMENTS—HYDROGEN AND GROUPS I–IV**
Professor M. F. Lappert, *University of Sussex*

2 **MAIN GROUP ELEMENTS—GROUPS V AND VI**
Professor C. C. Addison, F.R.S. and Dr. D. B. Sowerby, *University of Nottingham*

3 **MAIN GROUP ELEMENTS—GROUP VII AND NOBLE GASES**
Professor Viktor Gutmann, *Technical University of Vienna*

4 **ORGANOMETALLIC DERIVATIVES OF THE MAIN GROUP ELEMENTS**
Dr. B. J. Aylett, *Westfield College, University of London*

5 **TRANSITION METALS—PART 1**
Professor D. W. A. Sharp, *University of Glasgow*

6 **TRANSITION METALS—PART 2**
Dr. M. J. Mays, *University of Cambridge*

7 **LANTHANIDES AND ACTINIDES**
Professor K. W. Bagnall, *University of Manchester*

8 **RADIOCHEMISTRY**
Dr. A. G. Maddock, *University of Cambridge*

9 **REACTION MECHANISMS IN INORGANIC CHEMISTRY**
Professor M. L. Tobe, *University College, University of London*

10 **SOLID STATE CHEMISTRY**
Dr. L. E. J. Roberts, *Atomic Energy Research Establishment, Harwell*

INDEX VOLUME

Organic Chemistry
Series One
Consultant Editor
D. H. Hey, F.R.S.
Department of Chemistry
King's College, University of London

Volume titles and Editors

1 **STRUCTURE DETERMINATION IN ORGANIC CHEMISTRY**
Professor W. D. Ollis, F.R.S., *University of Sheffield*

2 **ALIPHATIC COMPOUNDS**
Professor N. B. Chapman, *Hull University*

3 **AROMATIC COMPOUNDS**
Professor H. Zollinger, *Swiss Federal Institute of Technology*

4 **HETEROCYCLIC COMPOUNDS**
Dr. K. Schofield, *University of Exeter*

5 **ALICYCLIC COMPOUNDS**
Professor W. Parker, *University of Stirling*

6 **AMINO ACIDS, PEPTIDES AND RELATED COMPOUNDS**
Professor D. H. Hey, F.R.S. and Dr. D. I. John, *King's College, University of London*

7 **CARBOHYDRATES**
Professor G. O. Aspinall, *Trent University, Ontario*

8 **STEROIDS**
Dr. W. F. Johns, *G. D. Searle & Co., Chicago*

9 **ALKALOIDS**
Professor K. F. Wiesner, F.R.S., *University of New Brunswick*

10 **FREE RADICAL REACTIONS**
Professor W. A. Waters, F.R.S., *University of Oxford*

INDEX VOLUME

Physical Chemistry
Series One

Consultant Editor
A. D. Buckingham

Consultant Editor's Note

The MTP International Review of Science is designed to provide a comprehensive, critical and continuing survey of progress in research. The difficult problem of keeping up with advances on a reasonably broad front makes the idea of the Review especially appealing, and I was grateful to be given the opportunity of helping to plan it.

This particular 13-volume section is concerned with Physical Chemistry, Chemical Crystallography and Analytical Chemistry. The subdivision of Physical Chemistry adopted is not completely conventional, but it has been designed to reflect current research trends and it is hoped that it will appeal to the reader. Each volume has been edited by a distinguished chemist and has been written by a team of authoritative scientists. Each author has assessed and interpreted research progress in a specialised topic in terms of his own experience. I believe that their efforts have produced very useful and timely accounts of progress in these branches of chemistry, and that the volumes will make a valuable contribution towards the solution of our problem of keeping abreast of progress in research.

It is my pleasure to thank all those who have collaborated in making this venture possible – the volume editors, the chapter authors and the publishers.

Cambridge A. D. Buckingham

Preface

Although the basic concept which underpins this series is to produce authoritative reviews of entire disciplines, the selection of topics for this particular volume presented a formidable problem because of the scope afforded by its title. It was decided to commission a relatively small number of individual reviews, to give the authors space to cover their topics in depth where this was appropriate, rather than to attempt a much wider, less selective and necessarily more superficial coverage.

The reviews were selected to provide vignettes of some areas of physical chemistry which have recently attracted renewed attention and others, which although classical, continue as areas of considerable activity. Thus it was considered timely to commission an article on magnetic susceptibilities because of recent developments in theoretical aspects of this subject. Molecular acoustics is included because only in the past few years has this technique begun to fulfil its early promise in the study of the dynamics of conformational change. The reviews on dielectrics and refraction demonstrate clearly that these topics still make major contributions to the understanding of structure–property relations. The article on high polymers is included to illustrate the progress made in unravelling the crystal structure of high polymers and to give a glimpse of how crystal dynamics lead to an understanding of the elastic constants of crystallites. Finally, there are two reviews dealing with purely structural techniques, for accurate structures are a prerequisite for the development of structure–property relations. There is a chapter on electron diffraction, which, for all but the simplest molecules, continues to be the major technique for determining the structure of molecules in the vapour state. Now that a high-flux beam-reactor has just been commissioned in Grenoble, and there are plans for one in the U.K. within 5 or 6 years, neutron diffraction is about to take a step forward. Chapter 2 outlines the special advantages of the technique, reviews the results obtained so far, and indicates the likely scope of both elastic and inelastic neutron scattering in the immediate future.

Manchester G. Allen

Contents

1
Techniques and Uses of Infrared Refractivity

J. HAIGH
Post Office Research Department

and

L. E. SUTTON, F.R.S.
University of Oxford

1.1 INTRODUCTION

1.1.1 General

Great advances in physical chemistry have been made in the past 40 years by measurement and interpretation of molecular properties, notably the characteristic energy levels, and this has tended to distract attention from the bulk properties. These are difficult to measure with high precision, they do not yield many units of information per measurement, and their interpretation is often complex and uncertain. Yet they are the properties that we experience directly and that we have to work and live with; so the study and discussion of them must be unremitting.

Recent advances in technique have opened up a new frequency range to investigation; moreover, the wealth of molecular information now accumulated should increase our power to understand bulk properties. This review is concerned mainly with the new techniques. They are too new for much chemically significant information to have emerged as yet, but they hold considerable promise.

1.1.2 Refractive and reflective properties of absorbing media

Monochromatic radiation propagates through empty space with a velocity $c_0 \doteqdot 2.998 \times 10^8$ m s^{-1} and a wavelength λ_0, wavenumber $\bar{\nu}_0$ and frequency ν_0 related by the equation $\lambda_0 \nu_0 = c_0$. As far as is known, c_0 is a constant for all values of ν_0. On encountering matter the radiation is partly reflected and partly transmitted; the transmitted part is characterised by an angle of refraction, a changed wavelength λ, and a rate of attenuation per unit path through the medium α. The angle of refraction and the wavelength λ, as well as the degrees of polarisation of the reflected and transmitted portions, depend upon the angle of incidence of the radiation and upon the optical

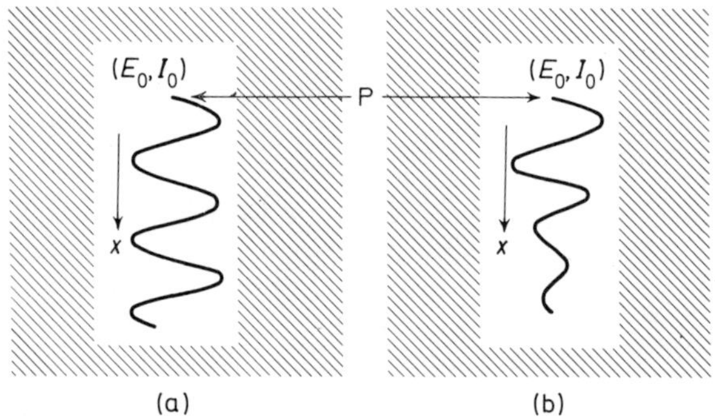

Figure 1.1 Radiation propagating in (a) non-absorptive and (b) absorptive media

constants of the material, which, as we shall see, are α and the refractive index n. The frequency remains unchanged at ν_0. To explore some aspects of the interaction of these optical constants we will consider monochromatic radiation propagating through two homogeneous media (a) and (b) (Figure 1.1). Consider the radiation passing fixed points P in the two media with synchronous phases and intensities I_0 in both media. One medium is totally non-absorbing. The other has an absorption coefficient α defined by Beer's law:

$$\frac{dI}{dx} = -\alpha I$$

so that if we write E as the amplitude of the electric vector corresponding to radiation intensity I, $dE/dx \doteq -\alpha E/2$. Thus at a distance x from P:

$$\text{(a)} \quad {}_{a}E_x = E_0 \exp 2\pi i(vt - x/\lambda) \tag{1.1}$$

$$\text{(b)} \quad {}_{b}E_x = E_0 \exp 2\pi i(vt - x/\lambda)\exp(-\alpha x/2)$$

Since $\lambda = \lambda_0/n$, we may write equation (1.1) as

$$\begin{aligned} {}_{a}E_x &= E_0 \exp 2\pi i(vt - nx/\lambda_0) \\ {}_{b}E_x &= E_0 \exp 2\pi i(vt - nx/\lambda_0[n - i(\alpha\lambda_0/4\pi)]) \end{aligned} \tag{1.2}$$

Thus we see that in medium (b) the expression $[n - i(\alpha\lambda_0/4\pi)]$, often written $n - ik$, replaces n, the simple refractive index of the non-absorbing medium. For this reason it is often called the complex refractive index and given the complex number symbol $\tilde{n}$, with its complex conjugate $(n + ik)$ being $\tilde{n}^*\tilde{n}$ replacing n in describing the other macroscopic radiation–matter interactions. For example, Maxwell's equations applied to non-absorbing matter give the reflection coefficient for a plane surface at normal incidence, the radiation passing from vacuum to the material:

$$\frac{I_{\text{reflected}}}{I_{\text{incident}}} = \rho = \left(\frac{n-1}{n+1}\right)^2$$

If now we make the medium absorbing, the equations yield

$$\rho = \left(\frac{\tilde{n}-1}{\tilde{n}+1} \cdot \frac{\tilde{n}^*-1}{\tilde{n}^*+1}\right) \tag{1.3}$$

This may be expanded by writing $n - ik$ for $\tilde{n}$ (i.e. $k = \alpha/4\pi\bar{v} = \alpha\lambda/4\pi$)†

$$\rho = \frac{(n-ik-1)}{(n-ik+1)} \cdot \frac{(n+ik-1)}{(n+ik+1)} = \frac{(n-1)^2+k^2}{(n+1)^2+k^2} \tag{1.4}$$

The reflected wave also suffers a phase change δ given by

$$\tan\delta = \frac{-2k}{(n^2-1+k^2)} \tag{1.5}$$

When $k \rightarrow 0$, we see that $\tan\delta \rightarrow 0$ from the negative side, i.e. the phase change tends towards π, which is the well-known result for reflection from a perfectly non-absorbing medium.

For a beam incident at an angle $\theta_{\text{inc.}}$ to the normal, Snell's law gives for the angle of refraction

$$\sin\theta_{\text{refr.}} = (1/\tilde{n})\sin\theta_{\text{inc.}} \tag{1.6}$$

This makes $\theta_{\text{refr.}}$ complex, a mathematical anomaly which arises because for the wave propagating through the absorbing medium at an oblique angle to the surface, planes of equal wave amplitude no longer coincide with planes of equal phase, as they do with non-absorbing media[1].

The reflection coefficient ρ at an oblique angle of incidence varies with the direction of polarisation of the incident radiation. If the polarisation vector

---

†Dropping the subscript $_0$ for λ, v and $\bar{v}$ in the context used so far; from now on all wavelengths etc. are considered as measured *in vacuo*.

is resolved into components in the plane of the reflection ($E_{\parallel}$) and perpendicular to it ($E_{\perp}$), these two components suffer different phase delays $\delta_{\parallel}$ and $\delta_{\perp}$, and have different reflection coefficients $\rho_{\parallel}$ and $\rho_{\perp}$. Measurements of these four quantities at different angles of incidence afford information which enables $\tilde{n}$ to be calculated; the calculation needs computer facilities. When radiation is reflected obliquely from a thin film of an absorbing medium deposited on another (substrate) medium, the equations are complicated by interference effects. They may be solved, however, and the technique of ellipsometry (see, for example, reference 2) has been developed to study the optical properties of such thin films. It has recently been extended into the far-infrared region[3,4].

Reflection of polarised light on the (high-n) side of a (high-n)/(low-n) boundary at an oblique angle of incidence constitutes the technique of attenuated total reflectance[5,6]. This technique has been used for studying intense infrared absorption bands in liquids, polymers and single crystals[5–8].

Reflection at normal incidence over a broad spectral range by Fourier interferometric techniques (see Section 1.2.3.2b) can give information on ρ and δ according to equations (1.4) and (1.5) and can be used to measure $\tilde{n}$[9], but this broad-band method is in its infancy.

The quantity $\tilde{n}$ characterises the properties of matter with respect to radiation at a particular wavelength. If, as is always the case, both parts of n vary with the wavelength λ_0 or wavenumber $\bar{\nu}_0$, the medium is said to be absorptive and dispersive. Taking this into account we may write $n(\lambda)$ for the value of n at wavelength λ, and, correspondingly, $n(\bar{\nu})$ for the value at wavenumber $\bar{\nu}$;

$$\tilde{n}(\lambda) = n(\lambda) - i\frac{\alpha(\lambda).\lambda}{4\pi}$$

$$\tilde{n}(\bar{\nu}) = n(\bar{\nu}) - i\frac{\alpha(\bar{\nu})}{4\pi\bar{\nu}} \tag{1.7}$$

Maxwell's equations show that $\tilde{n}(\bar{\nu})$ is related to $\tilde{\varepsilon}(\bar{\nu})$, the complex permittivity, as

$$[\tilde{n}(\bar{\nu})]^2 = \tilde{\varepsilon}(\bar{\nu})$$

(for an essentially non-magnetic material). $\tilde{\varepsilon}(\bar{\nu}) = \varepsilon'(\bar{\nu}) - i\varepsilon''(\bar{\nu})$ where $\varepsilon'(\bar{\nu})$ is the real permittivity or 'dielectric constant' and $\varepsilon''(\bar{\nu})$ the imaginary permittivity or 'dielectric loss'.

The motion of electrically-charged particles in the medium in response to the oscillating electromagnetic field is the factor which causes and determines the values of n and α. This causal relation also produces the Kramers–Kronig relations between the real and imaginary parts in the complex quantities

$$\varepsilon'(\bar{\nu}_1) - 1 = 2/\pi \int_0^\infty \frac{\varepsilon''(\bar{\nu}).\bar{\nu}.d\bar{\nu}}{(\bar{\nu}_1^2 - \bar{\nu}^2)}$$

$$\varepsilon''(\bar{\nu}_1) = 2/\pi \int_0^\infty \frac{\varepsilon'(\bar{\nu}).\bar{\nu}.d\bar{\nu}}{(\bar{\nu}_1^2 - \bar{\nu}^2)} \tag{1.8}$$

and, since $\tilde{n}^2 = n^2 + (\alpha/4\pi\bar{\nu})^2 - in\alpha/2\pi\bar{\nu} = \tilde{\varepsilon}$, we have a similar pair of relationships between n and α. One part of the complex permittivity at a particular wavenumber $\bar{\nu}_1$ is determined by an integral of the other part over the whole spectrum. The n–α relationship simplifies[10] to

$$n(\bar{\nu}_1)-n(\infty)=\frac{1}{2\pi^2}\int_0^{\infty}\frac{\alpha(\bar{\nu})\mathrm{d}\bar{\nu}}{\bar{\nu}_1^2-\bar{\nu}^2}$$

Thus we can see that, in principle, measurements of n or of ε' at well-separated spot frequencies enable one to derive integrals involving α or ε'' over wide frequency ranges. Also, as will be seen in Section 1.3.1, one may calculate the integrated absorption coefficient $A=\int_{\bar{\nu}_1}^{\bar{\nu}_2}\alpha\cdot\mathrm{d}\bar{\nu}$ by assuming a Lorentzian shape for the absorption between $\bar{\nu}_1$ and $\bar{\nu}_2$. One may not, however, except in special circumstances, obtain discrete α or ε'' values at particular frequencies from such spot values of n and ε'.

Figure 1.2 shows what corresponding n and α spectra look like. Passing downwards in the wavenumber scale, we see that each region of spectral

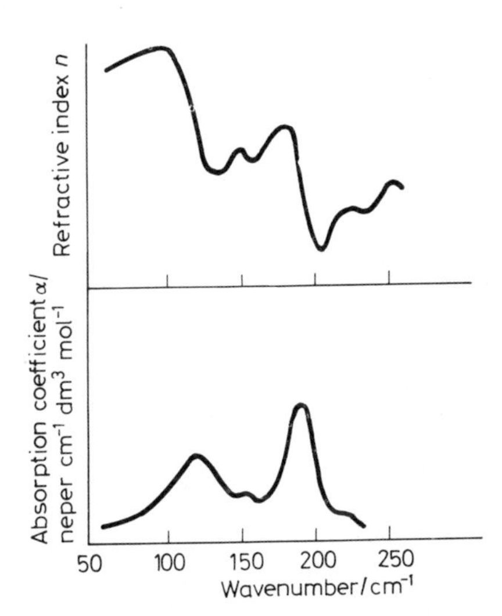

Figure 1.2 Absorption and refraction spectra of 1,2-dibromoethane in heptane (arbitrary vertical axes)

absorption adds a certain increment to the value of $n(\bar{\nu})$. From equations (1.8) we can see qualitatively that for a given size and shape of the ε'' profile, the increment to ε' increases as the reciprocal of the wavenumber. Thus an absorption feature that at higher frequencies would produce only a small increment in ε' will, at lower frequencies, produce a much larger change. There is a similar relation between n and α. This fact has important repercussions when we consider the optical constants of materials in the far-infrared and sub-millimetre regions (Section 1.3.2).

1.1.3 Choice of techniques for measurements in the infrared region

The measurement of the optical constants of materials which, as we have said, are invariably dispersive and absorptive somewhere in the spectrum, is spectrometry. Below about 10^{11} Hz, where the sizes of commonly used samples are comparable with the wavelength, the measurements become dielectric in character, although lumped-circuit measurements are not used much above 5×10^8 Hz. ε' and ε'' have been given equal emphasis in this region, partly because they can be measured with equal ease or difficulty, and partly because they are of equal practical importance in the traditional dielectric applications. In the millimetre and sub-millimetre regions up to *c.* 6×10^{12} Hz, ε' and ε'', or n and α, are, at the present date, more or less on an equal footing, largely because these spectral regions have been considered as adjuncts to the microwave region. Above 6×10^{12} Hz, a much greater emphasis has always been placed on α than on n. Part of the reason for this must be the fact that the equations linking optical constants to molecular behaviour at these frequencies are more easily formulated in terms of α. However, another part of the reason is the practical one that it is, on the whole, easier to make the fairly rough measurements of radiation intensities which give the usual semi-quantitative α spectra than to make the correspondingly much more precise length or angle measurements that are needed to get n spectra of equivalent quality. Nevertheless, when one needs to make *precise* measurements of spectral profiles, as for example when seeking to measure oscillator strengths or polarisability changes (see Section 1.3), one finds that even with sophisticated double-beam absorption spectrometers difficulties exist. One must know the intensity- and frequency-response characteristics of the detector, or, alternatively, have a precisely-calibrated attenuator of the shutter or variable-thickness type. The same difficulties arise with the methods, such as attenuated total reflectance and ellipsometry, which use reflected intensities.

Thus we find that direct measurement of the refractive index by a transmitted-radiation method is being consistently used, in the situations to which it is ideally adapted, to obtain information about microscopic or molecular properties of the transmitting medium. This ideal situation is that in which the absorption bands are strong (for example strong vibrational bands of liquids) but not so strong that transmitted intensities are too low and reflected intensities too high. This seems to rule out, for example ionic crystal absorptions, such as the far-infrared reststrahlen type[11] and semiconductor phonon processes. Refractivity measurements always depend on the precise determination of a length, in an interferometer, or an angle, in a dispersive system. They are nearly always, with the obvious exception of methods which measure n and α simultaneously, independent of precise knowledge of detector response characteristics. Length and angle[12] measurements can be extremely precise, and recently, several groups of workers have shown that refractivity measurements can quite practicably be undertaken in the far, middle and even near infrared regions. This chapter will survey these measurement methods as applied over the past few years, and discuss some of the chemically significant data obtained.

1.2 EXPERIMENTAL

1.2.1 Sources, detectors and filters

The prime requirements for refractivity measurements, as for absorption measurements, are sources, filters and detectors. In the near and middle infrared region, black-body and other heated-solid sources provide adequate broad-band radiation intensity. In the far-infrared and sub-millimetre regions the high-pressure mercury arc discharge lamp seems to be the most efficient broad-band source available, although rather poor. It is not a black-body source, since part of the emission arises from the bremsstrahlung effect in the discharge, and in particular it has an emission line due to the hot envelope (quartz) superimposed on it near 128 cm^{-1} (78 μm). The intensity of this line is highly temperature-sensitive and can fluctuate during a measurement if the lamp is old. Spot frequencies are provided by infrared lasers; only those in the far-infrared region seem to have been much used, however (Section 1.2.3.2).

Some broad-band methods need filters to remove unwanted radiation, especially for work at very low frequencies[68]. KBr and CaF_2 have been used to remove radiation above 50 cm^{-1} (200 μm) and 20 cm^{-1} (500 μm) respectively, and the Fourier transform spectrometers developed by Gebbie and co-workers use the interference effect produced by their thin polyethylene terephthalate beam-dividers[13] to perform most, if not all, of the necessary filtering.

The detectors used include thermopiles, bolometers and the Golay cell, which need no discussion here. Some applications demand fast detector response, which can be met by using photoconductive detectors; in particular in recent submillimetre work[14–17] the Putley indium antimonide detector[18], which responds adequately up to 0.5 MHz.

1.2.2 Prism refractometer methods

The simplest and oldest method of measuring refractive index in the visible spectral region uses the variation of the angle of refraction with wavelength, almost always with the sample in the form of a prism, and this method is applicable in the infrared. The equations linking the geometry of the prism and the angles of incidence and emergence of the radiation beam to the sample refractive index are simple, especially if the configuration giving minimum deviation[19] is used, or the beam is incident normally on one face of the prism[20, 21]. The sample prism is usually placed on a high-precision spectrometer table, and a source, collimator and monochromator system used to supply a nearly monochromatic variable-wavelength beam. The measurement of the angle of refraction can be very precise. Pereira and Hollis Hallett[12] describe an optical-lever system which can detect changes of angle corresponding, other things being equal, to a change of 0.000 01 in n. The limitation of the method becomes clear, however, when one realises that to utilise all the available precision on measuring angles one must have a perfectly parallel narrow beam of monochromatic radiation of usable intensity capable of being focused sharply on a detector of infinitesimally small receiving area. Feldman and Horowitz[22] have made measurements of n to ± 0.001 up to $\lambda = 1.2$ μm (8300 cm^{-1}) on a 90 degree prism of solid copper(I) chloride, accurately machined and inserted into a glass V-block, and an interesting variant on the angle of refraction method used at much lower frequencies has been described by Pilipenko and co-workers[23].

Lecomte[24] has discussed the limit on precision imposed by the absorption of the sample. Over most of the infrared region the overall precision is limited by the degree to which the radiation can be made truly monochromatic, and Dudermel and co-workers quote $10^{-3} \leqslant \delta\lambda/\lambda \leqslant 2 \times 10^{-3}$ in the range 1–10 μm (10 000–1000 cm^{-1}) for their Fabry–Perot monochromator, giving n to $\pm$0.000 03 at 7 μm (1430 cm^{-1}) and $\pm$0.000 07–0.0001 at longer wavelengths[20]. The refraction angle in their system is measured to 2 seconds of arc. At very long wavelengths the crucial problem becomes uncertainty in the angle of refraction caused by diffraction of the radiation at the monochromator or collimator slits. Thus, for a circular aperture radius R, the angular spread, as defined by the first dark ring in the Airy's disc diffraction pattern[1], is proportional to λ/R. Clearly on this basis, Dudermel's method would be much less precise at 100 μm (100 cm^{-1}) than at 10 μm (1000 cm^{-1}), and the diffraction effect would be the dominant source of error in n. A laser has been used[21] at 337 μm (29.7 cm^{-1}) in a prism refractometer. This disposes of the $\delta\lambda/\lambda$ uncertainty, but does not eliminate or even necessarily reduce the aperture diffraction problem which, with the forthcoming development of tunable infrared lasers, will presumably set the limit on the precision of this method and restrict it to the higher infrared frequencies.

1.2.3 Monochromatic interferometric methods

When monochromatic radiation encounters a change in the refractive index of the propagating medium, there is a wavelength change dependent (as shown in Section 1.1) on the change in n. By causing the several beams in an interferometer to traverse different path-lengths through a sample medium, before recombining them and observing interference fringes, we have a method of using an interferometer to measure the sample refractive index.

The Fabry–Perot interferometer, which produces an infinite number of interfering beams, and the Michelson and Mach–Zehnder interferometers which each produce only two, can all be used to produce interference patterns along the interferometer axis by moving one of the mirrors (Figure 1.3), with parallel incident radiation. All the work reviewed here has been concerned with axial interference patterns, since these are unaffected by aperture diffraction effects. This constitutes the great advantage of interferometric methods over methods which use prism dispersion and the production of source images at the detector (Section 1.2.2). Whereas the latter try to define precisely an angle of deviation of a beam, and are thereby limited by the factors discussed in Section 1.2.2, the only limitation on an interferometric method is that the radiation should traverse precisely-known paths through the sample. This is not a trivial requirement; it means in practice either that beams must be perfectly collimated, with or without focusing at the detector (Figure 1.4a) or, if sufficient intensity is available and the detector area is small enough, they may be divergent (Figure 1.4b), with the focusing omitted. These requirements cannot be met completely. Perfect collimation is impossible and good collimation becomes less feasible, because of diffraction, as the wavelength increases. Thus, unless the interferometer apertures can be made infinitely large, case (a) in Figure 1.4 becomes unattainable at long

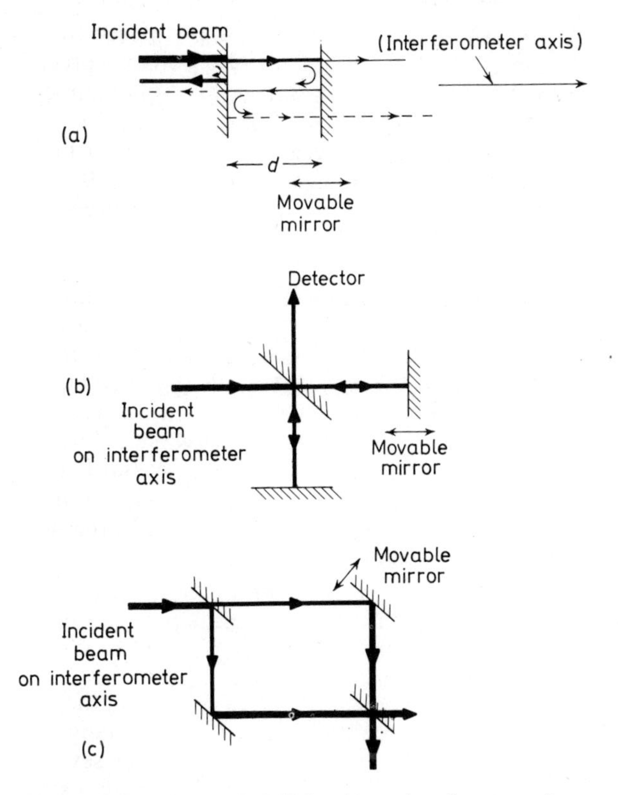

Figure 1.3 Scheme of (a) Fabry–Perot interferometer (beams shown separated from each other and from axis, for clarity) (b) Michelson and (c) Mach–Zehnder interferometers

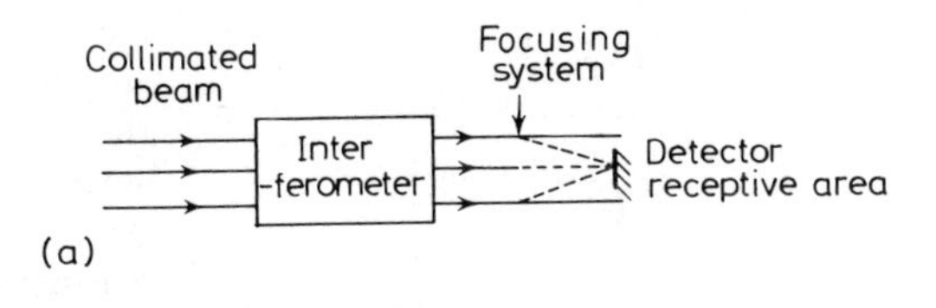

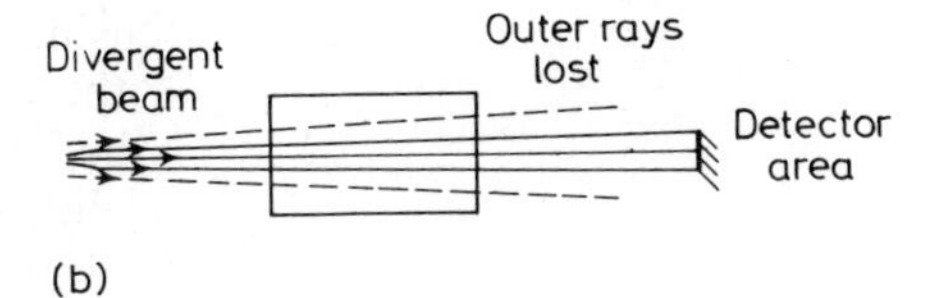

Figure 1.4

wavelengths. In case (b) the ideal detector would have infinitesimal receptive area, since otherwise it would pick up non-axial rays and thereby introduce uncertainty into the measured path length. Interferometers have recently been used in the middle- and far-infrared regions with precisions on the whole less than the near-infrared prism systems described in Section 1.2.2 (typically $\pm 10^{-3}$–10^{-4} compared with $\pm 10^{-5}$), but the longer wavelengths account for this.

1.2.3.1 Fabry–Perot interferometer

Some of the pre-1969 literature on infrared work with these interferometers has been briefly reviewed by Young and Jones[25], and articles by Wilhelmi[26, 27], Pratt and King[28], Kaliteevskii[29], Sztraka and co-workers[30] and McSweeney[31] have discussed techniques and theoretical aspects.

In the Fabry–Perot interferometer the multiple interfering beams are generated by partial reflections between parallel plates, which for work in the middle- and far-infrared, may be made of germanium[28, 32, 33] or silicon[33]. The outer surfaces may be bloomed to cut down unwanted reflections and improve the fringe contrast[34]. The radiation used is as monochromatic and parallel as possible, and usually incident normally to the plates. The wavelength λ_m at which a fringe maximum is observed is related to the plate separation, d, for normal incidence (see Figure 1.3a), as

$$\frac{2 \cdot d.n}{\lambda_m} + \phi = M$$

Here M is the 'fringe number', a positive integer equal to the number of cycles delay imposed on the radiation by each double traverse of the cell. ϕ is a number which takes into account the phase changes taking place at each reflection: it is equal to 2δ (equation (1.5)) and a function of n and α.

The 'finesse' of the interferometer, which for our purposes simply means the sharpness of the fringe maxima as a function of λ or d, is, roughly speaking, proportional to the number of interfering beams generated by the partially reflecting plates. Thus, for maximum precision, the incident radiation energy must be distributed as evenly as possible over as many as possible of the infinite number of beams which the interferometer is capable of producing. Hence the reflective properties of the plates are important.

Two methods of using the Fabry–Perot cell exist. In one, the wavelength of the radiation admitted to the cell is varied, and fringe maxima are recorded as a function of the wavelength. In the other, the wavelength is constant and fringe maxima are recorded as a function of the plate separation d.

If the first method is used, the cell (now strictly an 'étalon') is mechanically simple, as no provision is needed for varying the plate separation. The fixed plate separation must be known precisely, and the plates must be plane and parallel to a tolerance related to the wavelengths studied. Then, if adjacent fringe maxima are observed at wavelengths λ_1 and λ_2, we have

$$2d\left(\frac{n_1}{\lambda_1} - \frac{n_2}{\lambda_2}\right) = M_1 - M_2$$

n_1 may be known from an independent experiment and if ϕ is negligible this enables M_1 to be calculated. In addition, if it is known that n is a slowly-varying function of λ we may assume that $M_1\lambda_1 \approx M_2\lambda_2$; since M_1 and M_2 are integers this enables us to calculate the change in n. However, the assumption that n is slowly varying breaks down near absorption bands (Figure 1.2). In this case trial and error (computer-aided) is necessary[28]. This problem of obtaining a value for the fringe order number M in regions of rapidly varying n is general to all interferometric methods, and the need to assume that ϕ is either negligible or constant, i.e. α is small, is a further limitation on this refractivity method related to those discussed in Section 1.1.

The alternative method, recording fringe positions as a function of d at constant λ, requires a Fabry–Perot cell of variable thickness, with the mechanical complications that this implies and, because separate fringe plots are needed for each wavelength, greatly increases the time needed to build up the dispersion data over an appreciable wavelength range. However, the method does give absolute values of n directly[35].

Pratt and King[28] and Zelano and King[32] have used the first method, also known as the 'channelled-spectrum' method. To increase precision they recorded fringe patterns with the étalon stopped down to different apertures, and used the fringe frequencies obtained as functions of the aperture to extrapolate to zero aperture, thus obtaining frequency values undistorted by the effects of aperture diffraction and imperfect collimation. Their stated precision on n was ± 0.002.

Parsons and Coleman[36] have used a channelled-spectrum technique in the range 27–184 cm^{-1} (*c.* 300–50 μm) to obtain n for gallium phosphide to a precision claimed into the fourth or fifth significant figure.

Hoffman and co-workers[37] have used a variant of the technique to measure n for solid samples, using a flat plate made of the sample material (a polymer) as the étalon and working with the returned, rather than the transmitted, interfering beams.

Orville–Thomas and co-workers have made extensive use of the method[33, 38] (see reference 33 for bibliography of their earlier work). Yarwood and Orville–Thomas also describe a variable-thickness method, using a set of spacers of various thicknesses for their étalon, although they appear to have used it merely to check the fringe order.

Wilhelmi has used the variable-thickness method. He employed a variable-path cell, whose settings, checked with a visible-frequency laser, were reproducible to 0.05 μm. His precision on n was ± 0.001 over the wavenumber range 3300–1000 cm^{-1} (3–10 μm)[26].

If gases are being measured, an alternative to varying the path length is to vary the pressure. Bradley and Gebbie[39] have done this with measurements on atmospheric gases at maser spot frequencies in the sub-millimetre region.

1.2.3.2 Michelson and Mach–Zehnder interferometers

The division of labour between the Fabry–Perot, and the Michelson and Mach–Zehnder systems gives the middle infrared region almost entirely to the former, and the far-infrared and sub-millimetre regions (below 400 cm^{-1}) to the latter. The reasons for this may in part be rationalised, as follows:

(a) The reflectivity of the plates of a Fabry–Perot interferometer can be so adjusted as to give a large number of interfering beams and hence very sharp fringes (high finesse) (Figure 1.5), so that optical properties calculated from the fringe pattern can be very precise. In contrast, the two beams produced by both the Michelson and the Mach–Zehnder interferometers, on recombining in an interference pattern, give only a sinusoidal intensity variation (Figure 1.5). In general, this means that a measurement between two extrema is of low accuracy, unless the whole profile is well defined by many measurements. In the middle infrared this would normally be a tedious procedure, since, with wavelengths less than 10 μm (above 1000 cm^{-1}) in particular, the interferometer reflector movements need very precise calibration. In the far-infrared region, however, it is easier.

(b) Fabry–Perot interferometer cells are usually sufficiently small to fit into the sample space of a commercial infrared spectrometer[27], so that the apparatus for refractivity measurement is easily assembled. Michelson and Mach–Zehnder interferometers are not so amenable, so that their use is more practicable when the apparatus is more adaptable.

Spot-frequency work—In the far infrared and sub-millimetre regions monochromators for broad-band sources are so inefficient that for precise work laser sources become necessary, even though these provide only a

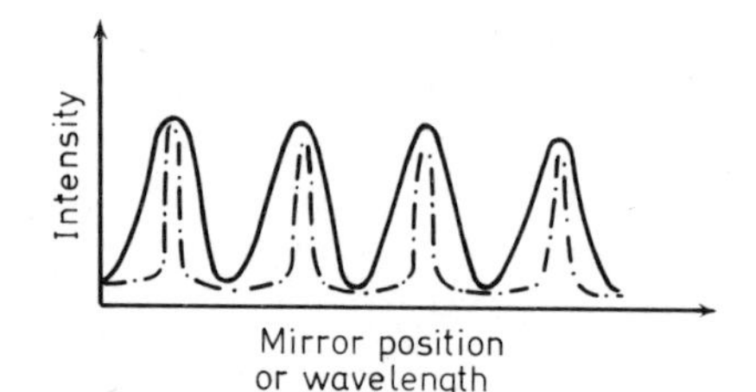

Figure 1.5 Fringe patterns from Michelson (continuous line) and Fabry–Perot (broken line) interferometers

scattering of spot frequencies and cannot usually define a spectrum adequately by themselves. The usual sources are the HCN gas laser for lines at 337, 311 and 128 μm[14,15,40–43] (the first of these is the strongest) and the H_2O gas laser at 119, 79 and 48 μm[14], either pulsed[14,43] or continuous-wave[15,40–42]. Plasmas[14,15,40], liquids[41,42] and solids[41] have been studied. Peterson and Jahoda[16] used a Michelson interferometer coupled to the laser cavity, that is, arranged so that the beam returned from the interferometer re-entered the laser cavity and altered the Q-factor or finesse of the system. The combined laser–interferometer system formed a higher-finesse system, giving sharper fringes (presumably) than those normally obtained with a Michelson interferometer. Other workers have used the Mach–Zehnder system, whose prime feature is that it does not return radiation to the source, principally in order to avoid such coupling[14,15,40,41,43]. Some discussion of this coupling effect has been given by Chamberlain and co-workers[41].

Fringes have been obtained for gases and plasmas by varying the density[14–16] or the length[40] of the sample. For liquids, only the length may be changed[41–43]. For solids one must resort to rotating plane-parallel slabs about an axis perpendicular to the beam from a position initially normal to the beam[41]. This produces a lateral beam displacement which may be disadvantageous, although Chamberlain[41] states that the accuracy is limited chiefly by the precision of measuring the angle of rotation. The measurements of these workers on solids and liquids[41, 42] using a continuous-wave 337 μm laser source are the most precise so far made in the infrared and sub-millimetre regions. They have used path-difference modulation in the interferometer to replace the amplitude modulation usually employed with continuous-wave laser or thermal sources, and this has been shown[44] to give a theoretical and practical signal-to-noise advantage, together with a drastic reduction in noise from source-intensity fluctuations. Perhaps its most important contribution, however, comes from the fact that it effectively differentiates

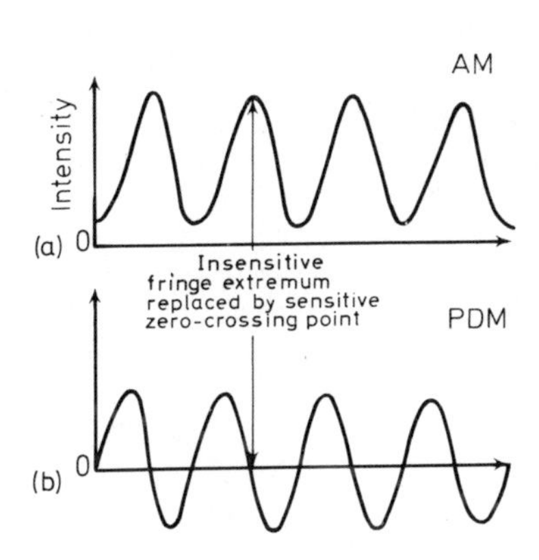

Figure 1.6 Corresponding amplitude-modulation (AM) and path-difference-modulation (PDM) interferograms

the interferogram curves shown in Figure 1.6a to produce Figure 1.6b. The conventional fringe extrema, points of slow variation of intensity in the normal interferogram, are represented in the PDM interferogram (Figure 1.6b) by turning points where the sensitivity of the fringe setting is a maximum. This gives a great increase in precision; n for solids, using the rotation technique[41], is quoted to $\pm 002_3$ in the most favourable case and ± 0.012 in the least favourable. The precision on liquid measurements is much better; using a commercial variable-thickness sample-holder not designed for the purpose, n has been measured to typically ± 0.0003 [42], and in favourable cases to ± 0.0001 or better[41]. Some of the advantages of the method may be seen by comparing the quoted precision for n from reference 42 with the precision on measurements on the same liquid by either amplitude-modulated[45] or pulsed-source[43] interferometer systems.

Table 1.1 Refractive index of chlorobenzene at 29.7 cm^{-1}

Reference	n	δn	*Temperature*/K
41	1.5248	±0.0003	298.2
43	1.535	±0.006	unstated
45	1.538	±0.01	*c.* 293

The precision is such that it becomes possible to determine the electric polarisation $\boldsymbol{P}$ of solute molecules at fairly high dilution in a solvent[46]. The refractive index and real permittivity of a two-component mixture, where there is no strong chemical interaction, is a linear function of solute weight-fraction, and the expression due to Halverstadt and Kumler[47], long used in high-precision work on dielectrics at radio-frequencies and refractivities at visible frequencies, may now be applied at a new intermediate frequency, at least in the case of solutions in the 0–10 wt % range.

Broad-band work (*Fourier-transform dispersion spectrometry*)—The technique of Fourier-transform absorption spectroscopy has been applied very successfully in the far infrared and sub-millimetre regions in recent years. Chamberlain, Gibbs and Gebbie[48] and Bell[49] have described methods by which both parts of the complex refractive index may be determined simultaneously over a wide spectral range, with all the advantages of light grasp and simplicity of optics that have already been demonstrated in the absorption apparatus[50]. The extended apparatus is slightly more elaborate than that used for α alone, and places somewhat higher demands on the stability of the radiation source and the detection and digitalisation electronics. Further, the α information, because of reflection losses and because only half the total detected radiation passes through the sample, is not as accurate as that obtained by the direct non-dispersive method. However, a commercial sub-millimetre dispersion spectrometer is being developed, and Chantry and co-workers have indicated that the method may be useful up to 1000 cm^{-1} (0.1 μm)[51]. In principle, the only difference from the simple absorption apparatus is that the sample is placed in only one beam of the Michelson interferometer, so that this beam suffers a phase dispersion relative to the other. This dispersion contains information about the dispersive characteristics of the sample. The same beam also suffers attenuation according to the absorption characteristics of the sample, and both n and α data may be extracted by Fourier transformation of the resulting interferogram[48].

The difficulties that arise with the method are analogous to those that we have already seen with the Fabry–Perot methods. Chamberlain[52] has discussed the corrections which must be applied when the radiation is not perfectly collimated. A much greater problem is that of knowing the sample thickness exactly; this may be of the order of 0.5–1 mm, and must be known to 0.1 % to give n to ± 0.001. An estimate of the thickness may be obtained by measuring the separation, on the interferogram record, of the points of maximum intensity of the transmitted and reflected structure[48] and assuming a value for the average refractive index, but this expedient is very approximate. This difficulty, trivial as it may seem, sets the limit on the precision of the method at present, although relative n values may be obtained to a much greater precision. For this reason the Fourier-transform method has often been combined with spot measurements at 337 μm, which are more precise in absolute terms[38].

When the samples to be studied have $n \gg 1$, a lot of radiation is reflected from the sample surface. This radiation adds both structure and noise to the interferogram. The structure may to a large extent be removed by editing, but the noise is a problem unless the source and detector are exceptionally stable. Chamberlain has introduced phase modulation[44] to remove this extra noise. Jinks[54] has shown how reflection losses may be estimated from the computed refractive index spectrum, thus enabling the α data to be improved, by a correction term which can be extremely important when working with dilute solutions.

1.3 CHEMICAL INFORMATION FROM INFRARED REFRACTIVITIES

1.3.1 Band-intensity measurements

The spectral region lying between 1 and 100 μm (10 000–100 cm^{-1}) contains, at least in the case of matter in rarified gas form, absorptions which are characteristic of the isolated molecule, and which arise through the interaction of radiation with those vibrational and rotational modes which have non-zero transition moments between their quantum states. For ordinary infrared studies at room temperature the lower state for a vibrational quantum transition is usually the ground state, but at long wavelengths, where modes with small interstate gaps absorb ($h\nu \doteqdot kT$, i.e. $\lambda \approx 50$ μm and $\bar{\nu} \approx 200$ cm^{-1}), appreciable numbers of molecules may be thermally excited to higher states, so that 'hot bands' arise. Because of anharmonicity, these higher transitions may lie at slightly different wavelengths from the ground-state transition, and hence far infrared bands may appear unsymmetrical.

The study of the absolute integrated intensities of these absorptions is of great interest because the intensities are proportional to the transition moment, or, classically, the value of $\partial\mu/\partial Q$ for the mode, $\partial\mu/\partial Q$ being the derivative of the molecular dipole moment with respect to the normal coordinate Q. A normal coordinate analysis of the molecular vibrations will tell us what Q is as a function of atomic coordinates; this may then enable us to calculate $\partial\mu/\partial r$, the derivative of the dipole moment with respect to a bond length or angle of interest, and to deduce information about the bond polarity. Such an analysis may be done rigorously or by guess-work, but in any case is rather easier in the far infrared because the vibrations observed there involve the movement of heavy atoms or groups, as rigid units, linked by weak bonds. This means that $\partial\mu/\partial Q$ and $\partial\mu/\partial r$ are related to each other more simply.

The absorptions observed are not infinitely sharp. This is fortunate because if they were their intensities could never be measured. They are broadened into Lorentz (or Cauchy[25]) line profiles by collisional interactions between molecules as well as by Doppler and natural-damping effects. The Lorentz line shape in terms of α is given by the equation

$$\frac{\alpha}{\alpha_{max}} = \frac{(\Delta\bar{\nu}_{\frac{1}{2}})^2}{(\Delta\bar{\nu}_{\frac{1}{2}})^2 + 4(\bar{\nu} - \bar{\nu}_0)^2}$$

where α_{max} is the intensity at the peak of the absorption at wavenumber $\bar{\nu}_0$ and $\Delta\bar{\nu}_{\frac{1}{2}}$ is the width of the peak at half-height. Then the integrated intensity $A_i = \int_0^\infty \alpha_i \, d\bar{\nu}$, where α_i is that part of the spectral absorption of the sample which belongs to the ith feature, i.e. with a correction for overlapping features.

Two problems which arise in finding A_i may have been brought out by the sketchy treatment given above. The first is that the data in the region around $\bar{\nu}_0$, which make important contributions to the integral, are the least likely terms to be known accurately since the spectral record will be most prone to error at this point of maximum sensitivity to experimental vagaries. The second problem is that of separating α_i from the total α at any wavenumber, where the tails of many broad absorption lines i, j, k etc. may contribute.

If the variation of refractive index in the region of the ith Lorentzian absorption (Figure 1.1) is examined by the Kramers–Kronig relations, it is found that it obeys the equation

$$n(\bar{\nu}) = n_{e,v}(\bar{\nu}) + \tfrac{1}{2}\pi^2 \frac{A_i}{\bar{\nu}_i^2 - \bar{\nu}^2 + \Gamma_i^2}$$

(reference 38 *et loc. cit.*), where Γ_i is written for

$$\frac{2\bar{\nu}}{\bar{\nu}_i^2 - \bar{\nu}^2} (\Delta\bar{\nu}_{1/2}),$$

and $n_{e,v}$ is the extrapolation to $\bar{\nu}_i$, the centre of the absorption of the curves of refractive index at higher frequencies, i.e. the contribution to the band-centre refractive index of electronic (e) and higher-frequency vibrational (v) transitions.

The estimation of $n_{e,v}(\bar{\nu})$ is difficult. The calculation of A is straightforward in cases where $n_{e,v}(\bar{\nu})$ can be found by extrapolation of the dispersion curves[55], which is possible when the refraction feature is not overlapped by others. In this case, and if we can neglect Γ_i [55] we obtain

$$n(\bar{\nu}) = n_{e,v}(\bar{\nu}) + \tfrac{1}{2}\pi^2 \frac{A_i}{\bar{\nu}_i^2 - \bar{\nu}^2}$$

A_i may be calculated by a least-squares plot of $n(\bar{\nu})$ against $(\bar{\nu}_i^2 - \nu^2)^{-1}$ [38, 55]. This method is ideal for isolated features, unresolved in the sense used by Chamberlain[55].

Where the band overlaps others of known strengths, the calculation is only slightly more complicated[55]. When one is faced with a group of overlapping features all of unknown strengths, however, difficulties increase. The two methods used, one of which is due to Schatz[53], have been described in Chamberlain's paper, although he considers only unresolved features and neglects the damping term Γ. Both methods involve iterative plotting repeated until the values for the absorption strengths remain constant; Chamberlain's method appears to give A to *c.* 1–2% accuracy.

Chamberlain has also described a method for the calculation of A for a resolved isolated Lorentzian refraction feature[55], i.e. one where an estimated value of Γ may be included in the calculation.

The only recent work on refractivity spectra of gases seems to be that of Chamberlain, Costley and Gebbie[56], on the $(J+1,\ J)$ rotational lines of NH_3 in the region 15–115 cm^{-1} (670–87 μm). Most of the reported work has been on liquids, in which the 1–100 μm region contains bands of the same Lorentzian profile type as in gases, but with larger half-widths.

The theoretical link between A and $\partial\mu/\partial Q$ is not so straightforward with liquids as it is with gases, since the closely-packed molecules have an appreciable dielectric effect and change the electromagnetic field vector experienced by an absorbing molecule. Various assumptions about the 'internal field', notorious in dielectric studies, lead to different equations linking A_{liquid} to A_{gas}, but the simplest is that of Polo and Wilson[57]:

$$\frac{A_{\text{liquid}}}{A_{\text{gas}}} = \frac{[n^2(\bar{\nu})+2]^2}{9n(\bar{\nu})}$$

Yarwood and Orville-Thomas[33], using this equation, found $\partial\mu/\partial Q$ to be different for liquid and gaseous phases of a polar species, indicating that there were further liquid interactions not described so simply. More recently, however, Thomas and Orville-Thomas[38] have found that the liquid-phase interaction in the non-polar liquid carbon tetrachloride is more nearly purely dielectric, at the frequencies studied, giving better agreement with the Polo–Wilson equation.

As already indicated, Orville-Thomas and his co-workers have made extensive studies of band intensities in liquids and, from them, have inferred the bond moments. A list of references to earlier work by this group is given in reference 33. More recently, Thomas, Orville-Thomas, Chamberlain and Gebbie have published dispersion curves for halides of Group IV elements C–Pb, with n given to ± 0.008 (the absolute level of the curves is found in many cases by maser-interferometric measurements at 29.7 cm^{-1} [38]). They have used Schatz' method of calculation of A for absorptions located in the region 1000–20 cm^{-1} (10–500 μm), of both skeletal stretching and bending types. From the way in which the intensities vary for the two types of vibration down the series C–Pb, they adduce evidence against the simple assumption that the intensity of the asymmetric bending modes should be proportional to the degree of polarity or the ionic character of the bonds. They rightly comment that factors such as steric hindrance may well swamp the bond-polarity effect, and a more thorough analysis of their results seems necessary. They calculate atom-polarisation contributions for the bands they measure. Methods of doing this in the case of samples of low refractive index may be arrived at by combining the Kramers–Kronig and Clausius–Mossotti equations (see, e.g. references 58 and 59).

1.3.2 Sub-millimetre absorption

Recently it has been demonstrated that a great many polar and non-polar organic liquids have a broad asymmetric absorption feature centred between 40 and 80 cm^{-1} (250 and 125 μm) in the sub-millimetre wavelength region. This process was first noted by Poley[60] in polar liquids, not by direct observation but by the discrepancy between the microwave and far infrared refractive indices of the liquids. For example, the real permittivity of nitrobenzene in the microwave region, i.e. at frequencies above the Debye absorption region, was 4.07, giving a refractive index of *c.* 2. The refractive index at *c.* 100 cm^{-1} (100 μm), as measured at an early date and with low precision, was 1.6, indicating that there was spectral absorption between these two reference points. In terms of n, this is a large discrepancy; the molar absorption, as later found for nitrobenzene in benzene solution[54], has a peak intensity of 7 neper cm^{-1} dm^3mol^{-1} at *c.* 30 cm^{-1} (330 μm), which is small compared with higher-frequency modes of the same molecule. Other polar molecules show absorption of the same magnitude with a peak between 30 and 80 cm^{-1}, and the $\Delta(n^2)$ across the band is roughly proportional to the square of the permanent dipole moment of the molecule, although no one as yet seems to have integrated the intensities to obtain polarisation increments.

Recently, Davies, Pardoe, Chamberlain and Gebbie have published spectra of α and n for polar and non-polar liquids, in some cases down to 2 cm^{-1} (5000 μm)[61–63, 68]. They have suggested that in non-polar liquids the phenomenon arises from the interconversion of translational and vibrational energy during molecular collisions. The molecules bend during the collisions and thus become transiently dipolar, and these dipoles form and decay at a frequency governed by the collisional frequency. In polar liquids the correlation between $\Delta(n^2)$ and μ^2 has led to the suggestion that in this case the absorption arises from the libration (incomplete rotation) of the dipole in the fluctuating cage formed by neighbouring molecules.

Chamberlain[64] and Hill[65] have shown that the spectra of both polar and non-polar liquids may be phenomenologically described by a superposition of several broad Lorentzian resonant absorption features.

In the case of non-polar liquids, the task of relating the sub-millimetre absorption/dispersion to chemical structure and properties has been made more difficult by the fact that few workers have studied a series of structurally similar molecules. In many cases the impression obtained is that the liquids studied were those that happened to be handy on the shelf. One attempt at a systematic study is that of Haigh and Sutton[42], who made measurements of refractive indices at 29.7 cm^{-1} (337 μm) on benzene solutions of metal chelate complexes of acetylacetone, using the phase-modulated Mach–Zehnder interferometer discussed in Section 1.2.3.2. The complexes were all non-ionised molecules of sufficient symmetry to make them non-polar if the chelate rings were planar. The chemical interest which prompted the measurements was twofold; firstly, it was known that there was a large discrepancy between radio-frequency and visible-frequency polarisation values, suggesting some absorption process of exceptional magnitude at low frequencies. Secondly, there was some suggestion from x-ray work on crystals codified later by Nelson and White[21] as an explanation of the anomalous polarisation, that the rings were not planar and could flip from one bent conformation to another. The 29.7 cm^{-1} measurements of n and polarisation together with n and α spectra in the sub-millimetre region[66], located an absorption centre at *c.* 60 cm^{-1} (170 μm). Complexes of several transition metals were studied and differences in the shape of the absorption profile were noted from one complex to another. It is not yet clear whether the absorption process is of the non-polar collisional type, if it has a ring-flipping origin as Nelson and White suggest, or if it has some other cause, but it seems to be mainly a solute property, not much affected by the nature of the solvent. The latest evidence suggests that there may be further absorption, as yet imperfectly delineated, at microwave frequencies.

Sutton and co-workers have very recently extended the work to cover other classes of compound which show anomalous polarisation, in particular quinones and donor–acceptor systems. An interesting example is furnished by a recent measurement of the polarisation of *p*-benzoquinone at 29.7 cm^{-1}. The value is 36.99 ± 0.51 cm^3 mol^{-1}. A recent radio-frequency value is 36.74 ± 0.24 cm^3 mol^{-1}. These values are the same within experimental error, although a slight uncertainty remains concerning an absorption feature at 108 cm^{-1} whose tail extends down to 29.7 cm^{-1} and which may perturb the values slightly. However, the near equality strongly suggests that there cannot be a dipolar complex formed between *p*-benzoquinone and the solvent, benzene, since such a complex would give rise to a microwave or radio-frequency absorption process and would cause the two values to differ. This contrasts with the findings of Crump and Price[67], that complexes exist between *p*-benzoquinone and mesitylene and *p*-benzoquinone and *p*-dioxan. It is also interesting that *p*-benzoquinone, in solution in either benzene or carbon tetrachloride, shows no absorption at 60 cm^{-1} which is significantly larger than (i.e. more than 3–5 times) the absorption by the solvent, despite its large quadrupole moment and the fact that it has a low bending force constant. Clearly the causes of the sub-millimetre absorption phenomenon are not yet fully known.

Acknowledgements

Acknowledgment is made to the Senior Director of Development of the Post Office for facilities to J. H. provided during the preparation of this review. The authors also thank Mr K. M. Jinks for providing the data on which Figure 1.2 is based.

References

1. Ditchburn, R. W. (1963). *Light*. (London: Blackie)
2. Archer, R. J. (1962). *J. Opt. Soc. Amer.*, **52,** 970
3. Jones, C. E. and Hilton, A. R. (1968). *J. Electrochem. Soc.*, **115,** 106
4. DeNicola, R. O. and Saifi, M. A. (paper given at the 1971 meeting of the Optical Society of America). *J. Opt. Soc. Amer.*, **61,** 1583 (abstract)
5. Gilby, A. C., Burr, J. and Crawford, B. (1966). *J. Phys. Chem.*, **70,** 1520
6. Tanaka, S. (1970). *Kagaku Kogyo*, **21,** 922 [*Chem. Abstr.*, **73,** 103 645]
7. Tsuji, F. and Yamada, H. (1968). *Bull. Chem. Soc. Jap.*, **41,** 1975
8. Kortüm, G. (1969). *Reflectance spectroscopy: Principles, Methods, Applications*. (trans. J. E. Lohr) (New York: Springer Verlag Inc.)
9. Parker, T. J. Chamberlain, J. and Burfoot, J. C. (1969). *J. Phys. Soc. (suppl.)*, **28,** 230
10. Chamberlain, J. E. (1965). *Infrared Physics*, **5,** 175
11. Mitra, S. S. (1970). *Far infrared properties of solids* (S. Nudelman and S. S. Mitra, editors) (New York: Plenum)
12. Pereira, F. N. D. D. and Hollis Hallett, A. C. (1971). *Rev. Sci. Instr.*, **42,** 490
13. Chamberlain, J. E., Chantry, G. W., Findlay, F. D., Gebbie, H. A., Gibbs, J. E., Stone, N. W. B., and Wright, A. J. (1966). *Infrared Physics*, **6,** 195
14. Turner, R. and Poehler, T. O. (1968). *J. Appl. Phys.*, **39,** 5726
15. Parkinson, G. J., Dangor, A. E. and Chamberlain, J. (1968). *Appl. Phys. Lett.*, **13,** 233
16. Peterson, R. W. and Jahoda, F. C. (1971). *Appl. Phys. Lett.*, **18,** 440
17. Olsen, J. N. (1971). *Rev. Sci. Instr.*, **42,** 104
18. Putley, E. H. (1965). *Applied Optics*, **4,** 649
19. Singh, R. N., Juyal, D. P., Bhargava, J. S., Mathur, K. M., Nagar, O. S. and Bhattacharyya, A. N. (1971). *J. Phys. E* **4,** 283
20. Dudermel, M. T., Corno, J. and Simon, J. (1969). *Optica Acta*, **16,** 587

21. Nelson, R. D. and White, C. E. (1969). *J. Phys. Chem.*, **73,** 3439
22. Feldman, A. and Horowitz, D. (1969). *J. Opt. Soc. Amer.*, **59,** 1406
23. Pilipenko, V. V., Polovnikov, G. G., Sologub, V. G. and Shestopalov, V. P. (1971). *Zhur. Tekh. Fiz.*, **41,** 201 [*Chem. Abstr.*, **74,** 92 417]
24. Lecomte, J. (1967). *Proceedings of the First Int. Conference on Spectroscopy, Bombay,* **2,** 273 [*Chem. Abstr.*, **69,** 23 200]
25. Young, R. P. and Jones, R. N. (1971). *Chem. Revs.*, **71,** 219
26. Wilhelmi, B. (1968). *Infrared Physics,* **8,** 157
27. Wilhelmi, B. (1969). *Jena Review,* **14,** 173
28. Pratt, H. A. and King, W. T. (1967). *J. Chem. Phys.*, **47,** 3361
29. Kaliteevskii, N. I. (1970). *Spektrosk. Gazorazryadnoi Plazmy,* 160 [*Chem. Abstr.*, **74,** 17591]
30. Sztraka, L., Grofcsik, A. and Jobbagy, A. (1970). *Perioda Polytechnica, Chem. Engineering,* **14,** 309
31. McSweeney, A. and Sheppard, A. P. (1970). *U.S. Clearinghouse Fed. Sci. Tech. Inform.* No. 709983 [*Chem. Abstr.*, **74,** 147738]
32. Zelano, A. J. and King, W. T. (1970). *J. Chem. Phys.*, **53,** 4444
33. Yarwood, J. and Orville-Thomas, W. J. (1966). *Trans. Faraday Soc.*, **62,** 3294
34. Nemes, L. (1968). *Instrument News,* **19,** 1;4. (Norwalk, Conn: Perkin Elmer Corp.)
35. Kagarise, R. E. and Mayfield, J. W. (1958). *J. Opt. Soc. Amer.*, **48,** 430
36. Parsons, D. F. and Coleman, P. D. (1971). *Applied Optics,* **10,** 1683
37. Hoffmann, V., Frank, W. and Zeil, W. (1970). *Kolloid-Z. Z. Polymere,* **241,** 1044
38. Thomas, T. E., Orville-Thomas, W. J., Chamberlain, J. and Gebbie, H. A. (1971). *Trans. Faraday Soc.*, **67,** 2710
39. Bradley, C. C. and Gebbie, H. A. (1971). *Applied Optics,* **10,** 755
40. Chamberlain, J., Gebbie, H. A., George, A. and Beynon, J. D. E. (1969). *J. Plasma Physics,* **3,** 75
41. Chamberlain, J., Haigh, J. and Hine, M. J. (1971). *Infrared Physics,* **11,** 75
42. Haigh, J. and Sutton, L. E. (1970). *Chem. Commun.*, 296
43. Allnutt, J. E. and Staniforth, J. A. (1971). *J. Phys. E,* **4,** 730
44. Chamberlain, J. E. (1971). *Infrared Physics,* **11,** 25
45. Chamberlain, J. E., Werner, E. B. C., Gebbie, H. A. and Slough, W. (1967). *Trans. Faraday Soc.*, **63,** 2605
46. Smyth, C. P. (1955). *Dielectric Behaviour and Structure.* (New York: McGraw-Hill)
47. Halverstadt, I. F. and Kumler, W. D. (1942). *J. Amer. Chem. Soc.*, **64,** 2988
48. Chamberlain, J. E., Gibbs, J. E. and Gebbie, H. A. (1969). *Infrared Physics,* **9,** 185
49. Bell, E. E. (1966). *Infrared Physics,* **6,** 57
50. Gebbie, H. A. and Twiss, R. Q. (1966). *Reports on Progress in Physics,* **29,** 729
51. Chantry, G. W., Evans, H. M., Chamberlain, J. and Gebbie, H. A. (1969). *Infrared Physics,*
52. Chamberlain, J. E. (1967). *Applied Optics,* **6,** 980
53. Lindsay, L. P. and Schatz, P. N. (1964). *Spectrochim. Acta,* **20,** 1421
54. Jinks, K. M. personal communication
55. Chamberlain, J. E. (1967). *J. Quant. Spectrosc. Radiat. Transfer,* **7,** 151
56. Chamberlain, J. E., Costley, A. E. and Gebbie, H. A. (1969). *Spectrochim. Acta,* **25A,** 9
57. Polo, S. R. and Wilson, M. K. (1955). *J. Chem. Phys.*, **23,** 2376
58. Whiffen, D. H. (1958). *Trans. Faraday Soc.*, **54,** 327
59. Rao, D.A.A.S.N. (1963). *Trans. Faraday Soc.*, **59,** 43
60. Poley, J. Ph. (1955). *Applied Sci. Res.*, **B4,** 377
61. Chamberlain, J. E., Gebbie, H. A., Pardoe, G. W. F. and Davies, M. M. (1968). *Chem. Phys. Lett.*, **1,** 523
62. Davies, M. M., Pardoe, G. W. F., Chamberlain, J. and Gebbie, H. A. (1968). *Chem. Phys. Lett.*, **2,** 411
63. Davies, M. M., Pardoe, G. W. F., Chamberlain, J. and Gebbie, H. A. (1968). *Trans. Faraday Soc.*, **64,** 847
64. Chamberlain, J. (1968). *Chem. Phys. Lett.*, **2,** 464
65. Hill, N. E. (1968). *Chem. Phys. Lett.*, **2,** 5
66. Haigh, J., Jinks, K. M. and Sutton, L. E., to be published
67. Crump, R. A. and Price, A. H. (1970). *Trans. Faraday Soc.*, **66,** 92
68. Davies, M. M., Pardoe, G. W. F., Chamberlain, J. and Gebbie, H. A. (1970). *Trans. Faraday Soc.*, **66,** 273

2
Neutron Diffraction Studies of Molecular Structures

G. E. BACON
University of Sheffield

2.1 GENERAL INTRODUCTION: X-RAYS AND NEUTRONS

It can be argued with some justification that the greatest practical contribution towards the determination of the shape and structure of molecules was the discovery by von Laue in 1912 that crystals could act as diffraction gratings for x-rays. Immediately afterwards, W. H. and W. L. Bragg demonstrated how this discovery provided a means for determining directly both the ionic and molecular details of the unit cell, which is the basic unit of the crystal of any chemical compound. Our task in this chapter is to recount and explain the contribution which *neutron* diffraction has made to the study of molecular structure. However, as will become clear, the use of x-rays and neutrons have proved to be complementary techniques, for both fundamental and circumstantial reasons, and our main topic will only be appreciated if we first outline briefly the way in which x-ray diffraction analysis has developed since 1912.

Early x-ray studies were made with simple compounds and metals which contained only a few atoms in the unit cell and these atoms could be precisely located from an application of space-group theory and a rudimentary knowledge of the diffraction patterns of the crystals. The greatest achievement by this approach was the determination of the structures of two relatively complicated, but highly symmetrical structures, namely, hexamethylenetetramine and urea; the former is cubic and the latter tetragonal. The 1930s saw the development of the Fourier method of approach, following the realisation that the diffraction pattern is the Fourier transform of the electron density in the unit cell, so that each spectral amplitude yields a term in a three-dimensional Fourier series by which the electron density at any point x, y, z in the cell can be expressed. Of the difficulties which beset this approach one was fundamental and the other practical. First, the experimenter can measure only the intensities of his reflections, whereas he needs to know amplitude and phase. Secondly, the labour involved in the separate summation of a series of at least 1000 terms for each point in the unit cell is prodigious. By 1940, procedures had been evolved for making the best of the fundamental difficulty by the use of indirect methods. particularly the Patterson synthesis, isomorphous replacement and 'heavy-atom' methods, and by *c.* 1950 the practical difficulty had been overcome by the development of electronic computers. Furthermore, with the use of electronic counters, rather than photographic films, for detecting the x-rays the intensity data were of much greater accuracy and could be analysed to give not merely the three-dimensional atomic coordinates of all the atoms but also to assess the thermal motion of the various atoms as well. By 1960 the technique of x-ray diffraction analysis had become so efficient that Robertson remarked that its success was in danger of proving a definite drawback to the development of chemistry. Organic chemists were no longer using the painstaking, but very informative, classical chemical methods of determining the shapes of their molecules, so that the wealth of by-product information which appeared in this process was now being forfeited!

The questions which we now have to ask are 'What are the limitations of the x-ray method, and to what extent can they be removed by using a beam of neutrons instead of x-rays?' In order to answer these questions we must look at the physical principles of the processes whereby atoms scatter, i.e. divert by re-radiation first x-rays and then neutrons.

2.2 PHYSICAL PRINCIPLES OF SCATTERING

When a beam of x-rays penetrates a solid, the electrons within the latter are set into oscillation by the electric field associated with the x-rays. As a result, the electrons themselves act as sources of x-radiation and they re-radiate this in all directions. These x-rays are identical in wavelength with the incident radiation and they are coherent in phase; this accounts for the fact that in certain directions, determined by the shape and size of the unit cell, constructive interference occurs between the contributions scattered by neighbouring cells, leading to the well-known Bragg reflections. Other processes, such as Compton scattering and the production of fluorescent x-rays, also occur but these do not contribute to the Bragg spectra. It is because of the electron dependence of this coherent scattering that we were able to say, in the introductory section above, that the diffraction pattern for x-rays was the Fourier transform of the *electron density*. It follows that the extent to which an atom is prominent in the electron-density picture which we find by experiment, or indeed whether the atom is visible at all, will depend on the number of electrons which it contains. Thus an atom such as hydrogen, which contains only one electron, is at a grave disadvantage compared with, say, an atom of iron, which contains 26 electrons, or an atom of uranium which contains 92. More important perhaps, this means that the *accuracy* with which atoms can be placed within the unit cell is very low for the light atoms which contain few electrons. This is a great limitation on the contribution which x-ray analysis can make to organic chemistry, where very accurate measurements of bond lengths and angles, dependent on very precise measurement of all the atomic positions, can be extremely valuable.

So far as the determination of molecular structure is concerned the advantage of using neutrons is that hydrogen atoms show no marked inferiority to other atoms and that, if the possibility of substituting deuterium for hydrogen can be countenanced, deuterium has a visibility which is equal to that of an 'average' element. Table 2.1 emphasises this point by comparing

Table 2.1 Neutron scattering length, *b*, for some typical atoms

Atom	10^{-12} *Scattering amplitude, b/cm*	
	x-rays	*Neutrons*
hydrogen	0.28	−0.38
deuterium	0.28	0.65
carbon	1.69	0.66
nitrogen	1.97	0.94
oxygen	2.25	0.58
manganese	7.0	−0.36
cobalt	7.6	0.25
bromine	9.8	0.67
iodine	15.0	0.52
lead	23.1	0.96

the 'scattering amplitudes' of a number of atoms for x-rays and neutrons. The scattering amplitude b, or more strictly the scattering length, is defined by saying that if a neutron beam of unit amplitude falls on an atom then the scattered beam which is observed at distance r from the atom has an amplitude equal to $-b/r$. The minus sign implies that for most atoms there is a change of phase of 180 degrees during the scattering process. For some atoms, such as hydrogen and manganese in Table 2.1, there is a phase change of zero and for these atoms the scattering amplitude is conventionally written as negative.

Figure 2.1 is a diagrammatic representation of the variation of the scattering amplitude with atomic mass and indicates particularly how there are seemingly random variations from element to element but, in the main, all the amplitudes are of the same order of magnitude. In fact, we find that the average value of the scattering amplitude for all elements is *c.* 0.60×10^{-12} cm and there are very few elements which have a value which is more than twice this or less than a half of it.

The pattern of variations which Figure 2.1 shows is accounted for by the fact that neutron scattering is a *nuclear* process, in contrast with the case of x-rays for which the electromagnetic field interacts with the *electrons*. The neutron, being uncharged, is virtually unaffected by the electrons in ordinary atoms and molecules, although a very important exception has to be made in the case of the unpaired-spin electrons in magnetic materials. The interaction of the neutrons with the nucleus depends on two factors, first the effective size of the nucleus (in fact its diameter increases as the cube root of the mass number) and, secondly on the energy levels of the nucleus or, more correctly, the energy levels of the compound nucleus formed by adding a neutron to the initial target nucleus. The contribution to the scattering from this second

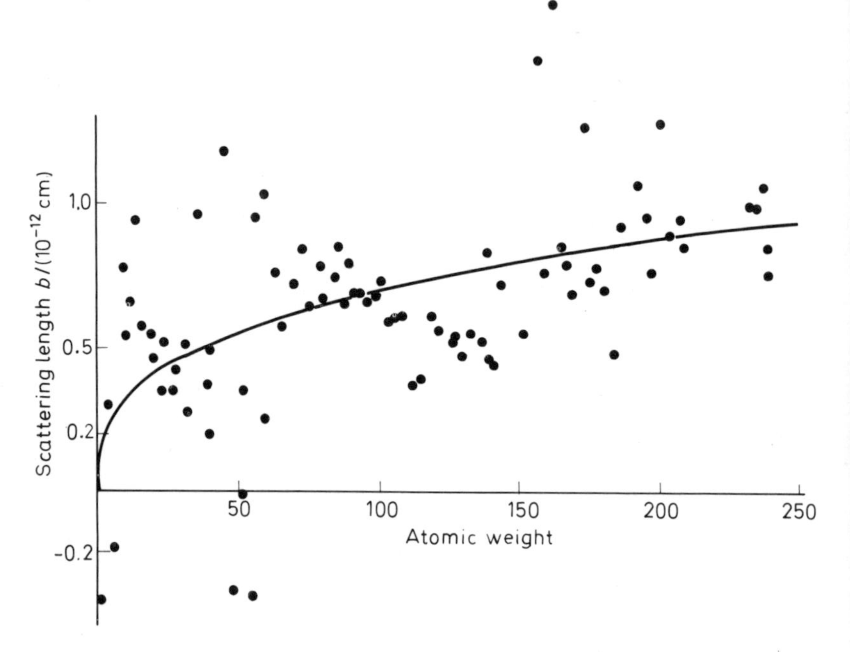

Figure 2.1 The variation of the neutron scattering length 'b' of the elements with atomic weight. The full line on the diagram represents $A^{\frac{1}{3}}$, where A is the mass number of a nuclide, and thus indicates the scattering in the absence of any nuclear resonance effects

effect varies in a quite random way from element to element in the Periodic Table and it is this feature which accounts for the very haphazard nature of b as displayed in Figure 2.1. It should be noted, in particular, that hydrogen and a few other elements show negative values of b.

We must refer to one or two further features of the scattering process for neutrons. We have already stated that the scattering amplitudes of hydrogen and deuterium are quite different. This is no more than a particularly good example of a general principle that the different isotopes of an element may have different scattering lengths, on account of the differing nuclear energy levels which the individual isotopes may possess. When these differences do occur, and particularly when they are substantial, there will be two noteworthy effects if a naturally-occurring element consists of several isotopes with substantial abundances. First, the effective scattering amplitude of the element is the algebraic weighted average of that for its individual isotopes, i.e. $\Sigma b_r \omega_r$, where b_r is the scattering length and ω_r is the fractional abundance of the rth isotope. A second consequence is that in addition to thus providing neutrons in the Bragg peaks in a diffraction pattern there is a contribution to the background intensity which is proportional to $\Sigma(\omega_r b_r^2) - (\Sigma \omega_r b_r)^2$. This feature is known as 'isotope incoherence', because this portion of the scattered radiation is incoherent from atom to atom and cannot contribute to the Bragg diffraction peaks.

Of more importance, because of its relevance to a technique of molecular spectroscopy using neutrons, which we shall describe later, is a rather related feature called 'spin incoherence'. Neutrons and the nuclei of many of the elements and isotopes possess a spin, and when the interaction between nucleus and neutron takes place it is possible for these two spins to line up in either the same $\uparrow\uparrow$ or opposite $\uparrow\downarrow$ fashion. It can be shown that the probabilities of these two arrangements are, respectively, $(I+1)/(2I+1)$ and $I/(2I+1)$ where I is the nuclear spin. Moreover, it is possible for the scattering lengths associated with these two processes to be different, and this gives the possibility of a further contribution to the background of the diffraction pattern, known as 'spin incoherence'. This effect is important to us simply because it is displayed by hydrogen to an enormous extent. In fact, although the ordinary coherent scattering cross-section of hydrogen, measured by $S = 4\pi b^2$, is only equal to 2×10^{-24} cm^2, nevertheless there is also a mammoth contribution to the background scattering which is 40 times as large, amounting to 81×10^{-24} cm^2. In some respects this background scattering is a great hindrance to the experimenter but in the technique of molecular spectroscopy it is put to advantage.

It will be evident that the ability to locate hydrogen atoms accurately is of great importance in organic chemistry. However, there are two factors worthy of mention which indicate that the gain with neutrons is more than simply an improved accuracy in the specification of the atomic coordinates. We have already made clear that x-rays locate the electrons in an atom, with the consequences that the atomic centre which they identify will be the effective centre of gravity of electron density. When atoms are involved in the formation of covalent bonds, or have lone-pair electrons, this centre of gravity may not be identical in position with the *nucleus* of the atom, which is identified incontravertibly by the neutrons. In such cases, therefore, the atomic positions found with neutrons may be judged to be more fundamental than the positions deduced with x-rays. The second factor of interest concerns thermal motion. Because of the thermal energy which atoms possess, they are not stationary in position but are in motion. This motion is compounded from many different vibrations, associated with the bending and stretching of individual bonds and with translations and oscillations of the molecules as a whole. In many cases a substantial portion of the atomic movements is associated with oscillations of the whole molecule as a rigid body, and it is important to be able to determine this motion. It very often happens that there are hydrogen atoms at the periphery of a molecule and these will necessarily be moving quite large distances if an oscillation of the molecule is taking place. The advantage of using neutrons is that they are able to determine with good precision the movements of the peripheral hydrogen atoms and from these measurements the oscillation of the molecule as a whole can be deduced more accurately.

2.3 EXPERIMENTAL METHODS OF NEUTRON CRYSTALLOGRAPHY

In order to make clear the extent to which it is practicable to carry out neutron-diffraction measurements we must give a short account of the experimental technique. The main requirement, and the one which limits the rate at which the use of the technique can grow, is that an intense source of neutrons is required and, from a practical point of view, only a nuclear reactor is adequate. Indeed, if really worth while and informative experimental work is to be done then it is necessary to have a high-flux reactor capable of providing a very intense beam. Most of the studies which have

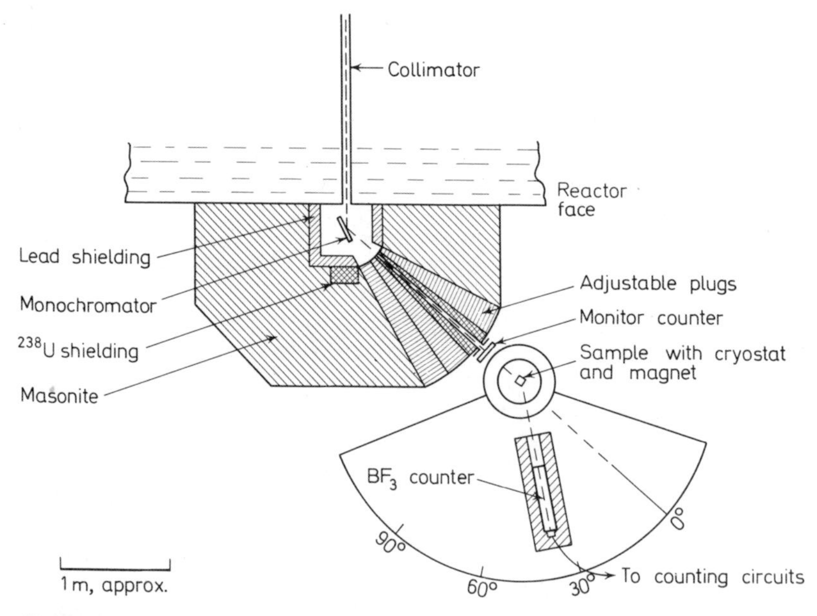

Figure 2.2 A sketch of the layout for a neutron-diffraction apparatus at a reactor. A well-collimated beam is monochromated by reflection at a crystal and then allowed to impinge on a sample; the diffraction pattern is recorded in a counter which traverses an angular range of *c.* 0–120 degrees

been made so far have been performed at reactors (now described as medium-flux reactors) which have a peak flux of a few times 10^{13} neutrons $cm^{-2}\ s^{-1}$, and there are 20–30 reactors of this type in the world. In Great Britain there are two reactors which satisfy the specification adequately and both are at the Atomic Energy Research Establishment, Harwell. University workers have access to the neutron equipment at these reactors through an arrangement with the Science Research Council. There remains a need for still higher fluxes and in the United States two reactors are in operation with fluxes of $c. 10^{15}$ and a similar Franco–German reactor is now almost available for use at Grenoble. There is a good chance that one will also be built in Great Britain during the next few years.

The apparatus used for the determination of structures by neutron diffraction is illustrated diagrammatically in Figure 2.2 and a photograph of an actual equipment is at Figure 2.3. It uses a counter method of detection and is very similar in principle to an x-ray counter diffractometer, except

Figure 2.3 An automatic diffractometer for examining and recording the diffraction patterns of single-crystals in three-dimensions (by courtesy of the United Kingdom Atomic Energy Authority)

for one important feature. The neutrons which emerge from a reactor have a wide distribution of wavelength, corresponding to a roughly Maxwellian distribution of velocities, and it is necessary to select from these a narrow band of wavelength, say 1.05 ± 0.02 Å, for use in the subsequent diffraction experiment. This selection is performed by a large single-crystal, usually a metal such as copper, lead or germanium, set at an appropriate angle. The 'monochromatic' neutrons which are reflected from this single-crystal then fall on the sample crystal, and their intensity at this point will be of the order of 5×10^5 neutrons cm^{-2} s^{-1} for a reactor of moderate flux. The need to employ a monochromator is to be contrasted with the situation in x-ray diffraction where the spectrum emitted from the x-ray tubes consists effectively of a few very narrow lines and can be converted readily into a closely-spaced doublet, the α_1,α_2 lines, by passing the beam through a thin foil which acts as a β-filter. In the apparatus shown in Figure 2.2 it is possible to choose one of several values for the operating neutron wavelength by suitable choice of the single-crystal and of the movable sections of shielding. In many neutron spectrometers the wavelength is maintained at a fixed value. The apparatus illustrated in Figure 2.3 requires that the material to be examined is available in the form of a single-crystal. It is indeed possible to make neutron-diffraction measurements with powdered or polycrystalline samples but, as for x-rays also, such measurements are unlikely in general to lead to detailed structural information of the type which we are seeking. The size of crystal required is a good deal larger than for x-rays; it varies with the material under study but a typical size for a medium flux would be a cube of side 2 mm or a spherical crystal of diameter 2.5 mm. It will be appreciated that this requirement is often a limitation, as many materials are not readily available in such large single-crystals. The principle of operation of the diffractometer is that the crystal and counter are suitably aligned, with the former in the neighbourhood of a position of Bragg reflection for a particular set of crystal planes. The diffraction peak is then scanned step by step at intervals of, for example, 0.04 degrees; each count is continued for a sufficient period of time to collect a constant number of counts in a monitor. This procedure serves to correct for any variations in the intensity of the incident neutron beam during the counting period. Each of the reflections to be measured has to be aligned and scanned in this way. The total number of reflections to be examined will depend on the nature of the material being studied and, in particular, the size of its unit cell; for a small organic molecule, of which we shall describe some examples later, it will be necessary to measure about 1000 reflections to give an accurate analysis of the structure in three-dimensions. It is evident that automation of the experimental work will be essential if it is to be completed in a reasonably short space of time – and this will be important not only from the experimenter's point of view but also to ensure that the utmost use is made of the nuclear reactor, which is very expensive both to construct and to operate. A common procedure is for the diffractometer to be controlled by a five-hole paper tape which carries coded instructions according to a predetermined programme, together with the possibility of inserting additional instructions via a teleprinter. Various aids are provided to enable the operator to maintain a watch on the progress of the experiment. A display of the diffraction curves on a cathode ray tube or a paper record provided by an automatic graph-plotter are possible alternatives. In this way it is possible to examine about four spectra per hour, or possibly six if they are intense reflections, so that a complete series of 1000 spectra could be accumulated in *c.* 2 weeks of continuous operation of the diffractometer.

The length of time involved in the experimental work, coupled with the fact that it can only be carried out in a very limited number of laboratories, usually means that an investigation with neutrons is only made as a sequel to an earlier analysis with x-rays, in particular to improve the precision of location of the hydrogen atoms and the details of the thermal motion of the molecule. This means that the fundamental 'phase problem' of the structure analysis will have already been solved, usually indirectly, by the x-ray investigator and a preliminary set of atomic coordinates and, at least, isotropic temperature factors for the heavier atoms will generally be available. A full analysis of the neutron data, specifying anisotropic thermal parameters for all the atoms, or an analysis in terms of rigid-body motion can then proceed by the standard use of least-squares methods. The results can be presented as tables of atomic coordinates and thermal parameters, supplemented by isometric views of the structure in which the atoms are indicated by the ellipsoids of thermal motion or by Fourier plots of the scattering density projected on to chosen planes.

2.4 SOME TYPICAL DIFFRACTION STUDIES

2.4.1 Hydrogen bonds

It is probably instructive at the outset to examine one or two of the earlier two-dimensional studies which were made before the present automated apparatus for working in three dimensions was developed. Most of this work was concerned with hydrogen-bonded substances, the relation between O—H and O—O bond lengths and study of the linearity or otherwise of the O—H · · · O chain.

A simple, but very direct example, is provided in Figure 2.4 which compares a two-dimensional projection of the anthracene molecule, determined with x-rays, with a two-dimensional projection of the molecule in solid benzene, secured with neutrons. As expected, the hydrogen atoms appear very insignificant in the x-ray projection but in the neutron picture, because of their relatively good scattering length which is *c.* 60% of that for carbon, they are well defined. The dotted, negative, contour lines for the hydrogen atoms result from the negative scattering length of hydrogen for neutrons, as previously discussed. Measurements were made at three different temperatures for solid benzene and it was possible to show that the molecules are executing oscillations with a root-mean-square amplitude of 8, 4.5 and 3 degrees of arc at −3, −55 and −135 °C respectively. It is worth commenting that the study of deuterated benzene would yield a considerable further improvement in the accuracy of location of the hydrogen atoms and in the assessment of the thermal motion.

There have been three separate diffraction studies of α-resorcinol, *m*-dihydroxybenzene, and a comparison of these provides an instructive example of the steadily-growing power of diffraction measurements. The classical x-ray study of resorcinol by Robertson[1] in 1936 gave a good picture of the planar molecules which linked their hydroxyl groups in such a way as to give continuous helices of hydrogen bonds which effectively held together the crystal. Twenty years later[2] the same substance provided the first example of a neutron diffraction study of an aromatic molecule, yielding the projection of molecular structure shown in Figure 2.5. The details of the hydrogen bonds are clearly revealed, there is evidence that one pair of bonds is substantially bent and detailed study of the thermal motion of the individual atoms suggested substantial thermal motion of the molecule as a whole. The conclusion was that carbon atoms C_D, C_F are relatively firmly anchored by the hydrogen bonds and that there is motion of the molecule as a whole about the

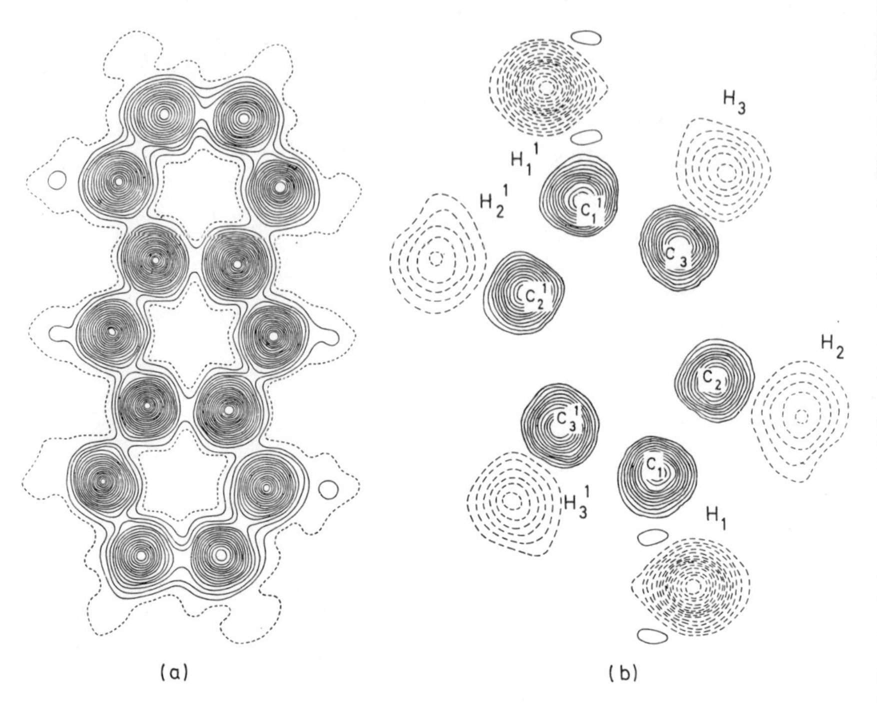

Figure 2.4 A comparison of (a) the poor detail of the hydrogen atoms revealed by x-rays in an anthracene molecule and (b) the full details provided by neutrons for the benzene ring in solid benzene. The former is based on the work of U. C. Sinclair *et al.* (1950). *Acta Crystallogr.* **3**, 251 and the latter is from G. E. Bacon *et al.* (1964). *Proc. Roy. Soc. A*, **279**, 98. The larger scattering density at H_1, H_1^1 in (b) occurs because pairs of hydrogen atoms overlap in the projection

axis which joins these pivots. This conclusion has been substantiated by a three-dimensional analysis which has recently been completed[3]. This latter analysis is typical of current work with a medium-flux reactor and used a crystal approximately 2 × 2 × 3 mm in size and measured 1300 reflections at a wavelength of 1.13 Å. Substantial corrections for secondary extinction had to be made for many of the reflections and this is characteristic of neutron measurements with crystals of this size. The data were analysed in two ways, first by a conventional least-squares analysis in which individual anisotropic temperature factors were postulated for each atom and, secondly, in terms of a rigid-body motion. The first analysis produced a discrepancy factor R of 3.6 % and standard deviations of the atomic coordinates amounting to 0.002 Å for the carbon and oxygen atoms and 0.005 for the hydrogen atoms. Figure 2.6 illustrates the results of the unconstrained analysis of the thermal vibrations by drawing the vibration ellipsoids. The pronounced motion of

C_A, C_B, C_C and their associated hydrogen atoms H_A, H_B and H_C in comparison with the restricted motion of C_D, C_E, C_F and H_E is well evident. The O—H distances in the two hydrogen bonds were 0.991 and 0.989 Å, with an estimated standard deviation of 0.006 Å. Both hydrogen bonds were significantly bent, one of them substantially to an angle of 166 degrees and the other only marginally to 176 degrees. It is worth while pointing out here that when the thermal motion of atoms is markedly anisotropic, as in this case, it can be shown that the measured centre of gravity of the atom does not give its equilibrium position. Thus all bond lengths measured to the atom are in error. A correction has to be first applied, amounting in the case of these hydrogen bonds to 0.004 Å, in order to arrive at the above values. In the

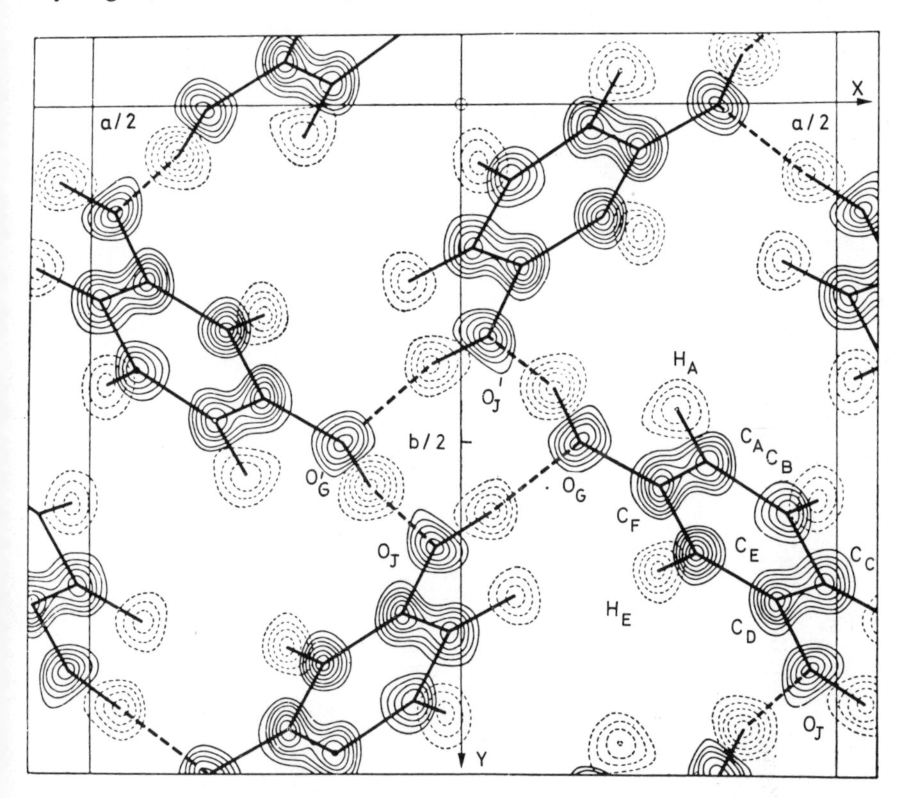

Figure 2.5 A projection on the (001) plane of the neutron scattering density in α-resorcinol. The hydrogen atoms are distinguished by the broken lines which indicate their negative scattering amplitudes. The hydroxyl groups link together neighbouring molecules by forming infinite helices of hydrogen bonds. In the two-dimensional projection these appear as the four-sided closed circuits $O_GO_JO_G'O_J'$. The molecules are inclined at 61 degrees to the plane of the projection

rigid-body analysis the starting point is the model illustrated in Figure 2.7 where translational and oscillatory motion is referred to three axes *L*, *M*, *N* through the centre of gravity of the molecule; the axis *M* is drawn through the point 0 and perpendicular to the diagram. In the refinement, the centre of oscillation was permitted to vary but it did not in fact move away significantly

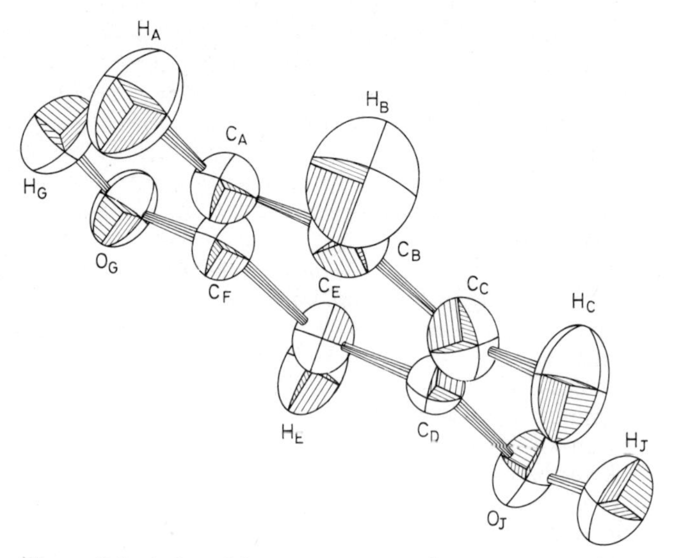

Figure 2.6 A view of the α-resorcinol molecule along the *c*-axis of the crystal, with ellipsoids drawn to indicate the thermal motion of the individual atoms as deduced from the three-dimensional neutron diffraction analysis. The relatively large motion of the hydrogen atoms H_A, H_B, H_C is very evident

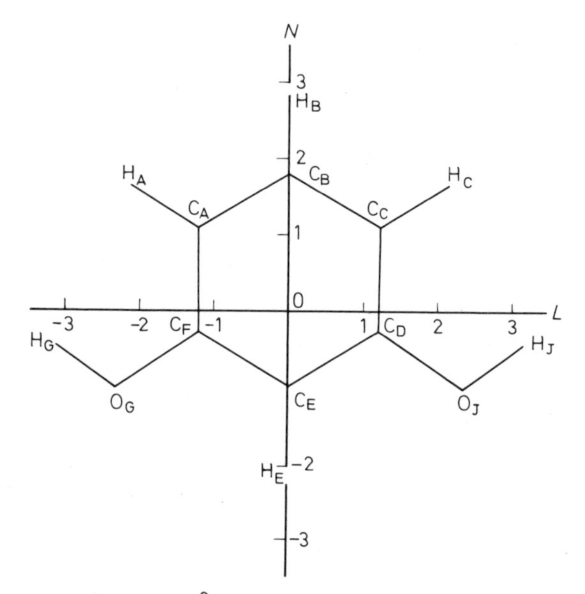

Figure 2.7 A plan of the molecule of α-resorcinol in the same orientation as Figure 2.6. The axis *L* is the line about which rigid-body oscillation is found to take place, thus accounting for the enhanced motion of atoms H_A, H_B, H_C in the previous figure

from the centre of gravity. The root-mean-square amplitudes of translation along the three axes were 0.19, 0.16 and 0.17 Å respectively and the amplitudes of libration were 4.37, 2.27 and 2.68 degrees; the largest amplitude was about *L*, the axis about which the moment of inertia is least. The discrepancy factor for the rigid-body analysis was 6.3 %, which is substantially worse than the 3.6 % given by the unconstrained refinement, and this indicates that there are significant internal vibrations within the molecule itself.

A further good example of neutron methods is provided by investigations of the structure of hexamethylenetetramine $C_6N_4H_{12}$, which is unusual for an organic compound by possessing cubic symmetry with an asymmetric unit of only three atoms. This feature limits the number of independent parameters which have to be determined and accordingly increases the accuracy of the conclusions which can be reached. Like resorcinol, hexamethylenetetramine is a much-studied compound, the investigations culminating in an x-ray study by Becka and Cruickshank[4] and a neutron study by Duckworth, Willis and Pawley[5,6]. In the neutron study, both unrestrained and rigid-body models were used for refining the thermal motion, giving *R* factors of 2.3 and 3.2 % respectively, and standard deviations of the atomic coordinates of 0.003, 0.002 and 0.004 Å for carbon, nitrogen and hydrogen respectively. The most interesting feature of this work however was that the authors also carried out a joint refinement[6] of their neutron data with the earlier x-ray data. The principle of this refinement is based on an acceptance of the fact that the *positional* parameters obtained with x-rays and neutrons may not be identical, because, as we have already indicated, x-rays locate the centroids of the electron clouds which surround the nuclei whereas neutrons locate the nuclei themselves. Accordingly the joint refinement is made in terms of separate scale factors for the neutron and x-ray data, a single set of thermal parameters applicable to both kinds of data and two sets of positional parameters – one for the neutron data and the other for the x-ray data. Table 2.2 shows the values of the position parameters for

Table 2.2 Position parameters in hexamethylenetetramine

	Neutron	*e.s.d.*	*x-Ray*	*e.s.d.*
v (N)	0.1221	0.0005	0.1237	0.0002
u (C)	0.2381	0.0008	0.2375	0.0003
x (H)	0.0899	0.0011	0.0802	0.0020
z (H)	−0.3258	0.0012	−0.3292	0.0030

e.s.d. = estimated standard deviation

the carbon and nitrogen atoms given by x-rays and neutrons. For the carbon atoms the two positions agree within one standard deviation, using the larger value of this deviation given by neutrons. This would be expected for carbon because the valence electrons are in four orthogonal hybrid orbitals, giving a resultant electron cloud which would be symmetrical about the carbon nucleus. On the other hand, for the nitrogen atom the difference between the two parameters is three times as great as the neutron standard deviation and indicates a displacement of the electronic centre of the atom by 0.018 Å from the position of the nucleus. Of the five electrons in the valence shell of nitrogen, three participate in bonding and the other two are in lone-pair orbitals which are displaced from the nucleus in a direction away from the centre of the hexamethylenetetramine molecule. Calculations by Coulson[7] have indicated that this displacement should be *c.* 0.016 Å, in remarkably good agreement with the above experimental observation. The same conclusion is illustrated rather strikingly by plotting the result of a difference Fourier synthesis in Figure 2.8. This plot has been obtained by synthesising the function $F^x - F'$ where F^x is the experimental

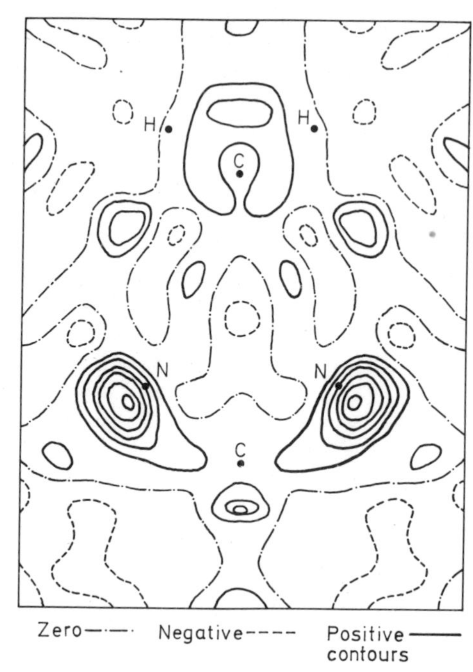

Figure 2.8 A difference synthesis for hexamethylenetetramine, produced by a combination of x-ray and neutron data, which indicates the asymmetry of electron distribution represented by the lone-pair electrons of the nitrogen atom. (From Duckworth *et al.*[5], by courtesy of Munksgaard). The diagram is a (110) section through the centre of the molecule showing two carbon, two hydrogen and two nitrogen atoms. The positions of the atomic nuclei, found by the neutrons, are marked by the dots

x-ray structure factor and F' is the calculated value of the x-ray structure factor *using the position coordinates given by neutrons.* The synthesis indicates therefore the extent to which the centroid of the electron density is displaced from the nuclear position. Thus the two large peaks in the projection, which is made on to the 110 plane of the crystal, represent the lone-pair electrons of the nitrogen atom and, as expected, they are displaced from the nucleus in a direction away from the centre of the molecule. Table 2.2 also indicates significant differences in the parameters of the hydrogen atoms found with x-rays and neutrons and although the standard deviations of the x-ray positions are rather large there is a significantly larger value for the C—H bond length given by neutrons; at present the best available values are 1.11 ± 0.01 Å for neutrons and 1.06 ± 0.01 Å for x-rays.

2.4.2 Biological materials

The potential advantages of neutrons for studying biological materials have been apparent for many years but it is only relatively recently, as higher neutron fluxes have become available, that much practical progress has been made. The main difficulty has been the provision of single-crystals of these substances which are large enough to study with a neutron flux between 10^{13} and 10^{14}, and this is undoubtedly a field which will be explored very rapidly when fluxes of 10^{15} become more widely available. The ability of neutrons to locate the hydrogen atoms accurately is of fundamental importance here but neutrons do have some other substantial advantages. With biological substances, because of the immense size of the molecules and the unit cells, the structural work is inevitably of lower resolution and reduced accuracy and with x-rays it is often quite difficult to distinguish nitrogen atoms from carbon and oxygen. Neutrons have a substantial advantage in this respect, for it happens, as is evident from Table 2.1, that nitrogen has an exceptionally large scattering amplitude and can be identified very readily. Advantage may also be taken of the completely different scattering amplitude of deuterium, compared with ordinary hydrogen, in order to identify the positions of particular atomic groupings by selective deuteration. The best idea of present progress is probably given by examining the work on the monocarboxylic acid derivative of vitamin B_{12}, $C_{63}H_{87}O_{15}N_{13}PCo{\cdot}16H_2O$, by Moore, Willis and Hodgkin[8]. The crystal has a monoclinic unit cell, measuring *c.* 15 Å in each direction. There are 228 atoms in the asymmetric unit and *c.* 1500 reflections, going down to an interplanar spacing of 1.3 Å, were measured with neutrons and the data were analysed to determine 820 parameters. The accuracy attained may be appreciated from Figure 2.9. At (a) in this figure is shown a plot of the neutron-scattering density over the corrin nucleus, and the hydrogen atoms are clearly distinguished and located; it is interesting to note that the central cobalt atom is a 'light' atom for neutrons, with a scattering amplitude of only 0.25×10^{-12} cm, whereas for x-rays it predominates as a heavy atom. At (b) is shown the detail of a methyl group whose possible rotation about the single bond linking it to the main molecule had been questioned; it is clear from the neutron contours that this group is not rotating. A further detail of the molecule, at (c), shows a group of atoms whose identification remained ambiguous from earlier x-ray work as either $CO{\cdot}NH_2$ or COOH. The neutron plot, with its ability both to see the separate hydrogen atoms and to distinguish the 'heavy' nitrogen atom identifies the group unquestionably as $CO{\cdot}NH_2$. This work has now been extended by Hodgkin, Moore, O'Connor and Willis[9] down to a spacing of 1.0 Å which involves the measurement of almost 3000 reflections. This has resulted in considerably improved resolution and some significant changes in some of the chemical details of the molecule. The data were refined by a succession of cycles of Fourier analysis and least-squares. It is noteworthy that limitation of data to 1.3 Å, rather than 1 Å, results in confusion between diffraction ripples, due to series termination, and the hydrogen atoms. If data only extend down to 2 Å, it becomes impossible to assign for certain the chemical atoms, although the course structure of the molecule

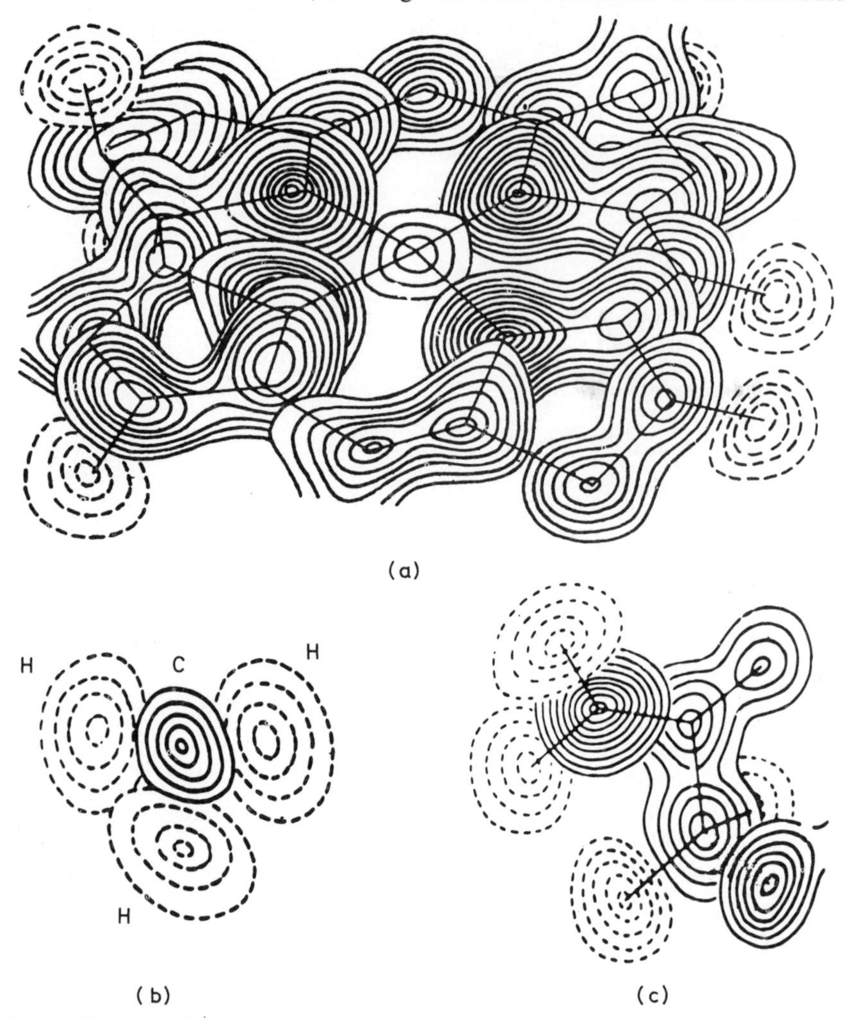

Figure 2.9 Some details of the first neutron diffraction contours of cyanocobalamin monocarboxylic acid (B_{12}). View (a) shows the corrin nucleus of the molecule, emphasising the 'light' cobalt atom at the centre. (b) Shows a methyl group, clearly not in rotation and (c) indicates a group of atoms which can be conclusively identified by neutrons, but not x-rays, as $CH_2{\cdot}CO{\cdot}NH_2$ (From Moore, Willis and Hodgkin[8] and reproduced by courtesy of Macmillan)

can be recognised. It is interesting to see that the great success of the 1 Å study is achieved in spite of the enormous incoherent scattering from the 119 hydrogen atoms in the molecule.

From a practical point of view it is worth noting that radiation damage of the biological materials while undergoing diffraction analysis is likely to be less serious with neutrons than with x-rays. On the other hand, it is only fair to point out that the neutron analysis of vitamin B_{12} is relatively simple compared with a classical protein analysis by x-rays. For example, in the x-ray analysis of myoglobin 17 000 reflections were measured, with the aim of determining the 21 500 parameters which were needed to define the structure completely, or 11 340 parameters if the hydrogen atoms were ignored altogether.

Finally, for these materials, there remains one advantage of neutrons which is possibly of much greater significance than any of those which we have previously mentioned. In our introductory remarks we stated that the fundamental drawback to a direct determination of structure by diffraction methods lay in the experimenter's restriction to measurements of intensity, rather than amplitude and phase. Neutrons, in certain circumstances, offer a way of overcoming this difficulty. We have already described how the extent to which a nucleus scatters neutrons depends partly on the energy levels within the compound nucleus of target-plus-neutron. This is a resonance process, but the resonant wavelength is usually so far from our thermal wavelengths of *c.* 1 Å that the outcome is no more than a numerical change in the value of scattering amplitude. However, for a few nuclei, notably ^{10}B, ^{113}Cd and ^{149}Sm, the resonance occurs near to thermal energy and the result of this is that 'anomalous scattering' occurs, anomalous in the sense that the scattering amplitude is not 'real' but 'complex', which means that the phase change on scattering is not just 0 or 180 degrees but can have an intermediate value. It is true that a similar effect occurs with x-rays in the neighbourhood of an absorption edge; the important feature is that the neutron effect is, quantitatively, of a different order of magnitude. Thus for ^{113}Cd at the resonant wavelength of 0.68 Å the imaginary term in the scattering amplitude is 7.6 times as large as the normal 'real' contribution, whereas x-ray anomalies amount only to a fraction of *c.* 0.2. The merit of the large imaginary term in these circumstances is that in crystals, which are not centrosymmetric, it leads to a method of phase determination based on comparison of the intensities of pairs of reflections hkl and $\bar{h}\bar{k}\bar{l}$. In the absence of anomalous scattering, such pairs of reflections are always of equal intensity for both neutrons or x-rays. An attempt is being made at Harwell to apply this technique to refine further the structure of insulin, first announced in 1969 by Hodgkin and her co-workers[10], by incorporating samarium into the molecule. The same project hopes also to identify the regions of the molecule which are occupied by water molecules, with the aid of the hydrogen–deuterium replacement technique.

2.5 MOLECULAR SPECTROSCOPY BY INELASTIC SCATTERING

In the applications which we have discussed so far, and which we may describe in broad terms as 'neutron crystallography' we have considered only one aspect of the scattering of the neutrons by the crystals. We have been solely concerned with those neutrons which have been scattered *elastically*, which means that they have undergone no change of energy in the scattering process and possess the same wavelength before and afterwards. Moreover, among these elastically-scattered neutrons we have been interested only in those for which constructive interference takes place between the contributions from neighbouring atoms, thus leading to the Bragg reflected beams which have been the basis of our interpretation of structures. Nevertheless we have indicated, in our discussion of spin and isotope incoherence, that there may also be other neutrons, incapable of causing constructive interference, which make up part of the background of the diffraction patterns.

Instead of being scattered elastically it is possible for neutrons to undergo a change of energy, leading to either an increase or a decrease of energy, during the scattering process. This same behaviour can occur with x-rays but we shall show that the *consequences* of the energy interchange, and the way in which we can utilise its study to further our knowledge, are quite unique to neutrons. This will be evident if we consider the process briefly in quantitative terms. The basis of this process is that energy is taken from the neutron to excite vibrations in the crystal or, alternatively, energy is taken away from the crystal's vibratory motion to increase the neutron's energy. These vibrations may be of many different kinds, for example vibrations of the crystal as a whole, oscillations of individual molecules or bending or stretching motions of some of the individual bonds within the molecules. The energies involved in these motions are about 0.01 eV for molecular oscillations, 0.1 eV for vibrations and roughly within the same range for acoustic vibrations. The important fact from our present point of view is that the energy of a thermal neutron of wavelength 1 Å is 0.08 eV and if the wavelength is 2 Å the energy is 0.02 eV. Thus the energies of our neutrons are quite similar to those of phonons of vibrations and oscillations. For the benefit of those readers who are more accustomed to thinking in terms of infrared wavenumbers, rather than energies in eV, we mention that the energy of a neutron of wavelength 1.5 Å corresponds to an optical wavenumber of 300 cm^{-1}. It follows, therefore, that if a neutron gains or loses a phonon of molecular energy then its wavelength will be altered substantially. Assuming that we can measure accurately the change in neutron energy, and this can indeed be done by time-of-flight techniques, then we shall have a method of measuring the energies of the molecular phonons, i.e. a technique of molecular spectroscopy. Before discussing the advantages which such a technique may possess, we must first make clear why x-rays could not be used in a similar way. This arises simply because the energy of an x-ray of wavelength 1.5 Å is *c.* 10^4 eV and any changes in wavelength brought about by inelastic scattering would be too small to detect.

Before describing the technique of neutron molecular spectroscopy in more detail, it will be worth while to point out the particular advantages which it possesses over other spectroscopic techniques. First, all molecular rotational and vibrational transitions contribute to the neutron spectrum and the latter is not restricted by the selection rules which operate for optical spectra. In a sense this might mean that the neutron spectrum would include an embarrassing wealth of information, particularly as the energy resolution in the spectrum is by no means so good as that in an optical spectrum. However, this conclusion is modified by the second distinctive feature of the technique, which is of over-riding importance. The quantitative expression for the intensity of the neutrons which have undergone a particular energy change includes both the amplitudes of vibration of the atoms concerned in the vibration and also their neutron-scattering amplitude: indeed, the intensity is proportional to the square of each of these factors. This means that the spectrum is heavily biased in favour of vibrations which involve hydrogen atoms, not only because the small mass of the hydrogen atom leads to a large amplitude of vibration but also because the scattering power of hydrogen is high. This latter statement will need a little explanation in view of the fact that the value of b for hydrogen in Table 2.1 is smaller than that for most other atoms. The values of b in this table relate to the coherent scattering which can produce interference effects, but in the case of the inelastic scattering it is the *total* scattering cross-section which is important

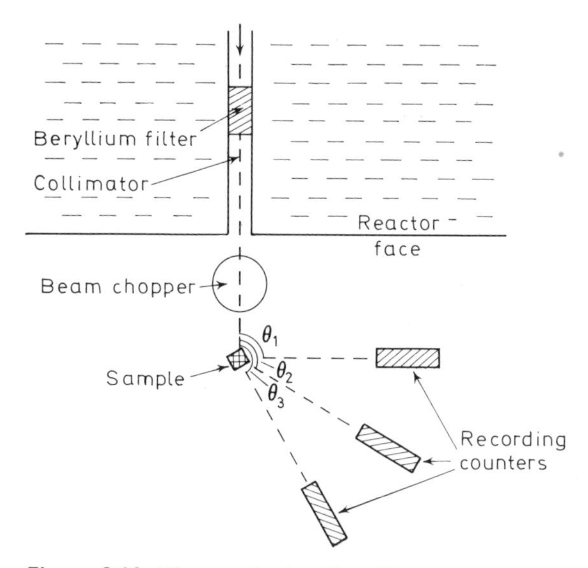

Figure 2.10 The use of a beryllium filter to separate the long-wavelength tail of the Maxwell spectrum, as a semi-monochromatic beam. The beam is then pulsed by a mechanical chopper and the neutrons scattered by a sample are collected, with energy analysis, in a group of counters at angles θ_1, θ_2, θ_3 . . . etc.

and this, as we have already seen, is extraordinarily large, amounting to 81 barns, in the case of hydrogen. As a result, not only do hydrogen modes figure particularly prominently in the inelastic spectra but they also can be identified quite conclusively by noting their drastic reduction when deuterium is substituted for hydrogen in the chemical compound under examination. The measurements are made with material in powdered or polycrystalline form and it is not necessary to have single-crystals.

A number of different experimental methods are available for observing the spectra. In the simplest, a monochromatic beam of neutrons is allowed to fall on the sample and the scattered neutrons are observed at a constant angle of scattering, which can conveniently be 90 degrees. The scattered neutrons are analysed into an energy spectrum, either by causing them to undergo Bragg reflection from a single crystal or, more commonly, by a time-of-flight technique, which requires the initial incident beam to be pulsed. The relative intensities of different bands in the neutron spectrum may vary considerably with the angle of scattering because of the significance of the Debye–Waller factor when the atomic displacements are of the same order as the neutron wavelength. If a series of counters is employed, then

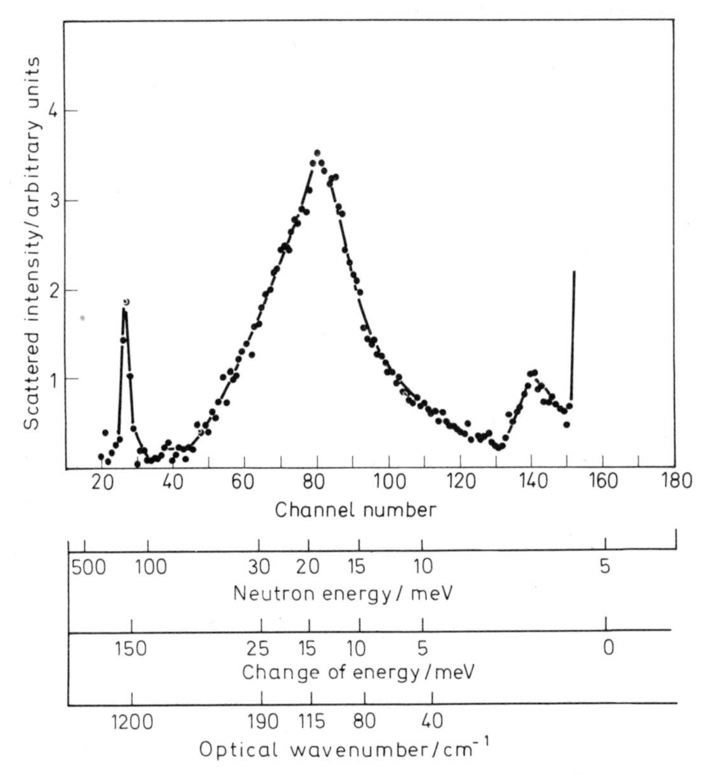

Figure 2.11 The spectrum of inelastically-scattered neutrons from KHF_2, for an incident neutron energy of 0.005 eV. The units on the axis of abscissae show (a) the neutron energy (b) the *gain* of neutron energy and (c) the wavenumber in cm^{-1} of the electromagnetic radiation whose energy corresponds to the energy change which the neutron has undergone
(From Boutin *et al.*[11], reproduced by courtesy of the American Institute of Physics)

data may be collected simultaneously at several different angles of scattering. Alternatively, as in Figure 2.10, the initial beam from the reactor may be passed through a block of polycrystalline beryllium, which acts as a low-pass filter, allowing passage only of the tail of the spectrum which comprises neutrons of wavelengths greater than 4 Å. This is, of course, not truly a monochromatic beam but has the advantage of giving greater intensities at the expense of some loss of resolution at low values of the transferred energy. The intensity of such a beam, of relatively long wavelength neutrons, may be increased still further by employing a 'cold source' within the reactor.

This cold source consists of a portion of moderator, of water or graphite, which is cooled by liquid hydrogen or helium. A third possible technique is the reverse procedure in which neutrons of all velocities fall on the sample and for each velocity, by applying a time-of-flight technique, we measure how many have had their energy reduced by the amount necessary to permit passage of a low-energy beryllium filter. In this method only processes in which there is a *loss* of energy by the neutron can be observed.

An early example of neutron molecular spectroscopy is illustrated in Figure 2.11 which shows the results of Boutin *et al.*[11] for polycrystalline KHF_2. The spectrum shows, in particular, a well-defined and intense peak which corresponds to an energy change of 147 meV (1180 cm^{-1}) and a very weak, but well-defined, peak at about 75 meV (600 cm^{-1}). These peaks are assigned respectively to the bending and stretching vibrations of the linear F—H—F ion. The large amplitude of the bending motion accounts for the intense peak. On the other hand, the weak peak is a consequence of the fact

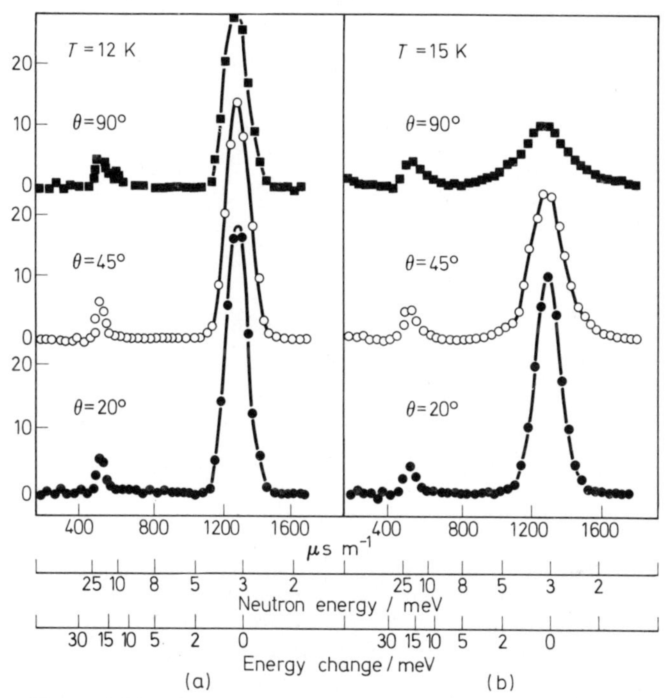

Figure 2.12 The spectra of scattered neutrons from (a) solid hydrogen at 12 K and (b) liquid hydrogen at 15 K measured in turn at scattering angles of 90, 45 and 20 degrees. The units on the abcissae indicate the time-of-flight, the neutron energy and the *change* of neutron energy.
(From Egelstaff *et al.*[12], reproduced by courtesy of the Physical Society)

that the hydrogen bond is very short, with the F—F distance equal to only 2.26 Å, and it is accordingly symmetrical and centred. As a result, the motion of the hydrogen atom in the stretching mode is insignificant and the neutron peak for this particular vibration is very small.

Figure 2.12 compares some results for solid and liquid hydrogen, at

12 and 15 K respectively, at several different angles of scattering, from the work of Egelstaff, Haywood and Webb[12]. Apart from the sharp elastic peak, corresponding to no change of energy by the neutron, the main feature of the solid spectrum is a sharp peak at an energy transfer of about 15 meV (120 cm^{-1}). The width of this is largely independent of the angle of scattering and the peak corresponds to a loss of one quantum of rotational energy by the hydrogen, from $J = 1$ to $J = 0$. Such a transition would not be observed in either infrared or Raman spectroscopy, as it is forbidden by the selection rules in each case. It is noteworthy that the rotational peak is still visible in the pattern for liquid hydrogen thus indicating that the rotations of the molecules are not significantly quenched by the intermolecular collisions in

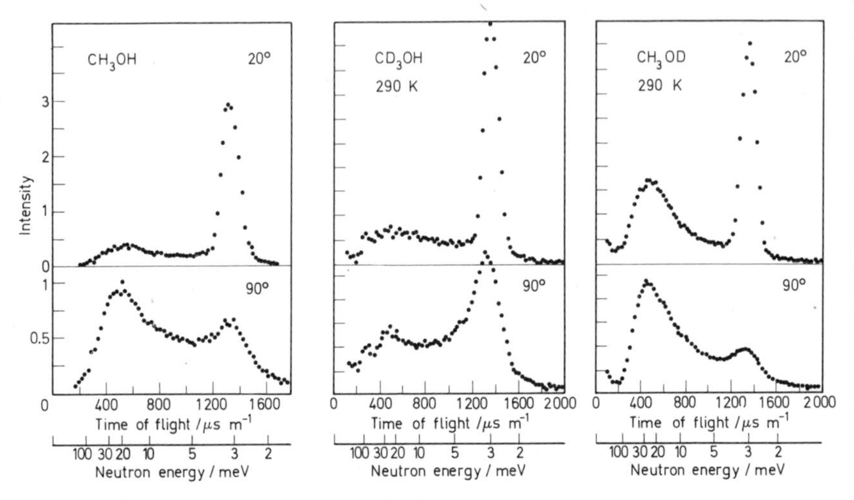

Figure 2.13 A comparison of the neutron spectra after inelastic scattering of neutrons of wavelength 5.3 Å (0.003 eV) from (a) CH_3OH (b) CD_3OH (c) CH_3OD, each at 290 K and for scattering angles of 20 and 90 degrees
(From Aldred *et al.*[13], reproduced by courtesy of the Faraday Society)

the liquid. On the other hand, the elastic peak is weaker in the liquid and becomes noticeably weaker and broader as the angle of scattering in-increases. It is, in fact, strictly only 'quasi-elastic' because the atomic motions give rise to Doppler shifts of the neutron's energy and it is possible to determine the diffusion behaviour of the atoms in the liquid by studying the variation of the width and intensity of this peak.

The virtue of the substitution of deuterium for hydrogen as a means of identifying particular features in a neutron spectrum is clearly shown by some measurements of Aldred *et al.*[13] on liquid samples of CH_3OH, CD_3OH and CH_3OD. The three sets of spectra are shown in Figure 2.13 and the most important feature is the variation in intensity of the broad peak at short times-of-flight, in relation to the size of the quasi-elastic peak; the changes are especially clear in the measurements at an angle of scattering of 90 degrees. This broad peak is much reduced in intensity when deuterium is inserted in the methyl group by the change from CH_3OH to CD_3OH. On the other hand, the intensity is little affected when deuterium is substituted in the hydroxyl group to give CH_3OD. It is deduced therefore that the peak centred on a time of flight of 500 μs m^{-1}, corresponding to an energy transfer of about 20 meV (160 cm^{-1}), is associated with the methyl group. In other applications, where the substitution of deuterium in molecules is difficult or not available, it is possible to compare the spectra of compounds of closely-related structure, one of which includes hydrogen atoms, but these are replaced in the other compound by atoms with small scattering cross-sections such as fluorine. Thus acetic acid shows a broad peak at about 500 μs m^{-1} (which is very similar to that of methanol which we showed in Figure 2.13) but this peak is completely removed when fluorine is substituted in the methyl group to give the molecule CF_3COOH. It is evident, therefore, that the peak is again associated with the methyl group.

The above examples will serve to show that neutron spectroscopy, by virtue of its partiality for hydrogen motions and with the aid of isotopic substitution, is a very valuable technique when used to complement spectroscopy with electromagnetic radiation. Together with the methods of neutron crystallography which we outlined earlier, it provides fundamental information concerning molecular structure. The use of these neutron methods to further chemical knowledge has spread rapidly during the past 10 years and now provides a generally-accepted technique. Future developments are likely to be extensive and to grow in significance as more intense neutron sources become available.

References

1. Robertson, J. M. (1936). *Proc. Roy. Soc. A.*, **157,** 79
2. Bacon, G. E. and Curry, N. A. (1956). *Proc. Roy. Soc. A.*, **235,** 552
3. Bacon, G. E. and Jude, R. J. (1972). In the press
4. Becka, L. N. and Cruickshank, D. W. J. (1963). *Proc. Roy. Soc. A*, **273,** 435
5. Duckworth, J. A. K., Willis, B. T. M. and Pawley, G. S. (1970). *Acta Crystallogr.*, **A26,** 263
6. Duckworth, J. A. K., Willis, B. T. M. and Pawley, G. S. (1969). *Acta Crystallogr.*, **A25,** 482
7. Coulson, C. A. (1970). *Thermal Neutron Diffraction*, 78, (B. T. M. Willis, editor). (London: Oxford University Press)
8. Moore, F. M., Willis, B. T. M. and Hodgkin, D. C. (1967). *Nature (London)*, **214,** 130
9. Hodgkin, D. C., Moore, F. H., O'Connor, B. H. and Willis, B. T. M. (1971). In the press
10. Adams, M. J. *et al.* (1969). *Nature (London)*, **224,** 491
11. Boutin, H. *et al.* (1963). *J. Chem. Phys.*, **39,** 3135
12. Egelstaff, P. A., Haywood, B. C. and Webb, F. J. (1967). *Proc. Phys. Soc.*, **90,** 681
13. Aldred, B. K., Eden, R. C. and White, J. W. (1967). *Discuss. Faraday Soc.*, **43,** 169

3
Structure of Crystalline Polymers

HIROYUKI TADOKORO
Osaka University, Japan

3.1 INTRODUCTION

In this article the structural studies of crystalline polymers reported in the past 5 years are briefly reviewed. The range of reviewing is limited to the developments of the methods of structure analysis, mainly x-ray diffraction, infrared and Raman spectroscopy, and to the results of analyses of synthetic polymers. The most important contributions in this field were made by the applications of new powerful instruments, especially the high-speed electronic computers, in addition to the development of the theoretical and experimental methods of analysis.

3.2 DEVELOPMENTS OF METHODS AND INSTRUMENTS

3.2.1 X-ray diffraction

The x-ray analyses of high polymers have been made mainly by trial-and-error procedures, and this is still the most useful method[1,2]. The reasons are: (a) it is difficult to obtain three-dimensional reflection data since the polymer samples are usually uniaxially orientated and partially crystalline, except for several special cases, e.g. polyoxymethylene prepared by solid-state polymerisation[3], (b) the number of independent observable reflections is small, at most about one hundred, (c) the accurate measurement of the reflection intensities is difficult because of the low intensity and broadening of the spots. However, several powerful methods have been developed and applied, giving successful results.

The molecular transform theory of helical molecules first given by Cochran *et al.*[4] and developed by Klug *et al.*[5] and Franklin and Holmes[6] have been extensively applied to many helical polymers. For non-helical polymer molecules, the molecular transforms have been calculated also and gave useful information[7,8].

For the case of complex uniform helical polymers in which plural chains are contained in a unit cell, it is more useful to calculate the lattice structure factor consisting of molecular transforms instead of the structure factor expressed by the fractional coordinates of the atoms.

For crystal structure analyses of low-molecular-weight compounds, the Patterson function is used as one of the most important procedures, but the synthesis of this function needs single-crystal diffraction data. The cylindrical Patterson function can be synthesised from fibre diffraction data alone, and gives information for the analysis without any assumption. This function was first proposed by MacGillavry and Bruins[9] and the example of the application will be shown in a later section.

The optical transform method of analysis[10] is based upon the similarity between the x-ray scattering by atoms and the visible-ray scattering by pin holes. Optical transforms have been used for (a) the study of disordered structures[11] and (b) as a convenient method for obtaining the molecular transforms[10].

In the commonly used least-squares method, the fractional atomic coordinates and temperature factors are used as the variable parameters. The least-squares method based on the molecular parameters has been developed by Arnott and Wonacott[12]. This method is most suitable for the analysis of high polymers where the number of available reflections is small. In their paper it was applied to polyethylene terephthalate and has since been used for many polymers[13,14,16].

A Weissenberg photograph of a uniaxially-orientated sample oscillating about an axis normal to the fibre axis contains the reflection data of the higher layer lines and also the meridian. This cannot be obtained from usual fibre diagrams taken with the cylindrical camera[18].

Further information is available in books[19, 20] and a review[21] on x-ray diffraction of high polymers, crystallographic data[22] and treatment of Lorentz and orientation factors in fibre diagrams[23].

3.2.2 Vibrational spectroscopy

The normal-coordinate treatment has become possible for many crystalline polymers, i.e. infinitely extended single molecular chains and also crystals because of the use of high-speed electronic computers. Laser Raman spectrophotometers, far-infrared spectrophotometers and microspectroscopic apparatus are commercially available and are very powerful instruments in this field. Many polymers have been synthesised[24] from selectively deuterated monomers and are very useful for making vibrational assignments.

Further information is available from infrared spectroscopy of high polymers[24–27], laser Raman spectroscopy[28–30] and standard data[31, 32].

3.2.3 Conformational analysis and other investigations

The energy calculations made for single molecules and crystals have the following three objectives. (a) The interpretation of the results of structure analyses. The most stable conformation may be found theoretically[33–37]. (b) The setting up of the probable models for structure analyses[38, 39]. (c) Provision of detailed information for the interatomic and intermolecular interactions[40].

Neutron inelastic scattering[41, 42], the measurements of elastic moduli of the crystalline regions[43] and high-resolution n.m.r.[44], etc. give important information from different points of view.

3.3 RESULTS OF ANALYSES

In the present review the *trans* and two types of *gauche* forms are denoted by T, G, and $\bar{G}$ (minus G), respectively[45]. Figure 3.1 shows the notation of the internal rotation angles, including C (*cis*) and S (*skew*). The conformations

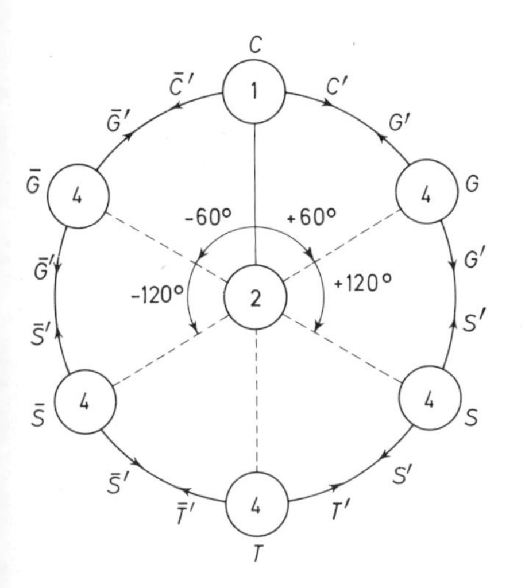

Figure 3.1 Notations of the conformation and the internal rotation angle. The sign of the angle is positive if, on viewing the atoms along the bond 2,3 with 2 nearer the observer than 3, the angle from the projection of 2,1 to the projection 3,4 is traced in the clockwise sense
(From Tadokoro[45], reproduced by courtesy of Interscience)

Table 3.1 Crystallographic data

Polymer	*Crystal system, space group, lattice constants and no. of chains per unit cell**	*Molecular conformation*	*Crystal density*/g cm^{-3}
Polyethylene[48] $(-CH_2-CH_2-)_n$	New modification, monoclinic, $C2/m\text{-}C_{2h}^3$, $a = 8.09$, b (f.a.) $= 2.53$, $c = 4.79$ Å, $\beta = 107.9°$, $N = 2$	Planar zig-zag (2/1)	0.998
it-Poly(3-methylbut-1-ene)[79] $(-CH_2-CH-)_n$ with side group $CH(CH_3)_2$	Monoclinic, $P2_1/b\text{-}C_{2h}^5$, $a = 9.55$, $b = 17.08$, c (f.a.) $= 6.84$ Å, $\gamma = 116°30'$, $N = 2$	Helix (4/1)	0.93
it-Poly(4-methylpent-1-ene)[80, 81] $(-CH_2-CH-)_n$ with side group $CH_2CH(CH_3)_2$	Tetragonal, $P\bar{4}b2\text{-}D_{2d}^7$ or $P\bar{4}\text{-}S_4^1$, $a = 18.63$, c(f.a.) $= 13.85$ Å, $N = 4$	Helix (7/2)	0.812
it-Poly((*S*)-4-methylhex-1-ene)[80] $(-CH_2-CH-)_n$ with side group $CH_2CH(CH_3)(C_2H_5)$	Pseudohexagonal, $P1\text{-}C_1^1$, $a = b = 19.85$, c(f.a.) $= 13.50$ Å, $N = 4$	Helix (7/2)	0.86
it-Poly((*S*),(*R*)-4-methylhex-1-ene)[80]	Tetragonal, $P\bar{4}\text{-}S_4^1$, $a = 19.85$, c(*f.a.*) $= 13.50$ Å, $N = 4$	Helix (7/2)	0.86
it-Poly(5-methylhex-1-ene)[84] $(-CH_2-CH-)_n$ with side group $(CH_2)_2-CH(CH_3)_2$	Monoclinic, $P2_1\text{-}C_2^2$, $a = 17.62$, $b = 10.17$, c(f.a.) $= 6.33$ Å, $\beta = 90°$, $N = 2$	Helix (3/1)	0.86
it-Poly((*S*)-5-methylhept-1-ene)[84] $(-CH_2-CH-)_n$ with side group $(CH_2)_2-CH(CH_3)(C_2H_5)$	Monoclinic, $P2_1\text{-}C_2^2$, $a = 18.40$, $b = 10.62$, c(f.a.) $= 6.36$ Å, $\beta = 90°$, $N = 2$	Helix (3/1)	0.90
it-Poly((*S*),(*R*)-5-methylhept-1-ene)[84]	Tetragonal, $P\bar{4}\text{-}S_4^1$ (probably), $a = b = 20.00$, c(f.a.) $= 38.76$ Å, $N = 4$	Helix (19/6)	0.91
it-Poly(4-methylpenta-1,3-diene)[85] $(-CH_2-CH-)_n$ with side group $CH{=}C(CH_3)_2$	Tetragonal, $I\bar{4}c2\text{-}D_{2d}^{10}$, $a = 17.80$, c(f.a.) $= 36.50$ Å, $N = 4$	Helix (18/5)	0.85
it-Poly(4-phenylbut-1-ene)[186]	Monoclinic, $Pa\text{-}C_s^2$, $a = 10.4$, $b = 18.0$,	Helix (3/1)	1.07

$SiH_2(C_2H_5)$			
it-Poly(vinyl methyl ether)[86] $(\text{—}CH_2\text{—}CH\text{—})_n$ OCH_3	Trigonal, $R\bar{3}$-C_{3i}^2, $a = 16.25$, c(f.a.) $= 6.50$ Å, $N = 6$	Helix (3/1)	1.168
st-Polypropylene[98] $(\text{—}CH_2\text{—}CH\text{—})_n$ CH_3	Orthorhombic, $C222_1$-D_2^5, $a = 14.50$, $b = 5.60$, c(f.a.) $= 7.40$ Å, $N = 2$	Helix (4/1) $(T_2G_2)_2$	0.93
Polyisobutylene[103] $[\text{—}CH_2C(CH_3)_2\text{—}]_n$	Orthorhombic, $P2_12_12_1$-D_2^4, $a = 6.94$, $b = 11.96$, c(f.a.) $= 18.63$ Å, $N = 2$	Helix (8/3)	0.962
Poly(vinylidene fluoride)[105, 106] $(\text{—}CH_2\text{—}CF_2\text{—})_n$	Modification (I)[105, 106] orthorhombic, $Cm2m$-C_{2v}^{14}, $a = 8.58$, $b = 4.91$, c(f.a.) $= 2.56$ Å, $N = 2$	Slightly distorted planar zigzag (1/0)	1.973
	Modification(II)[105], monoclinic, $P2_1/c$-C_{2h}^5, $a = 4.96$, $b = 9.64$, c(f.a.) $= 4.62$ Å, $\beta = 90°$, $N = 2$	Glide type $TGT\bar{G}$ (2/0)	1.925
	Modification(III)[105], monoclinic, $C2$-C_2^3, $a = 8.66$, $b = 4.93$, c(f.a.) $= 2.58$ Å, $\beta = 97°$, $N = 2$	Slightly distorted planar zigzag (1/0)	1.944
it-Poly(methyl methacrylate)[38] $[\text{—}CH_2\text{—}C(CH_3)\text{—}]_n$ $COOCH_3$	Orthorhombic?, $a = 21.08$, $b = 12.17$ c(f.a.) $= 10.50$ Å, $N = 4$	Helix (5/1)	1.23
st-Poly(α-methylvinyl methyl ether)[188] $[\text{—}CH_2\text{—}C(CH_3)\text{—}]_n$ OCH_3	Pseudohexagonal?, $a = 9.02$, c(f.a.) $= 16.6$ Å, $N = 1$	Helix (10/4)	1.02
trans-1,4-Polybutadiene[109] $[\text{—}CH{=}CH\text{—}(CH_2)_2\text{—}]_n$	Low temperature form, monoclinic, $P2_1/a$-C_{2h}^5, $a = 8.63$, $b = 9.11$, c(f.a.) $= 4.83$ Å, $\beta = 114°$, $N = 4$	(1/0) *trans*-$ST\bar{S}$	1.03
trans-1,4-Polyisoprene[14] $[\text{—}CH{=}C(CH_3)\text{—}(CH_2)_2\text{—}]_n$	α-form, monoclinic, $P2_1/c$-C_{2h}^5, $a = 7.98$, $b = 6.29$, c(f.a.) $= 8.77$ Å, $\beta = 102.0°$, $N = 2$	(2/0 (*trans*-$CT\bar{S}$) (*trans*-$CT\bar{S}$)	1.05
trans-Polypentenamer†[113] $[\text{—}CH{=}CH\text{—}(CH_2)_3\text{—}]_n$	Orthorhombic, $Pnam$-D_{2h}^{16}, $a = 7.28$, $b = 4.97$, c(f.a.) $= 11.90$ Å, $N = 2$	*Helix* (2/1) (*trans*-$ST_4\bar{S}$)$_2$	1.05
trans-Polyoctenamer‡[113] $[\text{—}CH{=}CH\text{—}(CH_2)_6\text{—}]_n$	Monoclinic, $P2_1/a$-C_{2h}^5, $a = 7.43$, $b = 5.00$, c(f.a.) $= 9.90$ Å, $\beta = 95°\ 10'$, $N = 2$	(1/0) *trans*-$ST_5\bar{S}$	1.00
	Triclinic, $P\bar{1}$-C_i^1, $a = 4.34$, $b = 5.41$, c(f.a.) $= 9.78$ Å, $\alpha = 64°\ 25'$, $\beta = 104°\ 50'$, $\gamma = 118°\ 35'$, $N = 1$		1.00

Table 3.1 (cont.)

Polymer	Crystal system, space group, lattice constants and no. of chains per unit cell*	Molecular conformation	Crystal density/g cm^{-3}
trans-1,4-Poly(2,3-dichlorobutadiene)[115] [—CCl=CCl—$(CH_2)_2$—]$_n$	Monoclinic, $P2_1/a$-C_{2h}^5, $a = 5.34$, $b = 9.95$, c(f.a.) = 4.80 Å, $\beta = 93.5°$, $N = 2$	(1/0) *trans*-$ST\bar{S}$	1.61
Polyallene[116, 117] (—CH_2—C(=CH_2)—)$_n$	Modification (I), orthorhombic, $Pnan$-D_{2h}^6, $a = 8.20$, $b = 7.81$, c(f.a.) = 3.88 Å, $N = 2$	Helix (2/1) G_4	1.07
	Modification (II), monoclinic, $P2_1$-C_2^2 or $P2_1/m$-C_{2h}^2, $a = 6.37$, b(f.a.) = 3.88, $c = 5.12$ Å, $\beta = 96.6°$, $N = 1$		1.06
	Modification (III), c(f.a.) = 3.88 Å, paracrystalline like		—
Polyoxacyclobutane[121, 137] [—$(CH_2)_3$—O—]$_n$	Modification (I) (hydrate), monoclinic, $C2/m$-C_{2h}^3, $a = 12.3$, $b = 7.27$, c(f.a.) = 4.80 Å, $\beta = 91°$, $N = 4$ with $4H_2O$	Planar zigzag (1/0)	1.18
	Modification (II), trigonal, $R3c$-C_{3v}^6, $a = 14.13$, c(f.a.) = 8.41 Å, $N = 9$	Glide type (2/0) $T_3GT_3\bar{G}$	1.19
	Modification (III), orthorhombic, $C222_1$–D_2^5, $a = 9.23$, $b = 4.82$, c(f.a.) = 7.21 Å, $N = 2$	Helix (4/1) $(T_2G_2)_2$	1.20
Poly(hexamethylene oxide)§[193] [—$(CH_2)_6$—O—]$_n$	Monoclinic, $C2/c$-C_{2h}^6, $a = 5.64$, $b = 8.98$, c(f.a.) = 17.32 Å, $\beta = 134.5°$, $N = 2$	Planar zigzag (2/1)	1.06
Poly(decamethylene oxide)[193] [—$(CH_2)_{10}$—O—]$_n$	Orthorhombic, $Pnam$-D_{2h}^{16}, $a = 7.40$, $b = 4.93$, c(f.a.) = 27.29 Å, $N = 2$	Planar zigzag (2/1)	1.04
Poly-1,3-dioxolane[189] [—OCH_2O—$(CH_2)_2$—]$_n$	Modification (II), orthorhombic, $Pbca$-D_{2h}^{15}, $a = 9.07$, $b = 7.79$, c(f.a.) = 9.85 Å, $N = 4$	Glide type $(G_2T\bar{G}\bar{S})(\bar{G}_2TGS)$	1.41
Poly-1,3-dioxepane[16] [—OCH_2O—$(CH_2)_4$—]$_n$	Orthorhombic, $P2_1cn$-C_{2v}^9, $a = 8.50$, $b = 4.79$, c(f.a.) = 13.50 Å, $N = 2$	Glide type $TGT\bar{G}TG_2T\bar{G}TGT\bar{G}_2$	1.23
it-Poly(propylene oxide)[132] (—CH_2—CH(CH_3)—O—)$_n$	Orthorhombic, $P2_12_12_1$-D_2^4, $a = 10.46$, $b = 4.66$, c(f.a.) = 7.03 Å, $N = 2$	Slightly distorted planar zigzag (2/1)	1.124
it-Poly(t-butyl ethylene oxide)[39]	Tetragonal, $P\bar{4}n2$-D_{2d}^8, $a = 15.42$, c(f.a.) = 24.65 Å,	Helix (9/4)	1.02

[—CH_2—C(CH_2Cl)$_2$—CH_2O—]$_n$	b = 8.16, c(f.a.) = 4.67 Å, N = 4	(1/0)	
	β-form, orthorhombic, $Pb2_1m$-C_{2v}^{12}, a = 13.01, b = 11.71, c(f.a.) = 4.67 Å, N = 4	Planar zigzag (1/0)	1.446
Poly(*trans*-but-2-ene oxide)[140] (—CH(CH_3)—CH(CH_3)—O—)$_n$	Orthorhombic, $P2_12_12_1$-D_2^4, a = 13.72, b = 4.60, c(f.a.) = 6.90 Å, N = 2	Slightly distorted planar zigzag (2/1)	1.10
Poly(3,3,3-trifluoro-1,2-epoxypropane)[144] (—CH_2—CH(CF_3)—O—)$_n$	Orthorhombic, $P2_12_12_1$-D_2^4, a = 11.42, b = 6.26, c(f.a.) = 6.26 Å, N = 2	Distorted planar zigzag (2/1)	1.67
Poly(*p*-phenylene oxide)[145] [—C_6H_4—O—]$_n$	Orthorhombic, $Pbcn$-D_{2h}^{14}, a = 8.07, b = 5.54, c(f.a.) = 9.72 Å, N = 2	Helix (2/1)	1.407
Poly(2,6-diphenyl-*p*-phenylene oxide)[146] [—C_6H_2(Ph)$_2$—O—]$_n$	Tetragonal, $P4_12_12$-D_4^4 or $P4_32_12$-D_4^8, a = 12.51, c(f.a.) = 17.08 Å, N = 2	Helix (4/1)	1.213
Polythiomethylene[147] (—CH_2S—)$_n$	Triclinic, $P1$-C_1^1, a = b = 5.07, c(f.a.) = 36.52 Å γ = 120°, N = 1	Helix (17/9)	1.60
Poly(ethylene sulphide)[148] [—(CH_2)$_2$—S—]$_n$	Orthorhombic, $Pbcn$-D_{2h}^{14}, a = 8.50, b = 4.95, c(f.a.) = 6.70 Å, N = 2	Glide type (2/0) $TG_2T\bar{G}_2$	1.41
Poly(pentamethylene sulphide)[17] [—(CH_2)$_5$—S—]$_n$	Monoclinic, $P2_1/a$-C_{2h}^5, a = 9.61, b = 9.78, c(f.a.) = 7.84 Å, β = 131°, N = 4	Planar zigzag (1/0)	1.218
Poly(propylene sulphide)[150] (—CH_2—CH(CH_3)—S—)$_n$	Orthorhombic, $P2_12_12_1$-D_2^4, a = 9.95, b = 4.89, c(f.a.) = 8.20 Å, N = 2	Slightly distorted planar zigzag (2/1)	1.23
Polyglycolide[151] (—CH_2COO—)$_n$	Orthorhombic, $Pcmn$-D_{2h}^{16}, a = 5.23, b = 6.19, c(f.a.) = 7.20 Å, N = 2	Planar zigzag (2/1)	1.65
Poly(β-propiolactone)[7] [—(CH_2)$_2$COO—]$_n$	Modification (II), a' = 7.76, b' = 4.50, c(f.a.) = 4.76 Å, N = 2	Planar zigzag paracrystalline like (1/0)	1.44
Poly(ε-caprolactone)[8] [—(CH_2)$_5$COO—]$_n$	Orthorhombic, $P2_12_12_1$-D_2^4, a = 7.47, b = 4.98, c(f.a.) = 17.05 Å, N = 2	Slightly twisted planar zigzag (2/1)	1.20
Poly(ethylene succinate)[157] [—O—(CH_2)$_2$—O—CO—(CH_2)$_2$—CO—]$_n$	Orthorhombic, $Pbnb$-D_{2h}^{10}, a = 7.60, b = 10.75, c(f.a.) = 8.33 Å, N = 4	Glide type (2/0) $T_3GT_3\bar{G}$	1.41

Table 3.1 (cont.)

Polymer	*Crystal system, space group, lattice constants and no. of chains per unit cell**	*Molecular conformation*	*Crystal density*/g cm^{-3}
Poly(ethylene oxalate)[157] $[-O-(CH_2)_2-O-CO-CO-]_n$	Orthorhombic, $Pbcn\text{-}D_{2h}^{14}$, $a = 6.44$, $b = 6.22$, c(f.a.) = 11.93 Å, $N = 2$	Glide type (2/0) $T_5GT_5\bar{G}$	1.61
Poly(tetramethylene succinate)[159] $[-O-(CH_2)_4-O-CO-(CH_2)_2-CO-]_n$	Monoclinic, $P2_1/n\text{-}C_{2h}^5$, $a = 5.21$, $b = 9.14$, c(f.a.) = 10.94 Å, $\beta = 124°$, $N = 2$	(1/0) $T_7GT\bar{G}$	1.32
Nylon 3[162] $[-(CH_2)_2-CONH-]_n$	Modification (I), monoclinic, $P2_1\text{-}C_2^2$, $a = 9.33$, b(f.a.) = 4.78, $c = 8.73$ Å, $\beta = 120°$, $N = 4$	Planar zigzag (1/0)	1.39
Nylon 4[163] $[-(CH_2)_3-CONH-]_n$	α-form, monoclinic, $P2_1\text{-}C_2^2$, $a = 9.29$, b(f.a.) = 12.24, $c = 7.97$ Å, $\beta = 114.5°$, $N = 4$	Planar zigzag (2/1)	1.37
Poly(*L*-proline)[13] $[-CH-CO-N-]_n$ with $(CH_2)_3$ ring	Modification (II), trigonal, $P3_2\text{-}C_3^3$, $a = 6.62$, c(f.a.) = 9.31 Å, $N = 1$	Left-handed helix (3/1)	1.37
Poly(*m*-xylene adipamide)[168] $[CH_2-C_6H_4-CH_2NHCO-(CH_2)_4-CONH-]_n$	Triclinic, $P\bar{1}\text{-}C_i^1$, $a = 12.01$, $b = 4.83$, c(f.a.) = 29.8 Å, $\alpha = 75.0°$, $\beta = 26.0°$, $\gamma = 65.0°$ $N = 1$	Slightly distorted planar zigzag (1/0)	1.251
Poly(dimethyl ketene)[170] $[-C(CH_3)_2-C(=O)-]_n$	α-form, orthorhombic, $P2_1cn\text{-}C_{2v}^9$, $a = 12.85$, $b = 6.53$, c(f.a.) = 8.80 Å, $N = 2$	Glide type (4/0) $TG_3T\bar{G}_3$	1.31
Fibrous sulphur[171] $(-S-)_n$	Monoclinic, $P2\text{-}C_2^1$, $a = 17.6$, $b = 9.25$, c(f.a.) = 13.8 Å, $\beta = 113°$, $N = 8$	Helix (10/3)	2.06
Poly(4,4′-isopropylidene diphenylene carbonate)[180]	Monoclinic, $Pc\text{-}C_s^2$, $a = 12.3$, $b = 10.1$, c(f.a.) = 20.8 Å, $\gamma = 84°$, $N = 4$	(2/1)	1.315

$[-C_6H_4-S-C_6H_4-O-CO-O-]_n$			
Poly(4,4′-methylene diphenylene carbonate)[180] $[-C_6H_4-CH_2-C_6H_4-O-CO-O-]_n$	Orthorhombic, $P2_1cn\text{-}C_{2v}^9$, $a = 5.0$, $b = 10.5$, c(f.a.) = 22.0 Å, $N = 2$	(2/1)	1.303
Poly(ethylene oxide)–$HgCl_2$ complex[181, 190]	Type (I), orthorhombic, $Ccmm\text{-}D_{2h}^{17}$, $Ccm2_1\text{-}C_{2v}^{12}$ $Cc2m\text{-}C_{2v}^{16}$ or $C222_1\text{-}D_2^5$, $a = 13.55$, $b = 8.58$, c(f.a.) = 11.75 Å, 16 (CH_2CH_2O) and 4 $(HgCl_2)$	Glide type (2/0) $T_5GT_5\bar{G}$	2.18
	Type (II), orthorhombic, $Pncm\text{-}D_{2h}^7$ or $Pnc2\text{-}C_{2v}^6$, $a = 7.75$, $b = 12.09$, c(f.a.) = 5.88 Å, 4 (CH_2CH_2O) and 4$(HgCl_2)$	Glide type (2/0) $TG_2T\bar{G}_2$	3.79
Poly(2,3-dichlorobutadiene)–thiourea complex[115]	Monoclinic, $P2_1/a\text{-}C_{2h}^5$, $a = 9.91$, $b = 15.85$, c(f.a.) = 12.5 Å, $\beta = 114.1°$, 5.2 monomeric units and 12 thiourea	*trans*-$ST\bar{S}$	1.44
trans-anti-trans-anti-trans-Perhydrotriphenylene–polymer complex[182]	Hexagonal, $P6_3/m\text{-}C_{6h}^2$, $a = 14.34$, $c = 4.78$ Å, 2 PHTP		

*f.a.: fibre axis.

†The structure of *trans*-polyheptenamer is similar to that of *trans*-polypentenamer with $c = 17.10$ Å.

‡The structures of *trans*-polydecenamer and polydodecenamer are similar to those of *trans*-polyoctenamer with different c values.

§$[-(CH_2)_8-O-]_n$ has a structure similar to $[-(CH_2)_6-O-]_n$ with $c = 22.44$ Å.

¶$[-(CH_2)_{12}-O-]_n$ has a structure similar to $[-(CH_2)_6-O-]_n$ with $c = 32.50$ A.

deviating appreciably from exact positions of *T*, *G*, *C*, *S*, $\bar{G}$, etc. are denoted by a prime. The notation (*u*/*t*) gives a conformation in which *u* chemical units turn *t* times in the fibre identity period. The non-helical polymers are indicated with $t = 0$. Table 3.1 summarises the crystallographic data concerning polymers reported since 1966.

3.3.1 Polyethylene

The crystal structure of orthorhombic polyethylene determined by Bunn[46] has not yet been essentially refined, but the structure of a new modification of polyethylene was analysed by Turner–Jones[47] and Seto *et al.*[48].

The normal modes of polyethylene crystal were calculated by Tasumi *et al.*[49, 50] and most of the vibrational assignments have been settled, including some Raman bands at 1440, 1418 and 1370 cm^{-1}. The first two were assigned to the crystalline field splitting of the CH_2 bending[51], and the last to the CH_2 wagging mode[51–55]. As a result, the frequency-phase relation has been investigated[50]. The far-infrared band at 73 cm^{-1} was assigned to the lattice vibration (B_{1u}, parallel to the *a* axis)[56, 57]. Based upon a normal coordinate treatments, the frequency distributions were calculated by Kitagawa and Miyazawa[58–60], and reasonable agreement with the results of neutron inelastic scattering[61, 62] was obtained.

The structure of the folded parts of a polyethylene single crystal has been studied by infrared spectroscopy. Koenig and Witenhafer[63] investigated the chain folding of linear polyethylene by utilising the 1350 and 1304 cm^{-1} infrared bands assigned to *gauche* methylene units in the disordered regions or in the folded parts. According to their results, the samples which were cooled slowly or crystallised from solution contain no appreciable disordered regions, while the rapidly-cooled samples contain anomalous structures which tend to disappear on annealing. Krimm *et al.*[64, 65] compared the infrared spectra of mixed crystals of normal and deuterated polyethylenes with the calculated results of several possible models, and suggested that the folding occurs along the (110) planes in the samples prepared from solution.

Raman scattering from the longitudinal acoustical vibration of polyethylene single crystals was observed by Peticolas *et al.*[66] and the frequencies in the 10–40 cm^{-1} range were found to be inversely proportional to the thickness of the single crystals within the range 90–250 Å. The extended chain length was estimated by assuming that the relation between frequency and chain length is the same as in the case of extended n-paraffin chains[67]. The results gave reasonable agreement with those estimated by long spacings in the x-ray data of the samples subjected to similar annealing[66].

According to Tasumi *et al.*[68, 69], the polymers prepared from (a) *cis*-dideuteroethylene and (b) *trans*-dideuteroethylene by a Ziegler catalyst show different infrared specta. The frequency distributions calculated for (i) the random copolymer with *erythro*-di-isotactic and disyndiotactic configurations and (ii) that with *threo*-di-isotactic and disyndiotactic ones were found to correspond to those spectra, respectively, and it was concluded that the *cis*-opening mechanism was operational.

Sakurada *et al.*[43, 70] measured the elastic moduli of the crystalline region of various polymers by using the meridional x-ray reflections. The force (in units of 10^{-5} dyne) required to stretch a molecule by 1% along the molecular axis is given as follows: (a) *planar zigzag type*; polyethylene: 4.3, poly(vinyl alcohol): 5.4, poly(vinylidene fluoride) modification (I): 3.7, nylon 66: 3.1, nylon 6: 3.0, (b) glide type; poly(vinylidene fluoride) modification (II): 1.5, poly(vinylidene chloride): 1.4, (c) *TG type helices*; isotactic (it)-polypropylene: 1.2, it-polystyrene: 0.81, it-poly(4-methylpent-l-ene): 0.58, (d) *other type helices*; polyoxymethylene: 0.91, polypivalolactone: 0.29, and poly(ethylene oxide): 0.21. The force required to stretch a molecule is sensitive to the molecular conformation.

Further references: Infrared studies of lamellar linking by cilia in polyethylene single-crystal mats[71], specific heat and root-mean-squared displacements[60], influence of defects on the infrared spectra[72] and Fourier analysis of polyethylene at low temperatures[73].

3.3.2 Vinyl polymers

Some preliminary data on the structure of it-poly(3-methylbut-1-ene) was first presented by Natta *et al.*[74] in 1955. Later, different structures, a monoclinic lattice with the double sized *a* and *b* axes[75,76] and a tetragonal lattice[77,78], were reported. The crystal structure was recently examined by

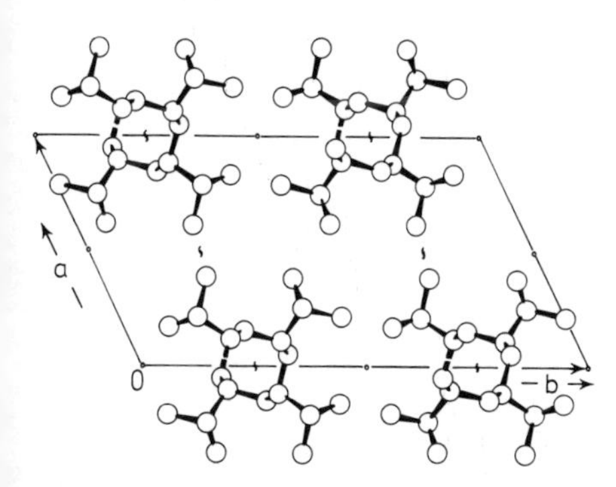

Figure 3.2 Crystal structure of it-poly-(3-methyl-but-1-ene)
(From Corradini *et al.*[79], reproduced by courtesy of Pergamon Press)

Corradini *et al.*[79] in detail and was determined as shown in Figure 3.2. It is interesting that the (4/1) helix molecules are packed in a monoclinic lattice instead of a tetragonal lattice.

The analysis of it-poly(4-methylpent-1-ene) was made by Corradini *et al.*[80] and Kusanagi[81] independently, but both the molecular conformations are essentially similar. The space groups are different; Kusanagi[81] introduced the disorder of anticlined chains ($P\bar{4}b2$), while Corradini *et al.*[80] did not ($P\bar{4}$). As reported by Griffith and Rånby[82], this polymer has the unusual property that the crystalline region is less dense at room temperature than the non-crystalline region ($d_{cr} = 0.812$ g cm^{-3}, $d_{noncr} = 0.838$ g cm^{-3}).

The usual crystal modification was found to transform into another modification, giving a different x-ray fibre diagram, by heat treatment under high pressure (200–270 °C, 4500 atm)[83]. The high-pressure modification has a density of 0.859 g cm^{-3}.

In optically active and racemic it-poly(4-methylhex-1-ene), four (7/2) helical chains, two right-handed and two left-handed, pass through a unit cell. Corradini *et al.*[80] proposed a $P\bar{4}$ unit cell for the racemic polymer, in which the positions of right-handed and left-handed helices are fixed, but a statistical distribution of the configurations (*R*) (rectus) and (*S*) (sinister) is allowed. For the optically active (*S*) polymer the skeletal atoms of the main chains have the $P\bar{4}$ symmetry in the lattice, while the side chains of all helices have only the (*S*) configuration, thus reducing the actual symmetry of the unit cell to *P*1.

The crystal structures of poly(5-methylhex-1-ene) and poly((*S*)-5-methylhept-1-ene) are similar, two anticlined (3/1) helices passing through a monoclinic cell with β = 90 degrees[84]. On the contrary, poly((*S*),(*R*)-5-methylhept-1-ene) is tetragonal, and the molecular conformation is different, a (19/6)

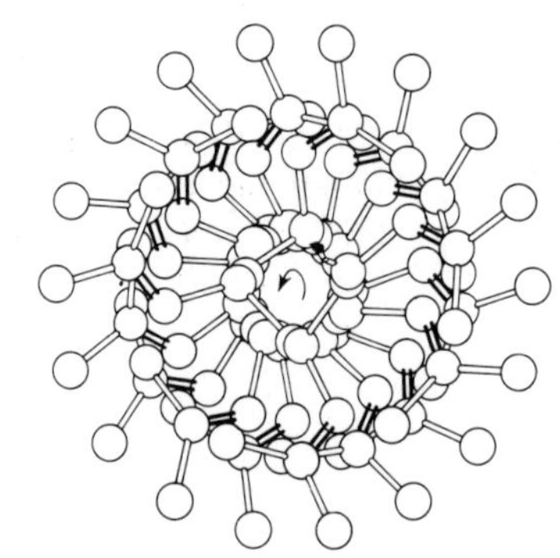

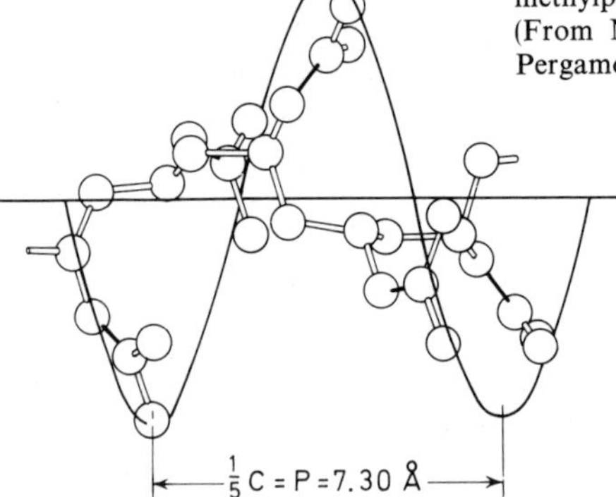

Figure 3.3 Molecular structure of it-1,2-poly(4-methylpenta-1,3-diene)
(From Natta *et al.*[85], reproduced by courtesy of Pergamon Press)

helix. The most probable space group is $P\bar{4}$, in which each up-right- handed helix is surrounded by four down-left-handed helices and vice versa. Owing to the few reflections, it was not possible to refine the structure and much conformational and configurational disorder of side groups may be present.

According to Natta *et al.*[85] the molecular structure of 1,2-it-poly(4-methylpenta-1,3-diene) is a (18/5) helix, as shown in Figure 3.3. The internal rotation angles along the main chain, however, correspond to a succession of nearly *gauche* ($\mp 78°\ 15'$) and nearly *trans* ($\pm 171°30'$). The calculated crystalline density of this polymer (0.85 g cm^{-3}) was found to be lower than the observed density[85], in the case of poly(4-methylpent-1-ene)[82].

Corradini and Bassi[86] determined the crystal structure of it-poly(vinyl methyl ether). The molecular conformation is a (3/1) helix and the internal rotation angles of the main chain are −80 and 160 degrees for the right-handed helix, deviating from $\bar{G}$ and T. Each polymer chain is surrounded by three enantiomorphous, anticlined chains. This mode of packing is similar to that of polyacetaldehyde (with four neighbours)[87], but different from those (enantiomorphous isoclined) of poly-α-butene (modification (I))[88], polypropylene[89], polystyrene[90], poly-*o*-fluorostyrene[91], and also the fourfold helices of poly(α-vinyl naphthalene)[92], and poly(*o*-methylstyrene)[93].

The conformational stability of four isotactic polymers have been analysed by calculating the intramolecular interactions, namely, polypropylene, poly(4-methylpent-1-ene), poly(3-methylbut-1-ene), and polyacetaldehyde[35–37]. These calculations were made without fixing the fibre identity period, using only the assumption that the chain forms a helical structure, i.e. the set of internal rotation angles repeats along the subsequent monomeric units of the chain. The potential energy was calculated for the internal rotation barriers (2 kcal mol^{-1} per bond for C—C and C—O), van der Waals interactions (Lennard–Jones type), and dipole–dipole interactions (point-dipole approximation for polyacetaldehyde only). Polypropylene and polyacetaldehyde are the simplest, having only two internal rotation

Table 3.2 Stable conformations[37]

Polymers	*Calculated*			*Determined by x-ray*		
	τ_1 (degrees)	τ_2	*N*	τ_1 (degrees)	τ_2	*N*
it-Polypropylene	179	−56	2.91	180	−60	3.00[189]
it-Poly(4-methylpent-1-ene)	155	−68	3.47	162	−71	3.50[81]
it-Poly(3-methylbut-1-ene)	158	−78	3.80	149	−80	4.00[79]
it-Polyacetaldehyde	131	−80	3.94	136	−83	4.00[87]

τ_1, τ_2 = internal rotation angles of main chain.

N = number of monomeric unit per turn.

angles of the main chain; τ_1 and τ_2. The other two polymers have additional rotation angles of the side chain; τ_3 and τ_4 for poly(4-methylpent-1-ene) and τ_3 for poly(3-methylbut-1-ene). For poly(3-methylbut-1-ene), the minimum was found in the three-dimensional plot. For poly(4-methylpent-1-ene), the stable conformation of the side chain was calculated with the fixed main chain conformation corresponding to the actual (7/2) helix and gave good agreement with the values determined by x-ray analysis[81]. The map was now plotted against τ_1 and τ_2 fixing the side chain to the actual one; the results are given in Table 3.2. For isotactic polypropylene, the result is essentially the same as those of Natta *et al.*[33] and Liquori *et al.*[34]. Good agreement was obtained between the calculated and x-ray structures for all four polymers, in spite of the assumption of only intramolecular interactions, suggesting that the helical structure of these polymers may arise from the intramolecular interactions, especially from the steric hindrance of the side chains.

Noether[78] determined unit-cell dimensions, space groups and helix types of various isotactic poly-α-olefins: Tetragonal, poly(vinylcyclohexane) (II): (24/7), poly(vinylcyclopentane) (I): (4/1), (II): (12/3), (III): (10/3), poly(vinylcycloheptane): (4/1), poly(vinylcyclobutane): (4/1), poly(allylcyclopentane): (24/7), poly(allylcyclohexane) (I): (10/3), poly(*m*-methylstyrene): (40/11), poly(4,4-dimethylpentene): (7/2), Hexagonal; poly(allylcyclohexane) (II): (3/1). It was deduced from these data and those of the literature that only (3/1) helices appear in the hexagonal (strictly speaking, trigonal) system. The tetragonal system seems to be able to accommodate any type of helix above (3/1), such as (4/1), (7/2), (10/3), (11/3), etc. between 3 and 4 monomers per turn. It was found in the case of tetragonal system that the body-centred space group $I4$ or $I\bar{4}$ requires an even number of monomeric units per fibre period [the molecular axis should coincide with the twofold screw axis; (4/1), (10/3), (24/7), (40/11), (18/5), etc.], while the primitive space group $P4$ or $P\bar{4}$ does not [(7/2), (11/3), (19/6), etc.].

Natta *et al.*[85] developed the close-packing consideration of the helical molecules with fractional values of (u/t) previously introduced by Frank *et al.*[94] and Noether[78]. When D_1 and D_2 are the helix-radii of the main chain and the side group, respectively, and the ratio D_1/D_2 lies between 0.3 and 0.8, the most efficient way of packing is that each right-handed helix is surrounded by four left-handed helices and vice versa. This rule was applied to the analyses of it-poly((*R*),(*S*)-5-methylhept-1-ene)[84] and it-1,2-poly(4-methylpenta-1,3-diene)[85].

Further references: Infrared studies of the conformation of isotactic vinyl polymers in solution[95,96], normal-coordinate treatment of syndiotactic 1,2-polybutadiene[97] and the crystal structure of syndiotactic polypropylene[98].

3.3.3 Vinylidene polymers

The structure of all vinylidene-type polymers, such as polyisobutylene, poly(vinylidene chloride), poly(vinylidene fluoride) and poly(methyl methacrylate), have long remained unresolved in spite of their basic and practical importance.

So far, the (8/5)[99–101] and (8/3) helices[102] have been proposed for the polyisobutylene molecule. A conformational analysis has been made for polyisobutylene by Allegra *et al.*[103] by taking into account the bond angle flexibility of the main chain. The most stable conformation was found as an (8/3) helix with $\angle CH_2—C(CH_3)_2—CH_2 = 110$ degrees, $\angle C—CH_2—C = 124$ degrees and dihedral angles of 156.7 and 47.5 degrees as shown in Figure 3.4. This model was further supported by comparing the molecular transform with the observed x-ray data.

Poly(vinylidene fluoride) gives three crystal modifications. Modification (I) is formed under special conditions, such as heat treatment under high pressure (4000–4500 atm at *c.* 290 °C) or under tension by drawing, etc.,

(II) is the most stable at atmospheric pressure and (III) may be an intermediate form between (I) and (II) with respect to the pressure treatment[83]. The crystal structures of these three modifications were determined (Figure 3.5)[105, 106]. The molecular structure of modification (II) is essentially $TGT\bar{G}$ and (I) and (III) may involve slightly distorted planar zigzag structures. In

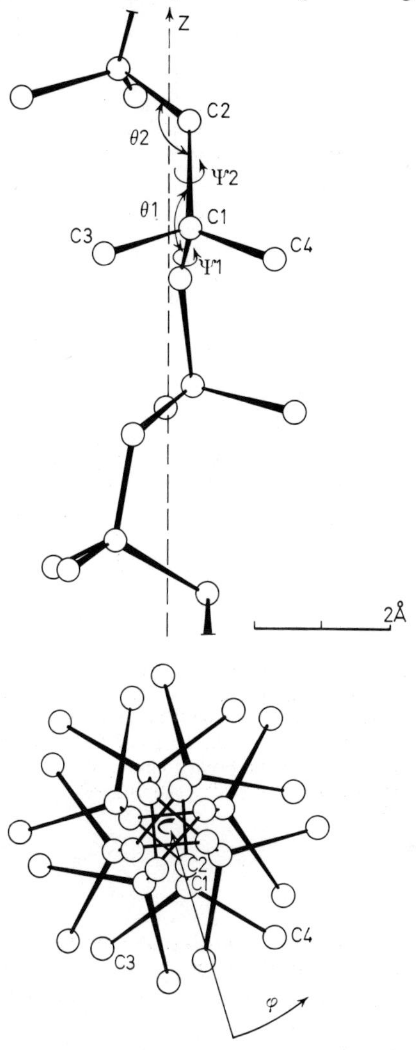

Figure 3.4 Molecular model of polyisobutylene
(From Allegra *et al.*[103], reproduced by courtesy of The American Chemical Society)

the latter two cases, statistically-disordered crystal structures were suggested[105].

The potential energies for intra- and inter-molecular interactions in the crystal were calculated for modifications (I) and (II)[105]. The result is given in

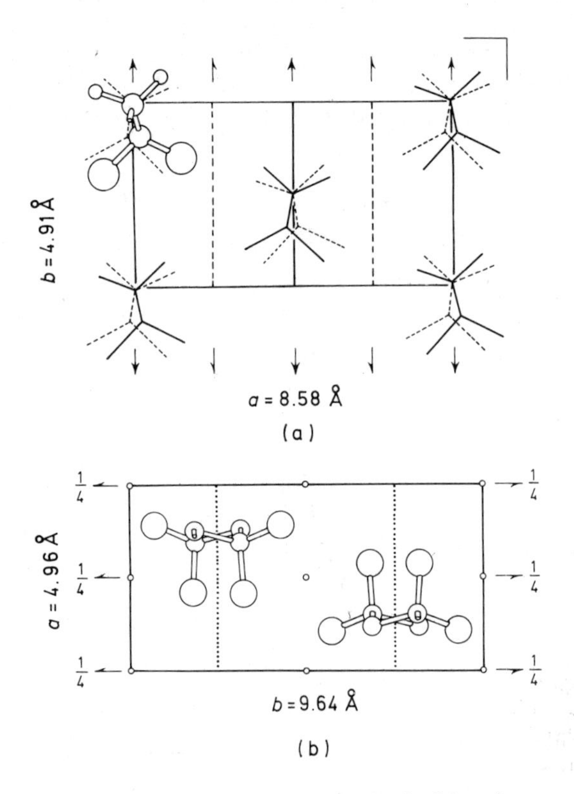

Figure 3.5 Crystal structure of poly(vinylidene fluoride); (a) modification (I) (Lando *et al.*[106]), and (b) modification (II)[105]. In modification (I), statistical deflections (7 degrees from the *bc* plane) are considered, as shown with solid and broken lines
(From Tadokoro *et al.*[105], reproduced by courtesy of IUPAC)

Table 3.3 Calculated potential energies of modifications (I) and (II) of poly-(vinylidene fluoride) (kcal mol^{-1} per monomeric unit)[105]

(From Tudokoro *et al.*[105], reproduced by courtesy of IUPAC)

Modification	(I) (*TT*)	(II) ($TGT\bar{G}$)
Intramolecular interaction	−0.48	−1.46
(van der Waals)	(−1.19)	(−1.57)
(Electrostatic)	(+0.71)	(+0.11)
Intermolecular interaction	−5.25	−4.57
(van der Waals)	(−5.06)	(−4.44)
(Electrostatic)	(−0.19)	(−0.13)
Total	−5.73	−6.03

Table 3.3. The intramolecular interaction energy of the planar zigzag type is -0.48 kcal mol^{-1} per monomeric unit, much higher than the value of the $TGT\bar{G}$ type, -1.46 kcal mol^{-1} per monomeric unit. The higher energy of planar zigzag type may be due to the steric hindrance between fluorine atoms, and also due to the parallel array of the CF_2 dipole moments. On the contrary, the F···F distance in the $TGT\bar{G}$ type is 2.70 Å, just twice the van der Waals radius of fluorine. In spite of the large difference between the intramolecular energies of the two molecular conformations, the stabilities of the crystal lattices of these two modifications are not very different, being respectively -5.73 and -6.03 kcal mol^{-1} per monomeric unit. This may be due to the more favourable intermolecular interactions in modification (I)

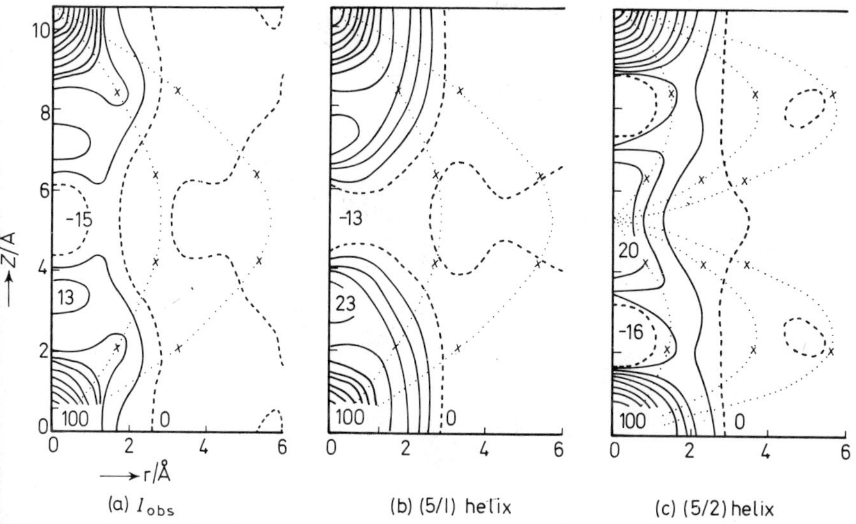

Figure 3.6 Cylindrical Patterson maps of it-poly-methylmethacrylate) synthesised from (a) the observed intensity, (b) and (c) the molecular structure factors calculated for (5/1) helix and (5/2) helix, respectively: ×, theoretical peaks for intrahelical $CH_3 \cdots CH_3$ (in the ester group) and C···C (in the skeletal chain) vectors
(From Tadokoro *et al.*[38] reproduced by courtesy of The American Chemical Society)

which are appreciably lower than those of (II). These values may cover the high energy of intramolecular interactions of the planar zigzag form, if certain conditions are satisfied.

The molecular structure of it-poly(methylmethacrylate) was analysed by use of x-ray diffraction and far-infrared spectroscopic methods[38]. Because of the broadness of reflections on the fibre photograph, application of the usual methods of x-ray analysis was difficult. Two helical conformations of the main chain, (5/1) and (5/2), were first assumed, and eight models were proposed from conformational analyses of the side chain. The comparison between the cylindrical Patterson map synthesised for the observed x-ray data and those for the (5/1) and (5/2) models excluded the (5/2) model (Figure 3.6). By comparing the observed reflection intensities with the molecular transforms for four (5/1) models, there remained two models each with the α-methyl groups pointing outward, but the selection of the final model was still difficult from the x-ray data alone. Then, model (I) was found to be

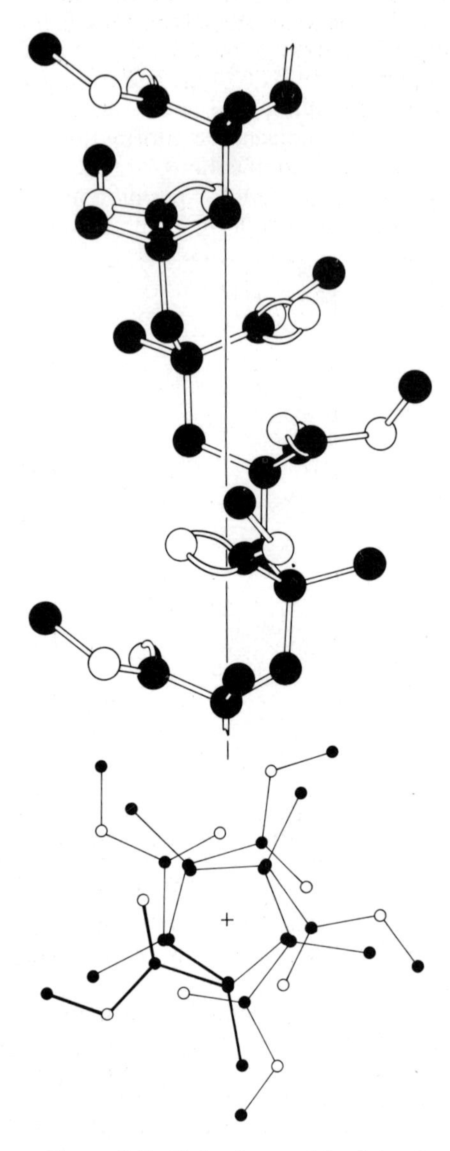

Figure 3.7 Molecular model of it-poly (methylmethacrylate)
(From Tadokoro *et al.*[38], reproduced by courtesy of The American Chemical Society)

the most reasonable from the results of normal-coordinate calculations for these two models which were compared with far-infrared spectra; the molecular model thus obtained is shown in Figure 3.7.

3.3.4 Polymers containing C=C double bonds

trans-1,4-Polybutadiene has two crystal modifications, a low-temperature form and a high-temperature form; the transition temperature is *c.* 75 °C. The low-temperature form was analysed by Iwayanagi *et al.*[109], the molecular form was of the *trans*-$ST\bar{S}$ type. For the high-temperature form, several papers[109–111] suggesting some kind of disordered structure have been published, but the definite structure has not yet been established.

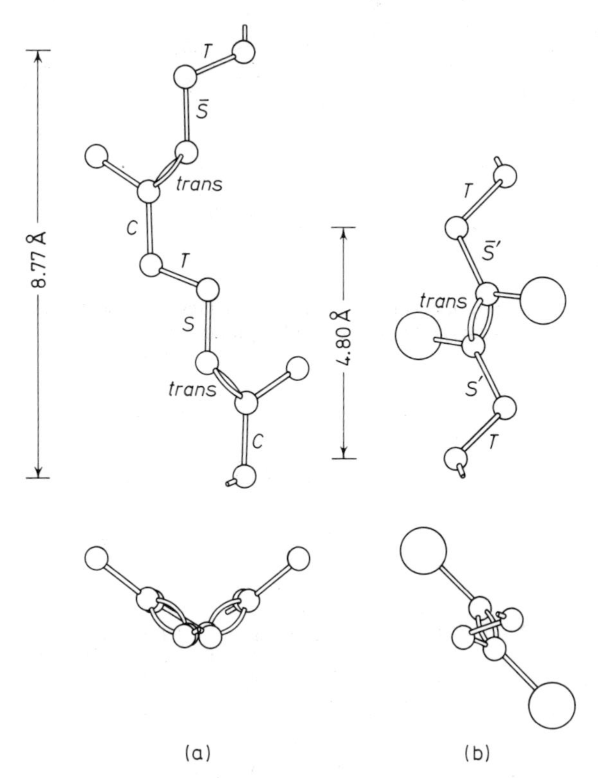

Figure 3.8 Molecular structures of (a) *trans*-1,4-polyisoprene (α-form)[14], and (b) *trans*-1,4-poly(2,3-dichlorobutadiene)
(From Tadokoro *et al.*[115], reproduced by courtesy of The American Chemical Society)

The crystal structure of the β-form of *trans*-1,4-polyisoprene (gutta-percha) was determined by Bunn[112] as early as 1942 giving a *trans*-$ST\bar{S}$ conformation, while the α-form was analysed recently[14] and the molecular conformation is a glide type (*trans*-CTS) (*trans*-$CT\bar{S}$) as shown in Figure 3.8(a).

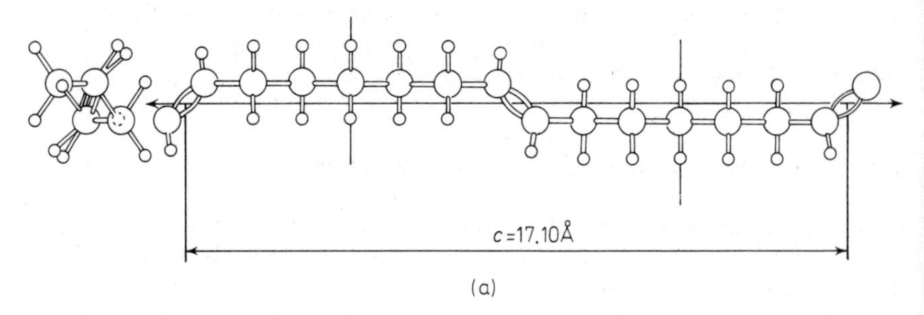

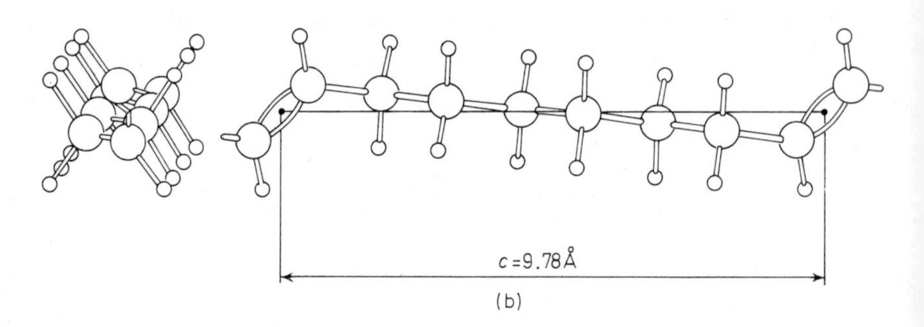

Figure 3.9 Molecular conformations of (a) *trans*-polyheptenamer and (b) *trans*-polyoctenamer (From Natta *et al.*[113], reproduced by courtesy of Pergamon Press)

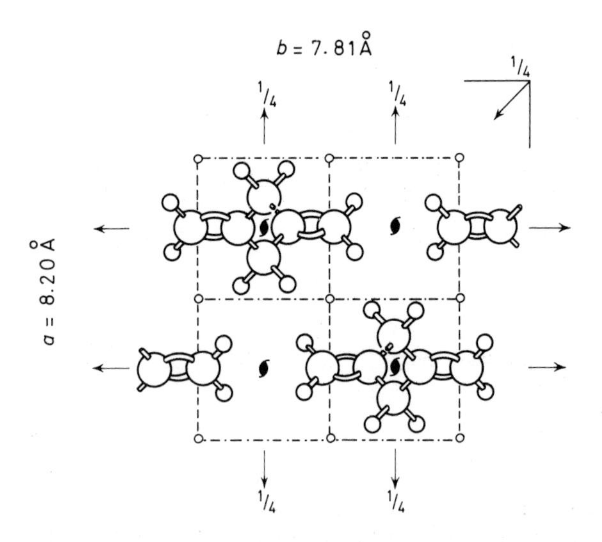

Figure 3.10 Crystal structure of polyallene (modification (I)) From Tadokoro *et al.*[117], reproduced by courtesy of Interscience)

Polyalkenamers [—CH=CH—$(CH_2)_m$—$]_n$ prepared from cyclo-olefins were analysed by Natta *et al.*[113,114]. Odd-numbered members, i.e. *trans*-polypentenamer and *trans*-polyheptenamer, have the same space group, *Pnam*, as orthorhombic polyethylene, and the molecular packing is also nearly identical; the two planes containing the methylene groups make an angle of 41 degrees with the *bc* plane. The molecular conformation is shown in Figure 3.9(a). The torsional angles around the single bonds adjacent to the double bonds are *c.* ± 120 degrees. Even-numbered members, i.e. *trans*-polyoctenamer, -polydecenamer, and -polydodecenamer, crystallise in monoclinic and also triclinic forms depending upon the conditions of preparation. In the monoclinic form, the molecular plane containing methylene groups inclines several degrees toward the *c*-axis as shown in Figure 3.9(b). The mode of molecular packing of monoclinic form in the lattice is also similar to that of the orthorhombic polyethylene. The molecular conformations of odd *trans*-polyalkenamers are essentially the same, *trans*-$ST_m\bar{S}$, in both monoclinic and triclinic forms, while the unit cells of triclinic forms are similar to the triclinic cell of polyethylene proposed by Turner-Jones[47].

Poly(2,3-dichlorobutadiene) obtained by radiation polymerisation of the 2,3-dichlorobutadiene–thiourea complex was found to have a high content of *trans*-1,4 enchainment, compared with the polymers by radical polymerisation[115]. Determination of the crystal structure gives the molecular conformation as sequences of *trans*-$S'T\bar{S}'$ as shown in Figure 3.8(b).

Polyallene polymerised by noble-metal catalysts forms three crystal modifications as shown in Table 3.1[116,117]. In modifications (I) and (II), the molecular structure is essentially the same, the G_4 type (2/1) helix shown in Figure 3.10. Modification (III) shows a characteristic x-ray pattern, suggesting a kind of paracrystalline disordered state. All the layer lines and also the equator are streak-like, exhibiting no distinct spots. The fibre period is 3.88 Å and is the same as those of the other modifications. Normal coordinate treatment of polyallene and its deutero derivatives was also reported[117].

Further references: Normal-coordinate treatment of 1,4-*trans*-polybutadiene[118], Raman spectra of polybutadiene[119] and stable conformation of 1,4-polybutadiene and the model compounds[15].

3.3.5 Polyethers

The structure of polyethers was reviewed by the present author in 1967[120] and so only further developments will be described here, to avoid overlap.

3.3.5.1 [—$(CH_2)_m$—O—$]_n$*-type polyethers*

This series of polyethers and also polythioethers show remarkable variations of physical properties as *m* increases. For example, Figure 3.11 shows how the melting points of polyethers and polythioethers depend upon *m*. These remarkable differences in properties cannot be explained solely in terms of different chemical structures, i.e. the one-to-one increase of methylene groups between oxygen (or sulphur) atoms. The molecular conformation and the crystal structure must be considered to be equally important.

Polyoxymethylene has two crystal modifications, a trigonal stable form[133] and an orthorhombic metastable form[134] with (9/5) and (2/1) molecular conformations, respectively. The molecular structure of the trigonal form was analysed in detail by use of three-dimensional Fourier syntheses with the data obtained from a sample of a single crystal of tetraoxane[3] prepared by solid-state radiation polymerisation. In Figure 3.12(a) a fibre diagram of an orientated sample of Delrin–acetal resin (du Pont) and (b) a rotation photograph

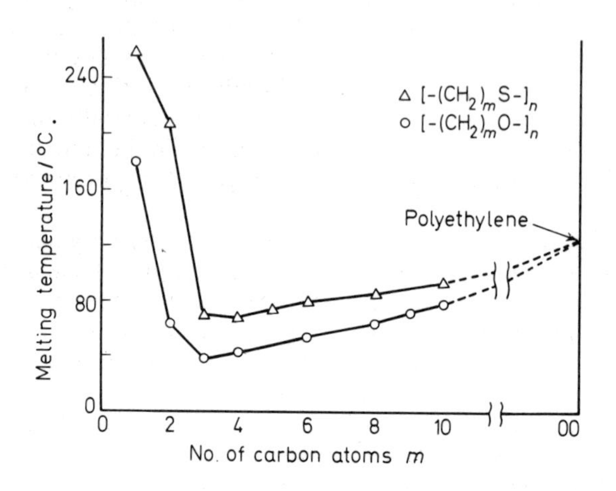

Figure 3.11 Melting temperature of polyethers and polythioethers (modified from the original by Lal and Trick[192])

around the *c* axis of a polyoxymethylene sample from a tetraoxane single crystal are shown. Figure 3.13 shows the electron-density distribution along the line intersecting the helix axis at a right angle and passing through (a) the carbon atom or (b) the oxygen atom. Molecular parameters thus obtained are: fibre identity period $c = 17.39$ Å, helix radius of the carbon atom $r(C) = 0.691$ Å, $r(O) = 0.671$ Å, bond distance C—O = 1.42_1 Å, $\angle$OCO = 110°49′, $\angle$COC = 112°24′ and dihedral angle of main chain = 78°13′.

The polymers prepared from *cis*- and *trans*-dideuteroethylene oxides by cationic, anionic and coordination catalysts showed different infrared spectra depending only on the configuration of the monomer[135]. If the inversion mechanism is the case, the *cis*-monomers should give polymers with *threo* monomeric units, and the *trans*-monomers should give *erythro* units. On the contrary, if the inversion does not occur, the relation should be opposed. However, the relation between the neighbouring monomeric units in polymers, i.e. isotactic or syndiotactic, may be random. The difference between the solution spectra of two polymers suggested that the spectrum in the 1500–1110 cm^{-1} region may reflect the characteristic features associated with the vibrational modes localised within a monomeric unit. Thus, the normal modes were calculated for the following five ideal models with

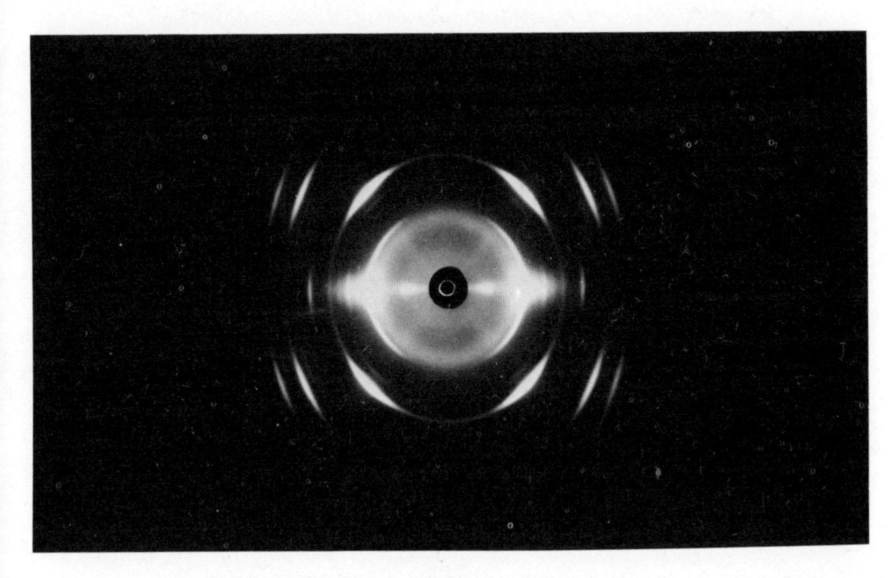

(a)

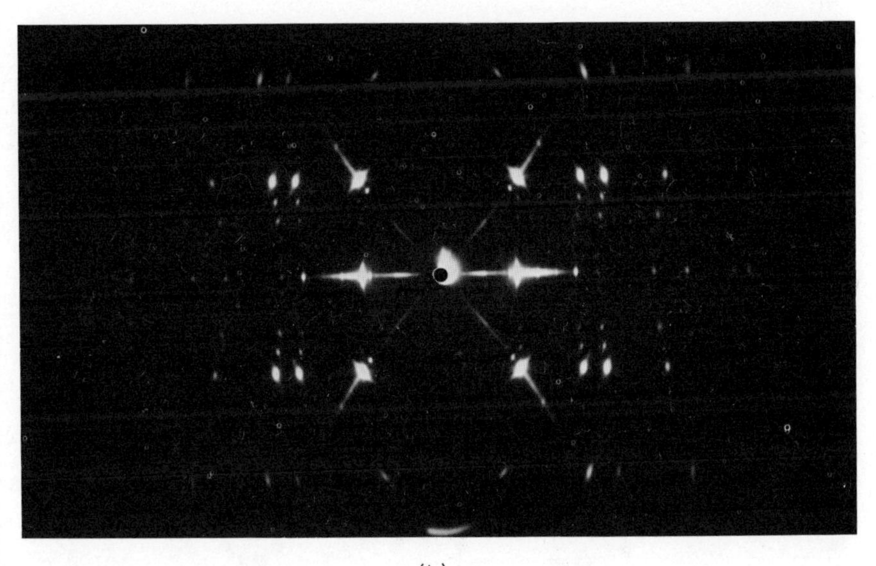

(b)

Figure 3.12 (a) Fibre diagram of an orientated sample of Delrin–acetal resin and (b) rotation photograph around the *c*-axis of a polyoxymethylene sample from a tetraoxane single crystal[3] (From Uchida and Tadokoro[3], reproduced by courtesy of Interscience)

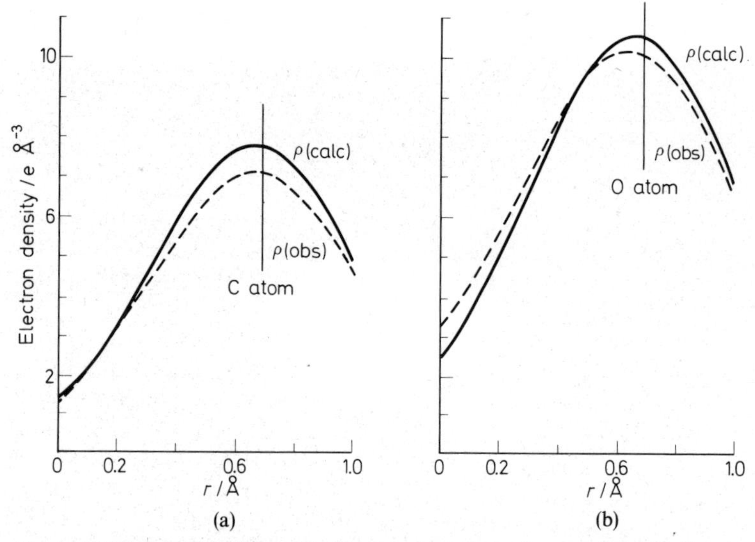

Figure 3.13 Electron-density distribution along the line intersecting the helix axis at a right angle and passing through: (a) the carbon atom or (b) the oxygen atom (From Uchida and Tadokoro[3], reproduced by courtesy of Interscience)

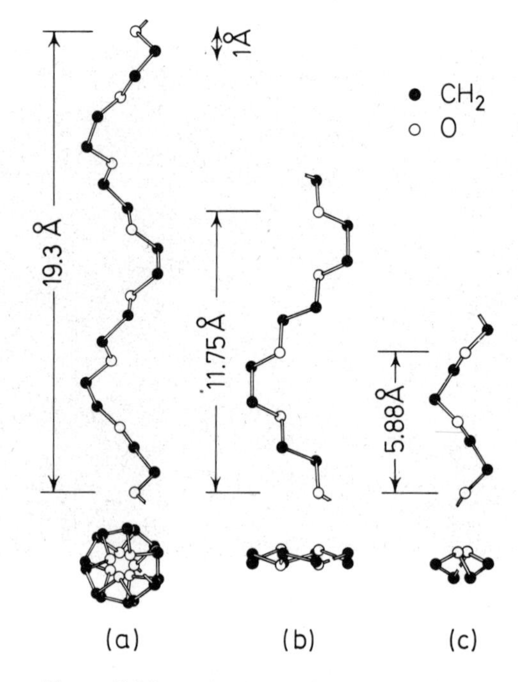

Figure 3.14 Molecular conformations of poly(ethylene oxide): (a) ordinary poly(ethylene oxide), $(T'G'_2)_7$; (b) $HgCl_2$ complex type (I), $T_5GT_5\bar{G}$; and (c) type (II), $T'G'_2\,\bar{T}'\bar{G}'_2$ [149, 181, 190]

the (7/2) helical conformation (Figure 3.14a); *threo*–di-isotactic (two optical antipodes), *threo*–disyndiotactic, *erythro*–di-isotactic, and *erythro*–disyndiotactic models. The two optical antipodes of *threo*–di-isotactic models give different conformations with respect to deuterium if the helix of the same sense is considered, namely, right-handed or left-handed. The comparison of the infrared spectra and the results of calculation indicated opening with inversion[136]. The vibrational localisation within a monomeric unit in the 1500–1100 cm^{-1} region was further confirmed by the potential-energy distribution and the flatness of the dispersion curves in this region.

Polyoxacyclobutane gives three crystal modifications. Modification (I) is stable only in the presence of water, (II) is obtained only as orientated samples and (III) is thermally the most stable form. The crystal structures of these three modifications were determined; the molecular conformations were as shown in Figure 3.15; (I) planar zigzag, (II) $T_3GT_3\bar{G}$ and (III) $(T_2G_2)_2$ [121, 122, 137].

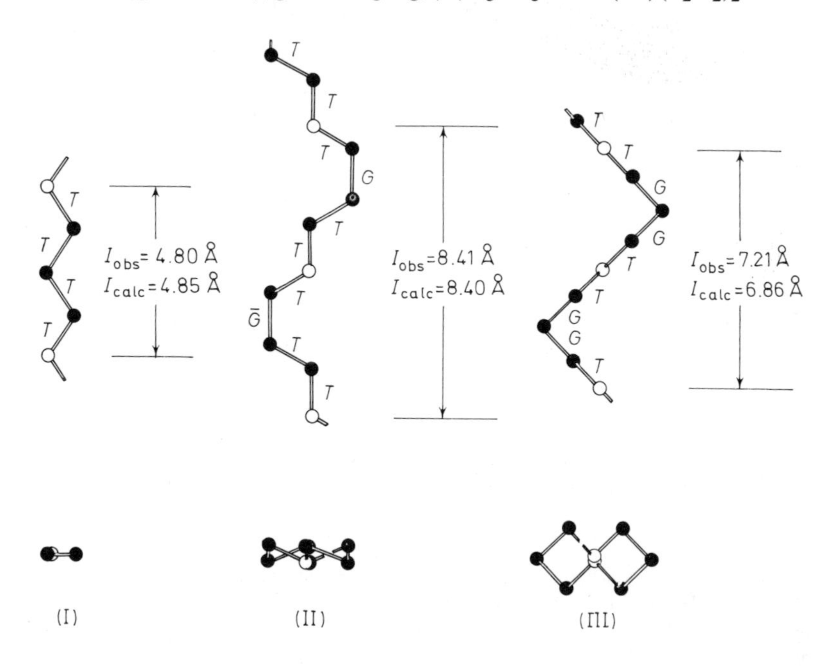

Figure 3.15 Molecular models of three modifications of polyoxacyclobutane (From Tadokoro *et al.*[137], reproduced by courtesy of Hüthig and Wepf Verlag)

Modification (I) was found to be a hydrate; the ratio $(CH_2)_3O:H_2O$ was stoichiometric, 1:1. In the crystal, two polymer molecules are connected by water molecules forming hydrogen bonds as shown schematically in Figure 3.16; the positions of the hydrogen atoms are based upon infrared studies[121, 122].

The members with $m = 4$, 6, 8–10, and 12 have been analysed, and all found to be planar zigzag[123, 124, 193] as in polyethylene type and polytetrahydrofuran type. Two types of molecular packing are found (Figure 3.17(a) and (b); In the polyethylene type (orthorhombic) (Figure 3.17(a)) the molecular plane makes an angle of 41 degrees with the *bc* plane. This type tends to appear in the cases of higher *m*. In the polytetrahydrofuran type (monoclinic) (Figure 3.17(b)) the molecular plane is parallel to the *bc* plane. This type appears in the lower *m* members.

Further references: Solid-state radiation polymerisation of ring oligomers,

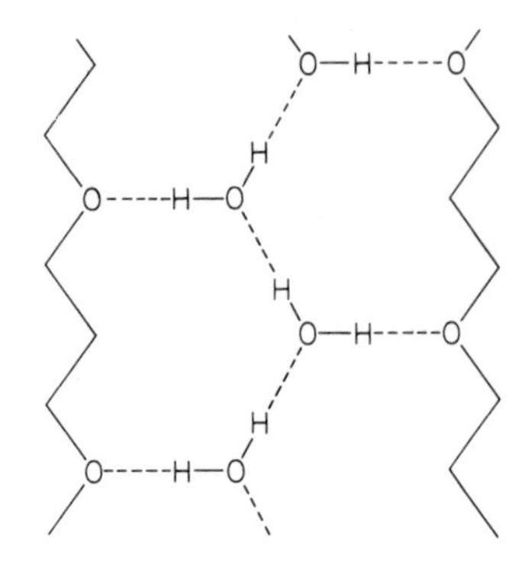

Figure 3.16 Ribbon-like chain in modification (I) of polyoxacyclobutane (From Tadokoro *et al.*[137], reproduced by courtesy of The American Chemical Society)

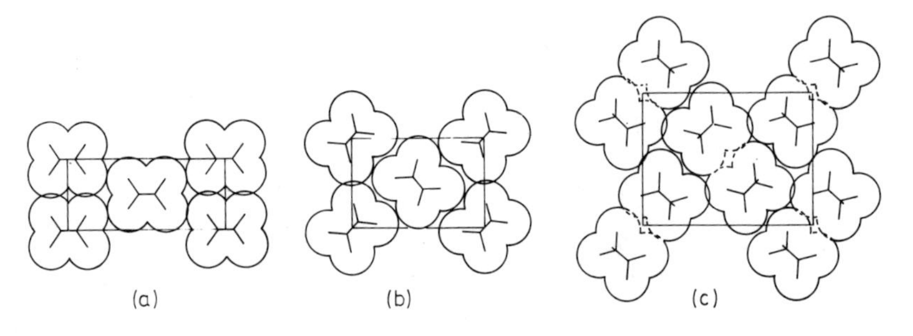

Figure 3.17 Crystal structures of (a) polyethylene[46] (b) polytetrahydrofuran[123] and poly(pentamethylene sulphide)[17]

trioxane and tetraoxane[125], pentoxane[126] and hexoxane[127], infrared studies of chain folding of poly(ethylene oxide)[128] and skeletal vibrations of planar zigzag polyethers[129].

3.3.5.2 *Polyformals* $[—OCH_2O—(CH_2)_m—]_n$

Poly(1,3-dioxolane) ($m = 2$) has three crystal modifications, among which modification (II) was recently analysed[189]; the molecular conformation is of the glide type (Figure 3.18(a)). The order of bonds is as follows:

$$—O—CH_2—O—CH_2—CH_2—O—)_n$$
$$G' \quad G' \quad T \quad \bar{G}' \quad \bar{S}' \qquad (\bar{G}'\bar{G}'TG'S')$$

The crystal structure of poly(1,3-dioxepane) ($m = 4$) also has been determined, giving a rather complicated glide type conformation, as shown in Figure 3.18(b)[16].

$$\text{—O—CH}_2\text{—O—CH}_2\text{—CH}_2\text{—CH}_2\text{—CH}_2\text{—O—)}_n\text{—}$$
$$G' \quad G' \quad T' \quad \bar{G}' \quad T' \quad G' \quad T' \qquad\qquad (\bar{G}'\bar{G}'T'G'T'\bar{G}'T')$$

In cationic ring-opening polymerisation of cyclic formals, the question arises whether the bond scission occurs exclusively at the same type of bond,

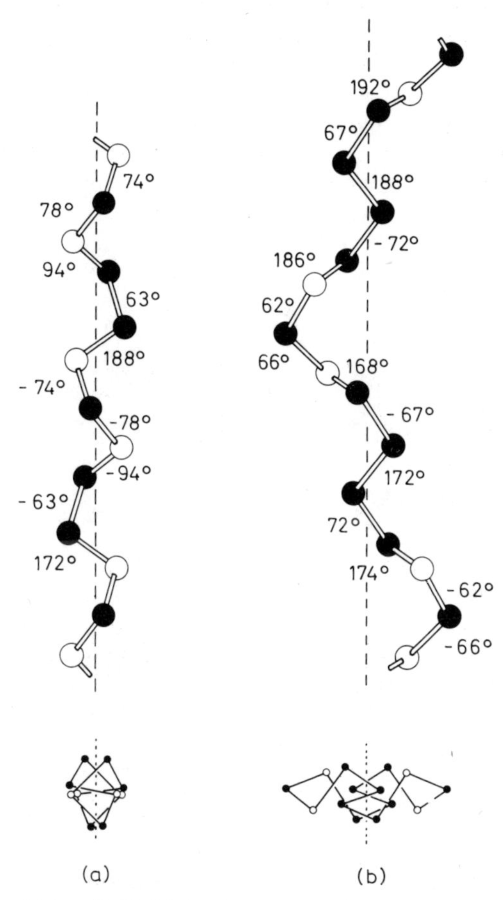

Figure 3.18 Molecular conformations of (a) poly-(1,3-dioxolane) and (b) poly(1,3-dioxepane)[16, 189]

for example, bond (1), or at both bonds (1) and (2) at random. If scission occurs at random, the resultant polymer should not have a well-defined crystal

$$\begin{array}{l} R^{+}\cdots\cdots O\overset{(1)}{\rule{2em}{0.4pt}}CH_2 \\ \quad\;\;(2)\,| \qquad\quad\;\; | \\ \qquad (CH_2)_m\text{—}O \end{array}$$

structure. Yamashita *et al.*[130] found, using a chemical method, that the ring opening occurs only at the bond (1). The above results of x-ray analyses shows the regular head-to-tail sequences and this fact represents additional confirmation that the ring opening occurs exclusively at the same type of bond, although it is not clear whether the scission occurs at bond (1) or (2).

3.3.5.3 Substituted polyethers

Corradini and Avitabile[131] calculated the lattice energy of it-polyacetaldehyde for various possible tetragonal space groups with (4/1) helices, by changing the angle of rotation of the chain about its axis and the level difference between two methyl groups of neighbouring chains. The result gave the lowest energy the space group actually found for $I4_1/a$.

The crystal structure of racemic poly(propylene oxide) was determined by Cesari *et al.*[132] and the molecule is a slightly distorted planar zigzag. Strictly speaking, there remains a problem whether the unit cell is optically

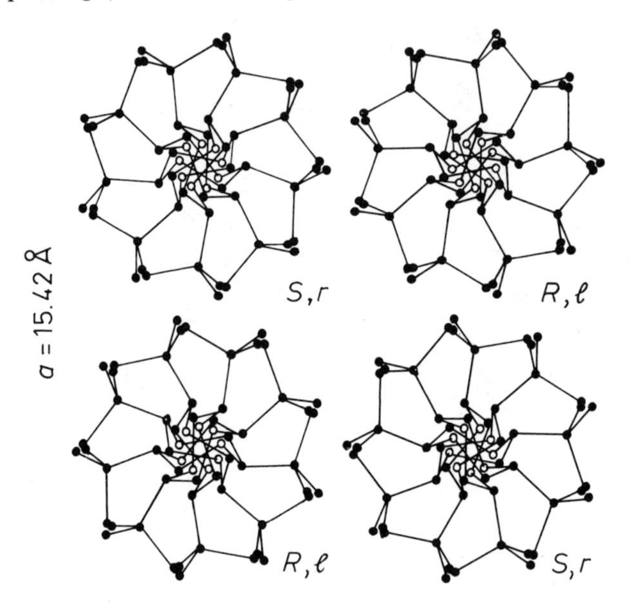

Figure 3.19 Crystal structure of poly(t-butyl ethylene oxide); *r*, right-handed; *l*, left-handed; *R*, rectus; *S*, sinister
(From Sakakihara *et al.*[39], reproduced by courtesy of the Society of Polymer Science, Japan)

active or racemic, i.e. whether the space group is $P2_12_12_1$ or *Pna*, since it is difficult to distinguish the possible difference due to the interchange of the oxygen and carbon atoms in the cell, because of the similarity of the x-ray scattering powers of oxygen atom and methylene group[138]. This problem will be discussed further in the section on poly(propylene sulphide).

The crystal structure (one of three crystal modifications) of poly(t-butyl ethylene oxide) was recently determined as shown in Figure 3.19[39]. The analysis was made on the racemic sample (polymerised with $ZnEt_2$–H_2O

catalyst) by conformational analysis and application of the lattice structure factor consisting of molecular transforms. The molecular structure is a (9/4) helix with the internal rotation angles;

$$—CH(Bu^t)—[—\overset{\tau_1}{O}\overset{\tau_2}{—}CH_2\overset{\tau_3}{—}\underset{\tau_4|\ Bu^t}{CH}—]_n$$

where $\tau_1 = -97$, $\tau_2 = 180$, $\tau_3 = 73$ and $\tau_4 = 64$ degrees for (*S*) polymer. According to the requirements of the space group $P\bar{4}n2$, the molecules at the upper-left and lower-right positions are right-handed (*S*) helices, and

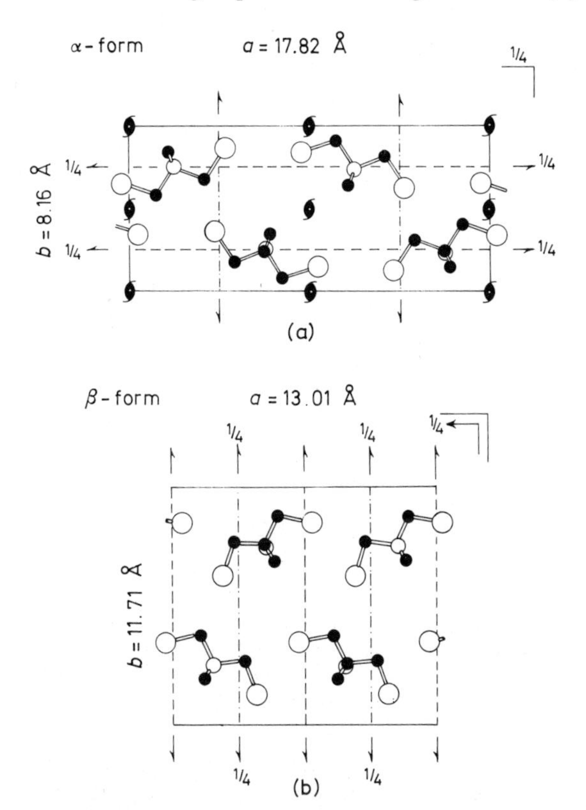

Figure 3.20 Crystal structures of (a) α-form and (b) β-form of poly 3,3(bischloromethyl) oxacyclobutane (From Takahashi *et al.*[139], reproduced by courtesy of the Society of Polymer Science, Japan)

those at the upper-right and lower-left positions are left-handed (*R*) helices. Thus two pairs of optical antipodes pass through definite positions in the unit cell, and as a result the unit cell is racemic. Although for simplicity Figure 3.19 shows only the upward helices, the actual structure is considered to be statistically disordered, consisting of a random mixture of upward and downward helices in 1:1 ratio. The result of the analysis clearly shows the formation of two types of isotactic polymers from the mixture of (*R*) and (*S*) monomers in 1:1 ratio, suggesting a stereoselective mechanism of polymerisation.

According to the x-ray analyses of poly[3,3-bis(chloromethyl) oxacyclobutane], the difference of the two crystal modifications is due to the difference between the modes of packing of the molecules having the same conformation, in which the main chain is a planar zigzag with the side chains of *TT* type (Figure 3.20)[139] (See also reference 191).

Barlow[140] analysed the structure of poly(but-2-ene oxide) prepared from the *trans* monomer and found it corresponded to an *erythro*–di-isotactic

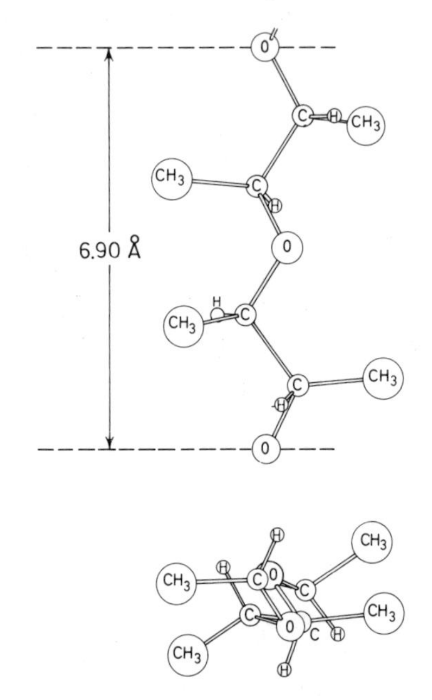

Figure 3.21 Molecular conformation of *erythro*–di-isotactic poly(*trans*-but-2-ene oxide)
(From Barlow[140], reproduced by courtesy of J. Wiley & Sons, Inc.)

polymer. This fact confirms the inversion-opening mechanism proposed by Vandenberg[141–143]. The molecular structure is a slightly distorted planar zigzag as shown in Figure 3.21.

Poly(3,3,3-trifluoro-1,2-epoxypropane) shows larger distortion than poly(propylene oxide) probably due to the bulky side chains, the fibre periods of these polymers being 6.26 and 6.92 Å, respectively[144].

3.3.5.4 *Aromatic polyethers*

The crystal structures of poly(*p*-phenylene oxide)[145] and poly(2,6-diphenyl-*p*-phenylene oxide[146]) were determined by Boon and Magré, giving (2/1) and

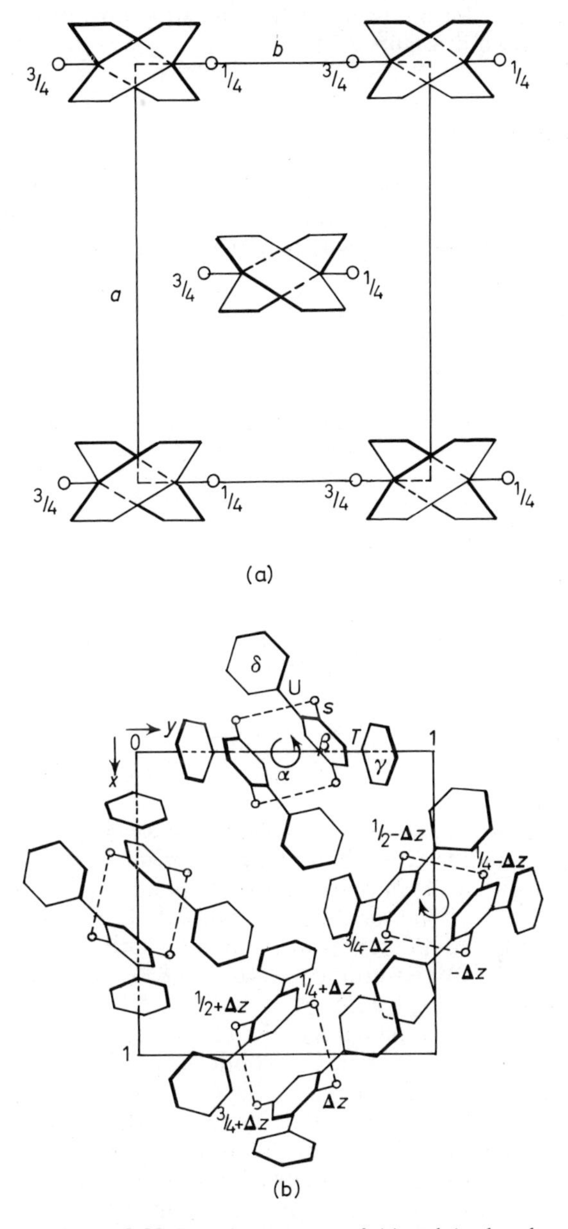

Figure 3.22 Crystal structures of (a) poly(*p*-phenylene oxide) and (b) poly(2,6-diphenyl-*p*-phenylene oxide) (From Boon and Magré[145,146], reproduced by courtesy of Hüthig and Wepf Verlag)

(4/1) helix conformations, respectively. The oxygen bond angles 124 and 127 degrees are larger than the usual value. The structures of these two polymers are reproduced in Figure 3.22.

3.3.6 Polythioethers

The melting points of polythioethers of $[-(CH_2)_m-S-]_n$ type have been compared already with those of polyethers in Figure 3.11. The molecular conformation of polythiomethylene is similar to that of polyoxymethylene,

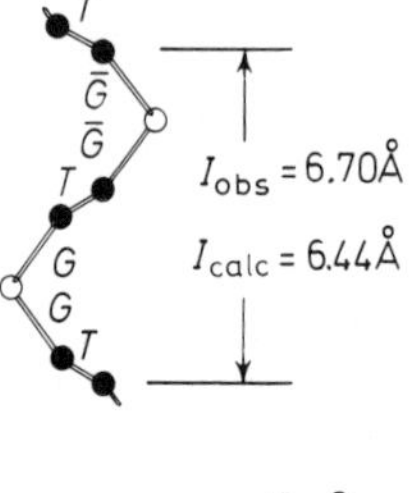

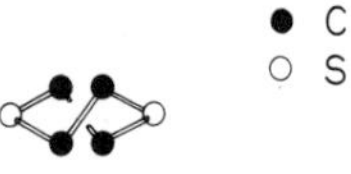

Figure 3.23 Molecular conformation of poly(ethylene sulphide)
(From Takahashi *et al.*[148], reproduced by courtesy of Marcel Dekker)

having a succession of nearly *gauche* forms[147]. On the other hand, the conformation of poly(ethylene sulphide)[148] is shown in Figure 3.23; the order of bonds is:

$$(-CH_2-CH_2-S-)_n$$
$$T \quad G \; G \qquad T\bar{G}\bar{G}$$

and is just opposite to that of poly(ethylene oxide), $T_2GT_2G\cdots$[149]. In poly(ethylene oxide), the C—C bond takes the *gauche* form and the C—O bond the *trans* form, while in poly(ethylene sulphide), the C—C bond takes the *trans* and the C—S bond the *gauche* conformation. Such a difference may be primarily due to the differences in the bond lengths (C—O, 1.43; C—S, 1.815 Å), the van der Waals radii (O, 1.41; S, 1.85 Å) and the dipole interactions (C—O, 0.74; C—S, 0.9 D).

Poly(pentamethylene sulphide) has the planar zigzag conformation, but the molecular packing is different from both the polyethylene type and the polytetrahydrofuran type as shown in Figure 3.17(c)[17]. This planar zigzag molecule is considered to be unstable because of the parallel array of the CSC dipole moments. Thus, a pair of chains become close together with the CSC dipoles antiparallel and such pairs of molecules, as a whole, pack in the lattice just as in the case of orthorhombic polyethylene.

Poly(propylene sulphide) has a true asymmetric carbon atom in each monomeric unit and so the isotactic polymer should be optically active. There are two kinds of optical isomers, (*R*) (rectus) and (*S*) (sinister) as shown in Figure 3.24. Actually, optically active and racemic polymers were obtained from optically active and racemic monomers, respectively. The crystal structure is similar to that of poly(propylene oxide), and the molecular chain is a (2/1) helix slightly distorted from a planar zigzag[150]. For poly(propylene sulphide) it was definitely confirmed that the unit cell was optically active (the space group: $P2_12_12_1$) even in the racemic polymer, because of the large difference between the x-ray scattering powers of the sulphur and carbon atoms, thus differing from the case of poly(propylene oxide). Since the

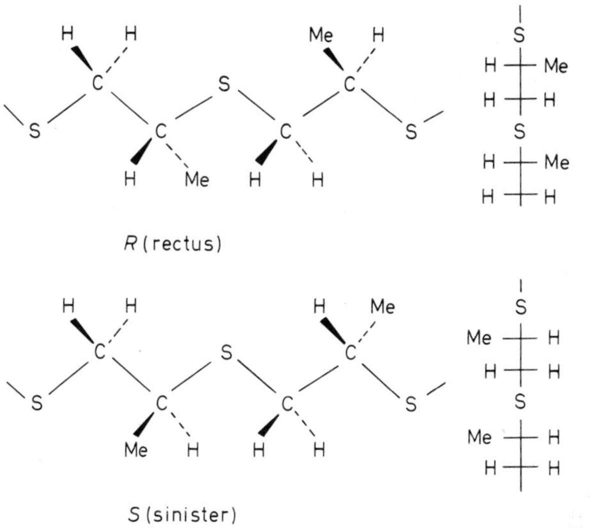

R (rectus)

S (sinister)

Figure 3.24 Absolute configurations of isotactic poly(propylene sulphide)

optically active and racemic polymers give essentially the same x-ray patterns and infrared spectra; it has been concluded that the molecular configuration is isotactic in both cases, and the molecular conformation and crystal structure are of the same type.

Three kinds of models may be considered for the crystalline racemic polymers; (a) racemic lattice, (b) racemic crystallite and (c) intercrystallite racemisation. it-Poly(t-butyl ethylene oxide) represents case (a), in which two pairs of optical antipodes are included in an ordered arrangement in the unit cell[39]. it-Poly(4-methylhex-1-ene) may probably be an example of case (b), in which an equal number of (*S*) and (*R*) polymers exist statistically in a crystallite[80]. it-Poly(propylene sulphide) is considered to be representative of case (c), where each crystallite is composed only of (*R*) polymers or only of (*S*) polymers[150]. As the result, the bulk polymer is optically inactive.

3.3.7 Polyesters

The first member of the series of aliphatic polyesters $[-(CH_2)_m-COO-]_n$ was analysed in detail by using three-dimensional Fourier syntheses[151], as shown in Figure 3.25. The planar zigzag chains form a sheet-structure parallel to the *ac* plane, differing from the polyethylene type-packing. The

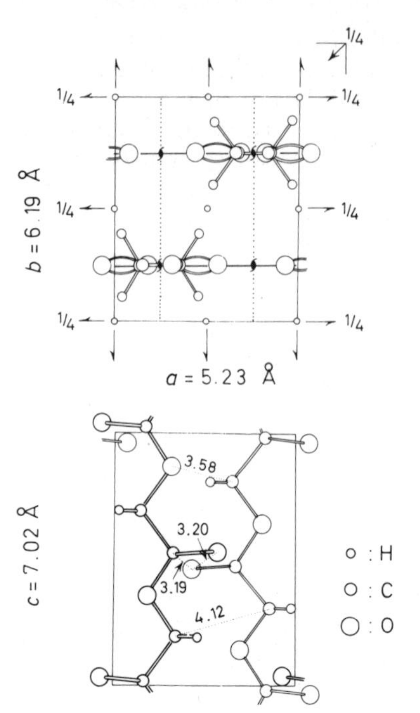

Figure 3.25 Crystal structure of polyglycolide
(From Chatani *et al.*[151], reproduced by courtesy of Hüthig and Wepf Verlag)

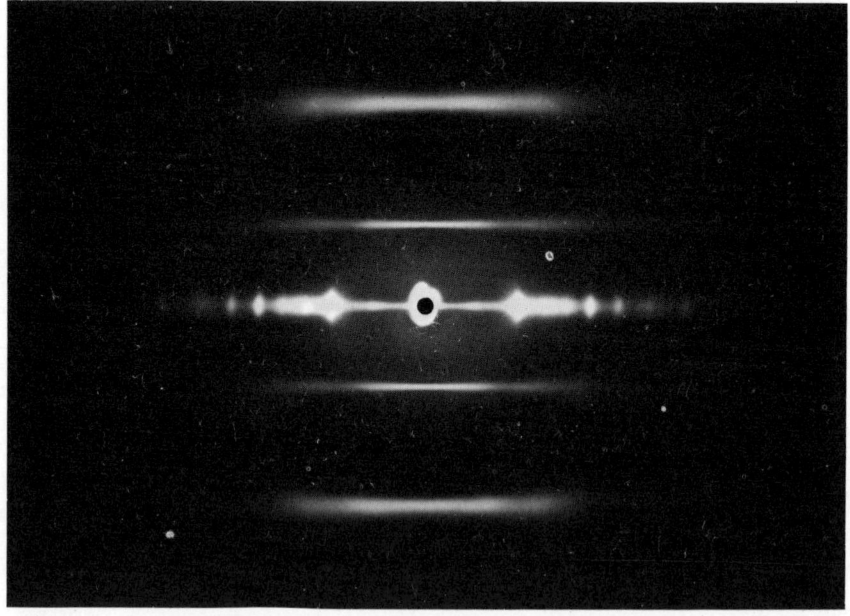

Figure 3.26 X-ray fibre diagram of poly-β-propiolactone

melting point of this polymer, 233 °C, is the highest in the series, and this may be due to the tight molecular packing (high density) and favourable position of approach of the ester groups.

Poly-β-propiolactone has two crystal modifications[152]. Modification (I) has a fibre period 7.02 Å, but the structure has not yet been clarified. Modification (II) gives a characteristic x-ray pattern as shown in Figure 3.26.

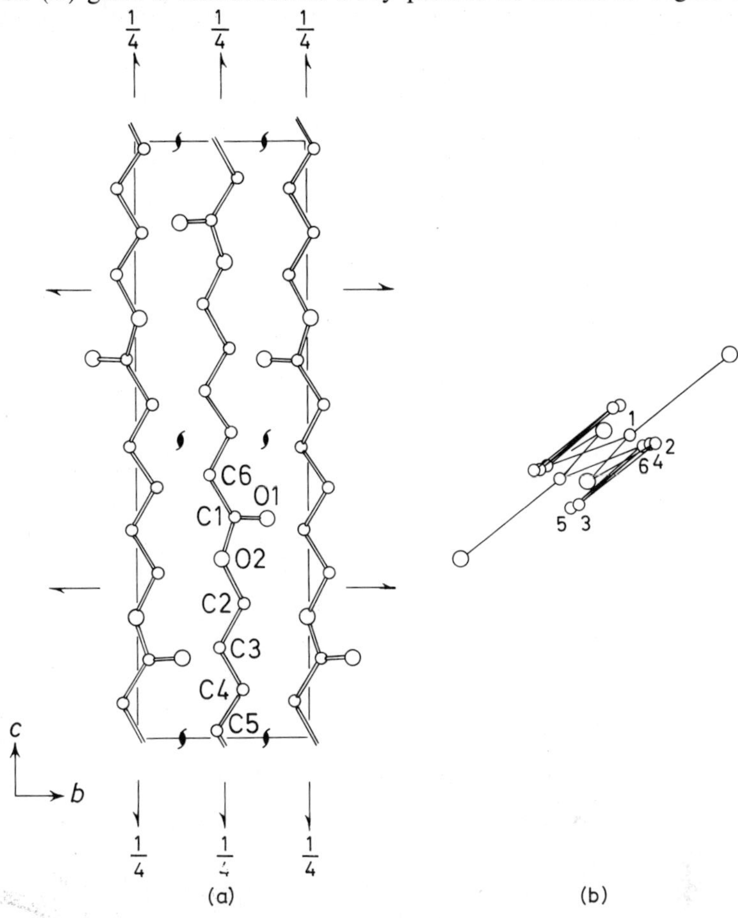

Figure 3.27 Crystal structure of poly-ε-caprolactone
(From Chatani *et al.*[8], reproduced by courtesy of The Society of Polymer Science, Japan)

The reflections on the equator are sharp and indexed with the lattice constants given in Table 3.1, but the layer lines are streaked, showing only intensity distribution along the lines. Such a feature could be interpreted by the molecular transform of a planar zigzag chain and disorder of the level of monomeric units[7].

The molecular packing of poly-ε-caprolactone is similar to that of polyethylene, but the molecule is slightly twisted and the carbonyl groups of the two chains in the unit cell are located at different levels as shown in Figure 3.27.[8] (See also reference 153.)

The far-infrared spectra of the series of polyesters $[-(CH_2)_m-COO-]_n$ (m = 1, 2, 4 and 5) have been analysed by the normal coordinate treatment[154]. The results of the calculations gave the assignments for the characteristic far-infrared bands of polyesters and polyamides.

For the structure of poly(ethylene succinate), the $(TGT\bar{G})_2$ [155] and $(TG)_2(T\bar{G})_2$ models[156] have been proposed, but the crystal structure was determined by the combination of x-ray and infrared studies as shown in

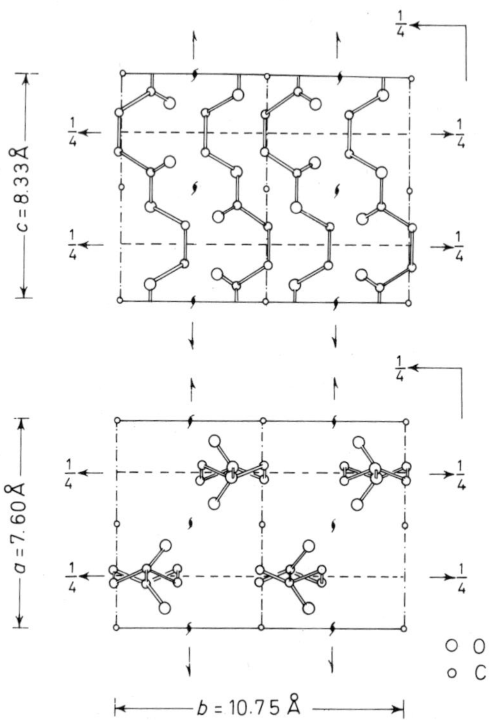

Figure 3.28 Crystal structure of poly(ethylene succinate)
(From Ueda *et al.*[157], reproduced by courtesy of The Society of Polymer Science, Japan)

Figure 3.28[157]. The molecule is a $T_3GT_3\bar{G}$ type, in which the two CH_2-CH_2 bonds in the chemical repeating unit are G and $\bar{G}$.

The molecular structure of poly(ethylene oxalate) is a $T_5GT_5\bar{G}$ type, in which the CH_2-CH_2 bonds are G and $\bar{G}$ [157]. There are some close similarities between the structures of the above two polyesters, which are quite different from the polyethylene-like structures of the higher members of the ethylene glycol series $[-O-(CH_2)_2-O-CO-(CH_2)_m-CO-]_n$, poly(ethylene adipate) (m = 4) and poly(ethylene suberate) (m = 6)[158].

The crystal structure of poly(tetramethylene succinate) was recently determined, giving the following molecular conformation[159]:

$$\begin{array}{cccccccccc} (-O- & C- & C- & C- & C- & O- & CO- & C- & C- & CO-)_n \\ T & G & T & \bar{G} & T & T & & T & T & T \quad T \end{array}$$

By summarising the skeletal conformations of the polyesters analysed so far, there seem to be some features concerning the internal rotation as shown below.

	—C	—C	—CO—	O—	C—
Poly-β-propiolactone[7]	*T*	*T*	*T*	*T*	*T*
Poly-ε-caprolactone[8]	*T*	*T*	*T*	*T*	*T*
Polypivalolactone[160]	*G*	*G*	*T*	*T*	*G*
Poly(ethylene adipate)[158]	*T*	*T*	*T*	*S*	*T*
Poly(ethylene succinate)[157]	*G*	*T*	*T*	*T*	$\bar{G}$
Poly(tetramethylene succinate)[159]	*T*	*T*	*T*	*T*	*G*

	—O—	CO—	CO—	O—	C—C—
Poly(ethylene oxalate)[157]	*T*	*T*	*T*	*T*	*G*

In these polyesters the CO—O bonds are all *trans*, and the CO—C bonds are also *trans* except for polypivalolactone in which two methyl groups instead of hydrogen atoms are attached to the main-chain carbon atom. Therefore, the CO—O and CO—C bonds are stable in the *trans* form so long as no large substituent groups are introduced. The O—C bonds take the *trans* or *skew*, and the C—C bonds take the *trans* or *gauche*.

Further references: Infrared studies of chain folding in polyethylene terephthalate[161].

3.3.8 Polyamides

Nylon 3 [162] and nylon 4 [163] have a few crystal modifications, among which the structures were determined giving crystal forms essentially similar to the α-form of nylon 6 [164].

Further references: Infrared studies of chain folding of nylon 66 [165], far-infrared spectra of polyamides[154, 166, 167], crystal structures of poly(L-proline)[192] and poly(*m*-xylene adipamide)[168].

3.3.9 Other polymers

Ganis *et al.*[169] proposed molecular conformations for two crystal modifications of poly(dimethyl ketene) $[\text{—C(CH}_3)_2\text{—}\underset{\underset{\text{O}}{\|}}{\text{C}}\text{—}]_n$; a G_4 model for the α-form and a $TG_3T\bar{G}_3$ model for the β-form. The latter was confirmed by x-ray analysis[170].

Fibrous sulphur has been studied by Lind and Geller[171] in detail by using x-ray diffractometer data. Four pairs of right-handed and left-handed (10/3) helices are included in a monoclinic unit cell. The molecular parameters are: S—S bond distance = 2.07 Å, ∠SSS = 106 degrees and the dihedral angle = 95 degrees. (Also see reference 172.)

For the structure of carbon fibres, see the papers written by Ruland (x-ray studies)[173–175] and Tuinstra and Koenig (Raman spectra)[176].

Further references: Infrared and Raman spectra of polytetrafluoroethylene[177–179] and crystal structures of polycarbonates[180].

3.3.10 Polymer complexes

Poly(ethylene oxide) forms crystalline complexes with urea, thiourea, mercuric chloride, etc.[120]. Mercuric chloride gives two types of complexes with different stoichiometric ratios of CH_2CH_2O and $HgCl_2$; type (I)(4:1) and type (II)(1:1). The crystal structures of these two types of complexes were determined.

The conformation of poly(ethylene oxide) in complex type (I)[181] was found to be $T_5GT_5\bar{G}$ as shown in Figure 3.17(b) compared with that of the ordinary poly(ethylene oxide), the (7/2) helix for comparison in Figure 3.17(a). These two conformations appear to be quite different, but are transformed by changing the conformations only of a limited number of bonds in the chain as shown below.

poly(ethylene oxide)

$$\begin{array}{l} \quad\;\; T\;\; T\;\; G\;\; T\;\; T\;\; G\;\; T\;\; T\;\; G\;\; T\;\; T\;\; G \\ -C-(-O-C-C-O-C-C-O-C-C-O-C-C-)_n \\ \quad\;\; T\;\; T\;\; T\;\; T\;\; T\;\; G\;\; T\;\; T\;\; T\;\; T\;\; T\;\; \bar{G} \end{array}$$

poly(ethylene oxide)–$HgCl_2$ complex type (I)

From the upper conformation when some of the Gs are transformed to T and $\bar{G}$, the lower conformation $T_5GT_5\bar{G}$ is obtained. In Figure 3.29 the bounded Fourier projection on the ac plane is reproduced. The figure shows very clearly the electron-density distribution corresponding to the $T_5GT_5\bar{G}$ conformation of the poly(ethylene oxide) molecule.

In the case of type (II), the contribution of the x-ray scattering of the Hg and Cl atoms is much larger than for type (I). For this reason the conformation of poly(ethylene oxide) was assumed so as to give a good agreement between the far-infrared spectra and the results of the normal-mode calculations, although the positions of $HgCl_2$ molecules were determined by the Patterson method and subsequent repetitions of Fourier analyses. The molecular conformation of poly(ethylene oxide) thus obtained is as follows:

$$\begin{array}{l} -O-(-CH_2-CH_2-O-CH_2-CH_2-O-)_n- \\ \qquad\quad T' \qquad G' \qquad G'\;\; \bar{T}' \qquad \bar{G}' \qquad \bar{G}' \\ \qquad\quad 167^\circ \quad 73^\circ \quad 81^\circ \end{array}$$

Figure 3.30 shows the crystal structure, in which the agreement between the observed and calculated reflection intensities was much improved relative to the case of mercuric chloride only. In Figure 3.14 three conformations of poly(ethylene oxide) were compared.

2,3-Dichlorobutadiene and 2,3-dimethylbutadiene form crystalline canal complexes with thiourea; the lattices are monoclinic, $P2_1/a$, deviating from rhombohedral symmetry (D_{3d}) which is usually found for thiourea complexes[115]. In the above-mentioned two complexes, the monomers polymerise by γ- or x-ray irradiation, to give polymer–thiourea complexes. Both the

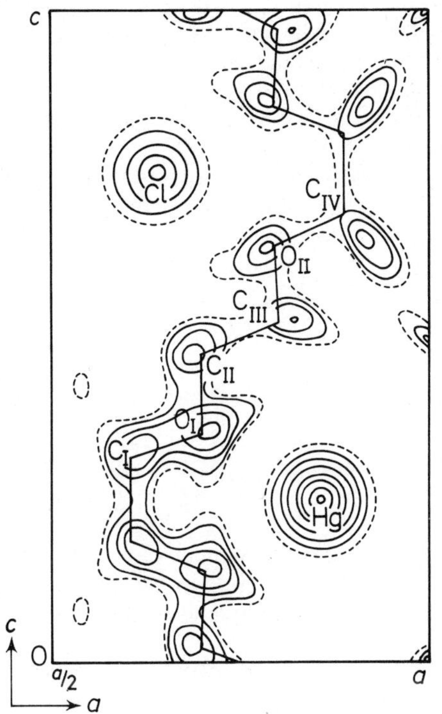

Figure 3.29 Bounded Fourier projection of poly(ethylene oxide)-$HgCl_2$ complex type (I) on the *ac* plane in the range from $y = -\frac{1}{4}$ to $+\frac{1}{4}$. The broken curves indicate the contour for 5 electrons $Å^{-2}$. The interval of the contours is 10 electrons $Å^{-2}$ for Hg and 2.5 electrons $Å^{-2}$ for Cl and poly(ethylene oxide)
(From Tadokoro *et al.*[181], reproduced by courtesy of Interscience)

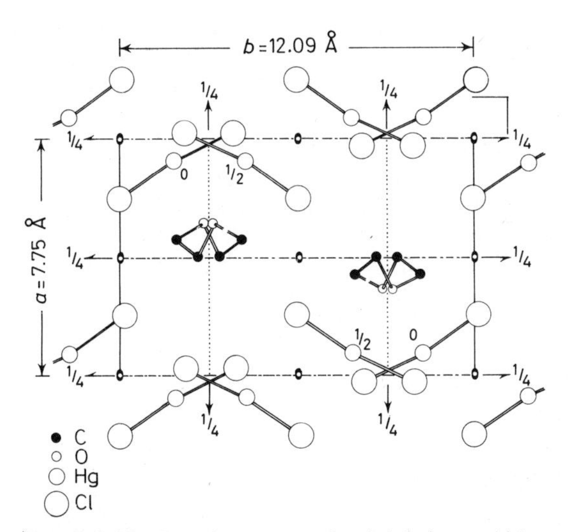

Figure 3.30 Crystal structure of poly(ethylene oxide)–$HgCl_2$ complex type (II)
(From Tadokoro *et al.*[190], reproduced by courtesy of The American Chemical Society)

monomer–thiourea and polymer–thiourea complexes have crystal structures very similar to the usual honeycomb-like structure of thiourea complexes, deforming slightly from hexagonal symmetry. The length of canal occupied by one monomer is 6.25 Å (two monomers in the period of the canal) in the monomer–thiourea complex, but is shortened to 4.80 Å in the polymer–thiourea complex. The polymer chain in the canal has the same conformation as that in the polymer crystal.

Several linear polymers, such as polyethylene, 1,4-polybutadiene and poly(ethylene oxide), form crystalline inclusion compounds with *trans-anti-trans-anti-trans*-perhydrotriphenylene (PHTP). Various monomers included in PHTP have been found by Natta *et al.*[182] to polymerise by γ-irradiation to give polymer complexes.

3.4 CONCLUDING REMARKS

The crystal and molecular structures of many polymers, including several polymers for which the structures have been unresolved for a long time, were determined in the past 5 years as described above. There still remain important problems:

(a) For several polymers, especially complicated helical polymers, the crystal structures have not yet been determined, although the molecular conformations were worked out, that is, for poly(ethylene oxide), it-poly(methylmethacrylate), polyisobutylene, etc.

(b) Structure analyses of the polymers with disordered molecular packing, irregular chemical structure, or low tacticity. Modification (III) of polyallene, modification (II) of poly-β-propiolactone, etc. may be examples of the first case. Of the second case, it has been recently found[183] that poly(methylene polysulphides) $(—CH_2—S_x—)_n$ with non-stoichiometric values of $x = 1.5$–3.0 give very sharp x-ray fibre diagrams. As an example of the third case, acryronitrile–butadiene alternate copolymer (first prepared by Furukawa *et al.*[184]) gives a fairly good x-ray fibre diagram irrespective of its possibly low tacticity[185].

(c) The treatment of intensities in infrared and Raman spectra.

(d) For the calculation of intra- and inter-molecular interaction energies, establishment of reliable potential functions, and parameters including electrostatic interactions, are most desirable.

(e) The relation of the crystal and molecular structure to the microtexture or morphological structure should be clarified, and may extend to the relation with the properties of polymers.

References

1. Bunn, C. W. (1963). *Chemical Crystallography*, 2nd edn., (Oxford: Oxford University Press)
2. Nyburg, S. C. (1961). *X-ray Analysis of Organic Structures*. (New York: Academic Press)
3. Uchida, T. and Tadokoro, H. (1967). *J. Polymer Sci. A-2*, **5**, 63
4. Cochran, W., Crick, F. H. C. and Vand, V. (1952). *Acta Crystallogr.*, **5**, 581
5. Klug, A., Crick, F. H. C. and Wyckoff, H. W. (1958). *Acta Crystallogr.*, **11**, 199

6. Franklin, R. E. and Holmes, K. C. (1958). *Acta Crystallogr.,* **11,** 219
7. Suehiro, K., Chatani, Y., Tadokoro, H., Kato, R. and Tanaka, A. (1966). *15th Annual Meeting Soc. Polymer Sci., Japan,* Nagoya, 220 (Tokyo: Soc. Polymer Sci., Japan)
8. Chatani, Y., Okita, Y., Tadokoro, H. and Yamashita, Y. (1970). *Polymer J.,* **1,** 555
9. MacGillavry, C. H. and Bruins, E. M. (1948). *Acta Crystallogr.,* **1,** 156
10. Taylor, C. A. and Lipson, H. (1964). *Optical Transform.* (London: G. Bell and Sons, Ltd.)
11. Hosemann, R. and Bagchi, S. N. (1962). *Direct Analysis of Diffraction by Matter,* (Amsterdam: North-Holland Publishing Co.)
12. Arnott, S. and Wonacott, A. J. (1966). *Polymer,* **7,** 157
13. Arnott, S. and Dover, S. D. (1968). *Acta Crystallogr.,* **B24,** 599
14. Takahashi, Y. and Tadokoro, H. (1970). *19th Annual Meeting Soc. Polymer Sci. Japan,* Tokyo, 244 (Tokyo: Soc. Polymer Sci., Japan)
15. Shimanouchi, T., Abe, Y. and Alaki, Y. (1971). *Polymer J.,* **2,** 199
16. Sasaki, S., Takahashi, Y. and Tadokoro, H. (1970). *23rd Annual Meeting Chem. Soc., Japan,* Tokyo, 1945. (Tokyo: Chem. Soc., Japan)
17. Gotoh, Y., Sakakihara, H. and Tadokoro, H. (1971). *24th Annual Meeting Chem. Soc. Japan,* Osaka, 2026 (Tokyo: Chem. Soc., Japan)
18. Norman, N. (1954). *Acta Crystallogr.,* **7,** 462
19. Vainshtein, B. K. (1966). *Diffraction of X-rays by Chain Molecules,* Translated into English by Express Translation Service (Amsterdam: Elsevier Publishing Co.)
20. Alexander, L. E. (1969). *X-ray Diffraction methods in Polymer Science,* (New York: Wiley-Interscience)
21. Corradini, P. (1966). *Advances in Structure Research by Diffraction Methods,* 141 (Braunschweig: Friedr. Vieweg & Sohn)
22. Miller, R. L. (1965). *Crystallographic Data for various Polymers,* Ed. by Brandrup, J. and Immergut, E. H., *Polymer Handbook,* p. III-1. (New York: Wiley-Interscience)
23. Cella, R. J., Lee, B. and Hughes, R. E. (1970). *Acta Crystallogr.,* **A26,** 118
24. Elliott, A. (1969). *Infrared Spectra and Structure of Organic Long-chain Polymers* (London: Edward Arnold)
25. Zbinden, R. (1964). *Infrared Spectroscopy of High Polymers,* (New York: Academic Press)
26. Tadokoro, H. and Kobayashi, M. (in press). 'Vibrational Analysis of Highly Ordered Polymers', *Polymer Spectroscopy,* Ed. by Hummel, D. O. (München: Carl Hanser Verlag)
27. Haslam, J. and Willis, H. A. (1965). *Identification and Analysis of Plastics.* (London: Ififfe Books Ltd.)
28. Koenig, J. L. (1971). *Appl. Spectrosc. Rev.,* **4,** 233
29. Hendra, P. J. (1969). *Fortshr. Hochpolm. Forsch.,* **6,** 151
30. Gilson, T. R. and Hendra, P. J. (1970). *Laser Raman Specroscopy.* (New York: Wiley-Interscience)
31. Hummel, D. O. (1966). *Infrared Spectra of Polymers in the Medium and Long Wave Length Regions.* (New York: Wiley-Interscience)
32. Hummel, D. O. (1968). *Atlas der Kunststoff-Analyse, Bd. I. Hochpolymere und Harze. Spectren und Methoden zur Identifizierung. Teil 1. Text und Teil 2. Spectren.* (München: Carl Hanser Verlag)
33. Natta, G., Corradini, P. and Ganis, P. (1960). *Makromol. Chem.,* **39,** 238
34. De Santis, P., Giglio, E., Liquori, A. M. and Ripamonti, A. (1963). *J. Polymer Sci. A,* **1,** 1383
35. Yokoyama, M. and Tadokoro, H. (1971). *Rep. Progr. Polymer Phys. Jap.,* **14,** 153
36. Tadokoro, H. (1971). *U.S.-Japan Seminar on Polymer Physics,* Amherst
37. Tadokoro, H., Kobayashi, M., Tai, K., Yokoyama, M., and Hasegawa, R. (1971). *20th Symposium on Macromolecules,* Tokyo, 531 (Tokyo: Soc. Polymer Sci., Japan)
38. Tadokoro, H., Chatani, Y., Kusanagi, H. and Yokoyama, M. (1970). *Macromolecules,* **3,** 441
39. Sakakihara, H., Takahashi, Y., Tadokoro, H., Oguni, N. and Tani, H. (1971). *20th Symposium on Macromolecules,* Tokyo, 537. (Tokyo: Soc. Polymer Sci., Japan)
40. Flory, P. J. (1969). *Statistical Mechanics of Chain Molecules.* (New York: Wiley-Interscience)
41. Danner, H. R., Safford, G. J., Boutin, H. and Berger, M. (1964). *J. Chem. Phys.,* **40,** 1417
42. Myers, W., Donovan, J. I. and King, J. S. (1965). *J. Chem. Phys.,* **42,** 4299
43. Sakurada, I. and Kaji, K. (1970). *J. Polymer Sci. C,* **31,** 57

44. Bovey, F. A. (1969). *Polymer Conformation and Configuration,* (New York: Academic Press)
45. Tadokoro, H. (1966). *J. Polymer Sci. C,* **15,** 1
46. Bunn, C. W. (1939). *Trans. Faraday Soc.,* **35,** 482
47. Turner-Jones, A. (1963). *J. Polymer Sci.,* **62,** 53
48. Seto, T., Hara, T. and Tanaka, K. (1968). *Japan. J. Appl. Phys.,* **7,** 31
49. Tasumi, M. and Shimanouchi, T. (1965). *J. Chem. Phys.,* **43,** 1245
50. Tasumi, M. and Krimm, S. (1967). *J. Chem. Phys.,* **46,** 755
51. Boerio, E. J. and Koenig, J. L. (1970). *J. Chem. Phys.,* **52,** 3425
52. Snyder, R. G. (1967). *J. Mol. Spectrosc.,* **23,** 224
53. Hendra, P. J. (1968). *J. Mol. Spectrosc.,* **28,** 118
54. Snyder, R. G. (1969). *J. Mol. Spectrosc.,* **31,** 464
55. Carter, V. B. (1970). *J. Mol. Spectrosc.,* **34,** 356
56. Krimm, S. and Bank, M. I. (1965). *J. Chem. Phys.,* **42,** 4059
57. Bank, M. I. and Krimm, S. (1968). *J. Appl. Phys.,* **39,** 4951
58. Kitagawa, T. and Miyazawa, T. (1967). *J. Chem. Phys.,* **47,** 337
59. Kitagawa, T. and Miyazawa, T. (1968). *J. Polymer Sci. B,* **6,** 83
60. Kitagawa, T. and Miyazawa, T. (1968). *Rep. Progr. Polymer Phys. Jap.,* **11,** 219
61. Myers, W., Summerfield, G. C. and King, J. S. (1966). *J. Chem. Phys.,* **44,** 184
62. Trevino, S. F. (1966). *J. Chem. Phys.,* **45,** 757
63. Koenig, J. L. and Witenhafer, D. E. (1966). *Makromol. Chem.,* **99,** 193
64. Tasumi, M. and Krimm, S. (1968). *J. Polymer Sci. A-2,* **6,** 995
65. Bank, M. I. and Krimm, S. (1969). *J. Polymer Sci. A-2,* **7,** 1785
66. Peticolas, W. L., Hibler, G. W., Lippert, J. L., Peterlin, A. and Olf, H. (1971). *Appl. Phys. Lett.,* **18,** 87
67. Schaufle, R. F. and Shimanouchi, T. (1967). *J. Chem. Phys.,* **47,** 3605
68. Tasumi, M., Shimanouchi, T., Kenjo, H. and Ikeda, S. (1966). *J. Polymer Sci. A-2,* **4,** 1011, 1023
69. Tasumi, M. and Zerbi, G. (1968). *J. Chem. Phys.,* **48,** 3813
70. Sakurada, I., Ito, T. and Nakamae, K. (1966). *J. Polymer Sci. C,* **15,** 75
71. Bank, M. I. and Krimm, S. (1969). *J. Chem. Phys.,* **40,** 4248
72. Opaskar, C. G. and Krimm, S. (1969). *J. Polymer Sci. A-2,* **7,** 57
73. Kasai, N. and Kakudo, M. (1968). *Rep. Progr. Polymer Phys. Japan,* **11,** 145
74. Natta, G., Corradini, P. and Bassi, I. W. (1955). *Rc. Accad. Naz. Lincei. Sez. 8,* **19,** 404
75. Turner-Jones, A. and Aizlewood, J. M. (1963). *J. Polymer Sci. B,* **1,** 471
76. Utsunomiya, H., Kawasaki, N., Niinouri, M. and Takayanagi, M. (1967). *J. Polymer Sci. B,* **5,** 907
77. Huguet, M. G. (1966). *Makromol. Chem.,* **94,** 305
78. Noether, H. D. (1967). *J. Polymer Sci. C,* **16,** 725
79. Corradini, P., Ganis, P. and Petraccone, V. (1970). *European Polymer J.,* **6,** 281
80. Bassi, I. W., Bonsignori, O., Lorenzi, G. P., Pino, P., Corradini, P. and Temussi, P. A. (1971). *J. Polymer Sci. A-2,* **9,** 193
81. Kusanagi, H. (1968). *Thesis,* Fac. of Sci., Osaka Univ.
82. Griffith, J. H. and Rånby, B. G. (1960). *J. Polymer Sci.,* **44,** 369
83. Hasegawa, R., Tanabe, Y., Kobayashi, M., Tadokoro, H., Sawaoka, A. and Kawai, N. (1970). *J. Polymer Sci. A-2,* **8,** 1073
84. Corradini, P., Martuscelli, E., Montagnoli, A. and Petraccone, V. (1970). *European Polymer J.,* **6,** 1201
85. Natta, G., Corradini, P., Bassi, I. W. and Fagherazzi, G. (1968). *European Polymer J.,* **4,** 297
86. Corradini, P. and Bassi, I. W. (1968). *J.'Polymer Sci. C,* **16,** 3233
87. Natta, G., Corradini, P. and Bassi, I. W. (1961). *J. Polymer Sci.,* **51,** 505
88. Natta, G., Corradini, P. and Bassi, I. W. (1960). *Nuovo Cimento, Suppl.,* **15,** 52
89. Natta, G. and Corradini, P. (1960). *Nuovo Cimento,* Suppl., **1,** 40
90. Natta, G., Corradini, P. and Bassi, I. W. (1960). *Nuovo Cimento, Suppl.,* **15,** 68
91. Natta, G., Corradini, P. and Bassi, I. W. (1960). *Nuovo Cimento, Suppl.,* **15,** 83
92. Corradini, P. and Ganis, P. (1960). *Nuovo Cimento, Suppl.,* **15,** 104
93. Corradini, P. and Ganis, I. W. (1960). *Nuovo Cimento, Suppl.,* **15,** 96
94. Frank, F. C., Keller, A. and O'Connor, A. (1959). *Phil. Mag.,* **4** (8), 200
95. Kobayashi, M., Tsumura, K. and Tadokoro, H. (1968). *J. Polymer Sci. A-2,* **6,** 1493

96. Kobayashi, M., Akita, K. and Tadokoro, H. (1968). *Makromol. Chem.*, **118,** 324
97. Zerbi, G. and Gussoni, M. (1966). *Spectrochim. Acta,* **22,** 2111
98. Corradini, P., Natta, G., Ganis, P. and Temussi, P. A. (1967). *J. Polymer Sci. C,* **16,** 2477
99. Liquori, A. M. (1955). *Acta Crystallogr.,* **9,** 345
100. De Santis, P., Giglio, E., Liquori, A. M. and Ripamonti, A. (1963). *J. Polymer Sci. A,* **1,** 1383
101. Bunn, C. W. and Holmes, D. R. (1958). *Discuss. Faraday Soc.,* **25,** 95
102. Wasai, G., Saegusa, T. and Furukawa, J. (1965). *Makromol. Chem.,* **86,** 1
103. Allegra, G., Benedetti, E. and Pedone, C. (1970). *Macromolecules.,* **3,** 727
104. Doll, W. W. and Lando, J. B. (1968). *J. Macromol. Sci. B,* **2,** 219
105. Tadokoro, H., Hasegawa, R., Kobayashi, M., Takahashi, Y. and Chatani, Y. (1971). *23rd International Congress of Pure and Applied Chemistry* (Boston), 865 (IUPAC)
106. Lando, J. B., Olf, H. G. and Peterlin, A. (1966). *J. Polymer Sci. A-1,* **4,** 941
107. Lando, J. B. and Doll, W. W. (1968). *J. Macromol. Sci. B,* **2,** 205
108. Cortili, G. and Zerbi, G. (1967). *Spectrochim. Acta,* **23A,** 285, 2216
109. Iwayanagi, S., Sakurai, I., Sakurai, T. and Seto, T. (1968). *J. Macromol. Sci.,* **B2,** 163
110. Natta, G. and Corradini, P. (1960). *Nuovo Cimento, Suppl.,* **1,** 9
111. Suehiro, K. and Takayanagi, M. (1970). *J. Macromol. Sci.,* **B4,** 39
112. Bunn, C. W. (1942). *Proc. Roy. Soc.,* **A180,** 40
113. Natta, G., Bassi, I. W. and Fagherazzi, G. (1969). *European Polymer J.,* **5,** 239
114. Natta, G. and Bassi, I. W. (1967). *J. Polymer Sci. C,* **16,** 2551
115. Chatani, Y., Nakatani, S. and Tadokoro, H. (1970). *Macromolecules,* **3,** 481
116. Tadokoro, H., Takahashi, Y., Otsuka, S., Mori, K. and Imaizumi, F. (1965). *J. Polymer Sci. B,* **3,** 697
117. Tadokoro, H., Kobayashi, M., Mori, K., Takahashi, Y. and Taniyama, S. (1969). *J. Polymer Sci. C,* **22,** 1031
118. Neto, N. and Di Lauro, C. (1967). *European Polymer J.,* **3,** 645
119. Cornell, S. W. and Koenig, J. L. (1970). *J. Polymer Sci. B,* **8,** 137
120. Tadokoro, H. (1967). 'Structure of Crystalline Polyethers', *Macromolecular Reviews,* **1,** Ch 4, 119 (New York: Wiley-Interscience)
121. Kakida, H., Makino, D., Chatani, Y., Kobayashi, M. and Tadokoro, H. (1970). *Macromolecules,* **3,** 569
122. Makino, D., Kobayashi, M. and Tadokoro, H. (1969). *J. Chem. Phys.,* **51,** 3901
123. Imada, K., Miyakawa, T., Chatani, Y., Tadokoro, H. and Murahashi, S. (1965). *Makromol. Chem.,* **83,** 113
124. Cesari, M., Perego, G. and Mazzei, A. (1965). *Makromol. Chem.,* **83,** 196
125. Chatani, Y., Uchida, T., Tadokoro, H., Hayashi, K., Nishii, M. and Okamura, S. (1968). *J. Macromol. Sci.,* **B2,** 567
126. Chatani, Y., Kitahama, K., Tadokoro, H., Yamauchi, T. and Miyake, Y. (1970). *J. Macromol. Sci.,* **B4,** 61
127. Chantani, Y., Ohno, T., Tadokoro, H., Yamauchi, T. and Miyake, Y. (1970). *19th Symposium on Macromolecules,* Kyoto, 753. (Tokyo: Soc. Polymer Sci., Japan)
128. Angood, A. C. and Koenig, J. L. (1968). *J. Appl. Phys.,* **39,** 4985
129. Makino, D., Kobayashi, M. and Tadokoro, H. (1970). *Rep. Progr. Polymer Phys. Japan,* **13,** 169
130. Okada, M., Yamashita, Y. and Ishii, Y. (1964). *Makromol. Chem.,* **80,** 196
131. Corradini, P. and Avitabile, G. (1968). *European Polymer J.,* **4,** 385
132. Cesari, M., Perego, G. and Marconi, W. (1966). *Makromol. Chem.,* **94,** 194
133. Tadokoro, H., Yasumoto, T., Murahashi, S. and Nitta, I. (1960). *J. Polymer Sci.,* **44,** 266
134. Carazzolo, G. and Mammi, M. (1963). *J. Polymer Sci. A,* **1,** 965
135. Price, C. C. and Spector, R. (1966). *J. Amer. Chem. Soc.,* **88,** 4171
136. Ochi, H., Yokoyama, M., Tadokoro, H. and Price, C. C. (1971). *20th Annual Meeting Soc. Polymer Sci., Japan,* Tokyo, 418 (Tokyo: Soc. Polymer Sci., Japan)
137. Tadokoro, H., Takahashi, Y., Chatani, Y. and Kakida, H. (1967). *Makromol. Chem.,* **109,** 96
138. Stanley, E. and Litt, M. (1960). *J. Polymer Sci.,* **43,** 453
139. Takahashi, Y., Tsugaya, H. and Tadokoro, H. (1970). *19th Symposium on Macromolecules,* Kyoto, 743 (Tokyo: Soc. Polymer Sci., Japan)
140. Barlow, M. (1966). *J. Polymer Sci. A-2,* **4,** 121
141. Vandenberg, E. J. (1960). *J. Polymer Sci.,* **47,** 489

142. Vandenberg, E. J. (1961). *J. Amer. Chem. Soc.*, **83,** 3538
143. Vandenberg, E. J. (1964). *J. Polymer Sci. B,* **2,** 1085
144. Kumpanenko, I. V., Kazanskii, K. S., Ptitsyna, N. V. and Kushnerev, M. Ya. (1970). *Polymer Sci. USSR,* **12,** 930
145. Boon, J. and Magré, E. P. (1969). *Makromol. Chem.*, **126,** 130
146. Boon, J. and Magré, E. P. (1970). *Makromol. Chem.*, **136,** 267
147. Carazzolo, G. A. and Valle, G. (1966). *Makromol. Chem.*, **90,** 66
148. Takahashi, Y., Tadokoro, H. and Chatani, Y. (1968). *J. Macromol. Sci.*, **B2,** 361
149. Tadokoro, H., Chatani, Y., Yoshihara, T., Tahara, S. and Murahashi, S. (1964). *Makromol. Chem.*, **73,** 109
150. Sakakihara, H., Takahashi, Y., Tadokoro, H., Sigwalt, P. and Spassky, N. (1969). *Macromolecules,* **2,** 515
151. Chatani, Y., Suehiro, K., Okita, Y., Tadokoro, H. and Chujo, K. (1968). *Makromol. Chem.*, **113,** 215
152. Okamura, S., Higashimura, T., Tanaka, A., Kato, R. and Kikuchi, Y. (1962). *Makromol. Chem.*, **54,** 226
153. Bittiger, H., Marchessault, R. H. and Niegisch, W. D. (1970). *Acta Crystallogr.*, **B26,** 1923
154. Tadokoro, H., Kobayashi, M., Yoshidome, H., Tai, K. and Makino, D. (1968). *J. Chem. Phys.*, **49,** 3359
155. Fuller, C. S. and Erickson, C. L. (1937). *J. Amer. Chem. Soc.*, **59,** 344
156. Bunn, C. W. (1952). *Proc. Roy. Soc.*, **A180,** 67
157. Ueda, A. S., Chatani, Y. and Tadokoro, H. (1971). *Polymer J.*, **2,** 387
158. Turner-Jones, A. and Bunn, C. W. (1962). *Acta Crystallogr.*, **15,** 105
159. Chatani, Y., Hasegawa, R. and Tadokoro, H. (1971). *20th Annual Meeting Soc. Polymer Sci., Japan,* Tokyo, 420 (Tokyo: Soc. Polymer Sci., Japan)
160. Carazzolo, G. (1964). *Chim. Ind.* (Milano), **46,** 525
161. Koenig, J. L. and Hannon, M. J. (1967). *J. Macromol. Sci.*, **B1,** 119
162. Masamoto, J., Sasaguri, K., Ohizumi, C. and Kobayashi, H. (1970). *J. Polymer Sci. A-2,* **8,** 1703
163. Fredericks, R. J., Doyne, T. H. and Sparague, R. S. (1966). *J. Polymer Sci. A-2,* **4,** 899
164. Holmes, D. R., Bunn, C. W. and Smith, D. J. (1955). *J. Polymer Sci.*, **17,** 159
165. Koenig, J. L. and Agboatwalla, M. C. (1968). *J. Macromol. Sci.*, **B2,** 389
166. Matsubara, I., Itoh, Y. and Shinomiya, M. (1966). *J. Polymer Sci. B,* **4,** 47
167. Matsubara, I. and Magill, J. H. (1966). *Intern. Symp. Macromol. Chem.*, Tokyo-Kyoto, VII-10 (Tokyo: Soc. Polymer Sci., Japan)
168. Ota, T., Yamashita, M., Yoshizaki, O. and Nagai, E. (1966). *J. Polymer Sci. A-2,* **4,** 959
169. Ganis, P. and Temussi, P. A. (1966). *European Polymer J.*, **2,** 401
170. Bassi, I. W., Ganis, P. and Temussi, P. A. (1967). *J. Polymer Sci. C,* **16,** 2867
171. Lind, M. D. and Geller, S. (1969). *J. Chem. Phys.*, **51,** 348
172. Tuinstra, F. (1966). *Acta Crystallogr.*, **20,** 341
173. Ruland, W. (1967). *J. Appl. Phys.*, **38,** 3585
174. Ruland, W. and Tompa, H. (1968). *Acta Crystallogr.*, **A24,** 93
175. Ruland, W. (1969). *J. Polymer Sci. C,* **28,** 143
176. Tuinstra, F. and Koenig, J. L. (1970). *J. Chem. Phys.*, **53,** 1126
177. Koenig, J. L. and Boerio, F. J. (1969). *J. Chem. Phys.*, **50,** 2823
178. Hannon, M. J., Boerio, F. J. and Koenig, J. L. (1969). *J. Chem. Phys.*, **50,** 2829
179. Koenig, J. L. and Boerio, F. J. (1970). *J. Chem. Phys.*, **52,** 4170
180. Bonart, R. (1966). *Makromol. Chem.*, **92,** 149
181. Iwamoto, R., Saito, Y., Ishihara, H. and Tadokoro, H. (1968). *J. Polymer Sci. A-2,* **6,** 1509
182. Farina, F., Natta, G., Allegra, G. and Löffelholz, M. (1967). *J. Polymer Sci. C,* **16,** 2517
183. Sakakihara, H., Takahashi, Y. and Tadokoro, H., to be published.
184. Furukawa, J., Iseda, Y., *et al.* (1969). *J. Polymer Sci. B,* **7,** 561
185. Gotoh, Y., Sakakihara, H., Tadokoro, H. and Yamamoto, K., to be published
186. Golemba, F. J., Guillet, J. E. and Nyburg, S. C. (1968). *J. Polymer Sci. A-1,* **6,** 1341
187. Bassi, I. W. and Chioccola, G. (1969). *European Polymer J.*, **5,** 163
188. Chen, V. Y., Allegra, G., Corradini, P. and Goodman, M. (1970). *Macromolecules,* **3,** 274
189. Sasaki, S., Takahashi, Y. and Tadokoro, H., to be published
190. Yokoyama, M., Ishihara, H., Iwamoto, R. and Tadokoro, H. (1969). *Macromolecules,* **2,** 184

191. Enomoto, S., Opasker, C. G. and Krimm, S. (1967). *J. Polymer Sci. C,* **16,** 2263
192. Lal, J. and Trick, G. S. (1961). *J. Polymer Sci.,* **50,** 13
193. Kobayashi, S., Tadokoro, H. and Chatani, Y. (1968). *Makromol. Chem.,* **112,** 225

4
Magnetic Susceptibilities of Diamagnetic Molecules

R. DITCHFIELD
Bell Telephone Laboratories, Murray Hill, New Jersey

4.1 INTRODUCTION

When a molecule is placed in a uniform magnetic field $\boldsymbol{H}$, its electrons move in a way which corresponds to an intramolecular flow of current. If there are no unpaired electrons, this is the dominant effect of the field and the orbital electronic currents induced by the magnetic field give rise to the diamagnetic polarisation of the molecule. The magnetic moment, $\boldsymbol{\mu}$, resulting from these induced currents is given by

$$\boldsymbol{\mu} = \boldsymbol{\chi}\boldsymbol{H} \tag{4.1}$$

where $\boldsymbol{\chi}$ is the molecular diamagnetic susceptibility. These induced currents will also give rise to secondary magnetic fields at the nuclei in the molecule. For example, the screening field at nucleus A, $\boldsymbol{H}^{\mathrm{A}}_{\mathrm{ind}}$, is given by

$$\boldsymbol{H}^{\mathrm{A}}_{\mathrm{ind}} = -\boldsymbol{\sigma}^{\mathrm{A}}\boldsymbol{H} \tag{4.2}$$

where $\boldsymbol{\sigma}^{\mathrm{A}}$ is the magnetic shielding constant for nucleus A. The total field experienced by nucleus A, which determines the n.m.r. frequency, is then

$$\boldsymbol{H}^{\mathrm{A}} = \boldsymbol{H}(1-\boldsymbol{\sigma}^{\mathrm{A}}) \tag{4.3}$$

Thus the magnetic susceptibility $\boldsymbol{\chi}$ and the magnetic shielding constant $\boldsymbol{\sigma}$ both arise from the same origin, namely, the orbital electronic currents induced by the external magnetic field. A complete understanding of these diamagnetic currents, therefore, should lead to a full interpretation of the diamagnetic susceptibility and the magnetic shielding constants of all nuclei in a molecule.

Although the magnetic susceptibility has been considered above at the molecular level, it is really a macroscopic property and involves a large number of molecules. For example, the total magnetic moment $\boldsymbol{I}$ of an assembly of non-interacting molecules placed in a uniform magnetic field $\boldsymbol{H}$ is given by

$$\boldsymbol{I} = \boldsymbol{\chi}_{\mathrm{s}}\boldsymbol{H} \tag{4.4}$$

where $\boldsymbol{\chi}_{\mathrm{s}}$ is defined as the magnetic susceptibility of the sample under consideration. Thus, $\boldsymbol{I}$ will depend on the size of the sample and we may define $\boldsymbol{\chi}_{\mathrm{s}}$ for unit volume ($\boldsymbol{\chi}_{\mathrm{v}}$), for a gramme ($\boldsymbol{\chi}_{\mathrm{g}}$) or for a mole ($\boldsymbol{\chi}_{\mathrm{m}}$). In the following discussion we shall only be concerned with molecular susceptibility $\boldsymbol{\chi}$ or molar susceptibility $\boldsymbol{\chi}_{\mathrm{m}}$. These are simply related by equation (4.5)

$$\boldsymbol{\chi}_{\mathrm{m}} = N\boldsymbol{\chi} \tag{4.5}$$

where N is Avogadro's number (6.0238×10^{23}).

Since, for a molecule of low symmetry, the induced magnetic moment and the secondary fields at the nuclear positions need not be parallel to the primary field, $\boldsymbol{\chi}$ and $\boldsymbol{\sigma}$ are tensor quantities. Thus the induced diamagnetic moment is a vector with components given by

$$\mu_{\alpha} = \sum_{\beta} \chi_{\alpha\beta} H_{\beta} \qquad \alpha,\beta = x,y \text{ or } z \tag{4.6}$$

The set of quantities $\chi_{\alpha\beta}$, of which six are independent, form the diamagnetic susceptibility tensor. If the axes are rotated so that χ_{xy}, χ_{xz} and χ_{yz} all vanish (principal axes), the three principal values χ_{xx}, χ_{yy} and χ_{zz} may be different if the molecule is of sufficiently low symmetry. The differences $\chi_{zz}-\chi_{xx}$, $\chi_{zz}-\chi_{yy}$ are the anisotropies of the susceptibility. For molecules with three-fold or higher symmetry there is only one unique anisotropy given by

$$\Delta\chi = \chi_{||}-\chi_{\perp} \tag{4.7}$$

where $\chi_{\|}$ is the susceptibility along the axis of highest symmetry (principal axis) and $\chi_{\perp}$ is the susceptibility perpendicular to this axis. The average susceptibility χ and the average shielding constants σ are just the average values of these tensors and are given by equation (4.8)

$$\begin{aligned} \chi &= (\chi_{xx}+\chi_{yy}+\chi_{zz})/3 \\ \sigma &= (\sigma_{xx}+\sigma_{yy}+\sigma_{zz})/3 \end{aligned} \tag{4.8}$$

There have been a large number of empirical studies[1] of the bulk diamagnetism of molecules without unpaired electrons. The best known scheme for interpreting diamagnetic susceptibilities χ_m on an atomic or group basis is probably the set of Pascal constants[2,3]. These permit the calculation of χ_m as a sum of atomic contributions together with a sum of constitutive corrections for the various types of bond or ring present. This scheme has been successful in coordinating a large body of experimental data for χ_m within a simple framework.

Studies of diamagnetic anisotropy have been somewhat smaller in number. For many years most direct experimental data on $\Delta\chi$ have been derived from measurements on molecular crystals[4]. In the early years[5,6] magnetic birefringence studies (Cotton–Mouton effect) were also used to obtain $\Delta\chi_m$ and there has been recent renewed interest in this technique[7]. However, the most important source of accurate $\Delta\chi$ values in the last few years has been high-field Zeeman microwave studies[8,9]. These have provided a wealth of information on diamagnetic anisotropy for many types of molecule. Since the magnetic anisotropy of individual electron groups plays an important part in predictive theories of magnetic shielding[10,11], there is considerable general interest in this area.

The aims of this review are threefold;

(a) to survey the theoretical methods that have been used to calculate magnetic susceptibilities. An attempt will be made to present these in such a way as to illustrate the fundamental relationship between molecular electronic structure and diamagnetism.

(b) to survey the experimental data on magnetic susceptibilities. Here, particular emphasis will be placed on those aspects of susceptibility which are important to an understanding of n.m.r. chemical shifts.

(c) to survey the research which has used a knowledge of magnetic anisotropy to interpret n.m.r. chemical shifts.

4.2 THEORETICAL METHODS

4.2.1 General theory and the Van Vleck formula

Several types of theory have been used to treat the general problem of molecular diamagnetism. These range from the early formal treatment of Van Vleck[12] using the Rayleigh–Schrödinger perturbation theory to the recent perturbation theories[13] based on self-consistent wave functions.

The electronic Hamiltonian describing a singlet-state molecule in a uniform magnetic field $\boldsymbol{H}$ has the form (in atomic units)

$$H(\boldsymbol{H}) = \tfrac{1}{2} \sum_j \left\{ \left(\boldsymbol{p}_j + (1/c)\boldsymbol{A}(\boldsymbol{r}_j) \right)^2 - 2 \sum_{\mathrm{B}} Z_{\mathrm{B}} r_{j\mathrm{B}}^{-1} \right\} + \sum_{j \neq l} \sum r_{jl}^{-1} \tag{4.9}$$

Here c is the velocity of light, $\sum_j$ is a sum over all electrons, and $\sum_{\mathrm{B}}$ is a sum over all nuclei. The last two terms represent the electron–nucleus and electron–electron contributions to the potential energy. $\boldsymbol{p}_j$ is the quantum-mechanical momentum which in the Schrödinger representation is replaced by

$$\boldsymbol{p}_j = -i\boldsymbol{\nabla}_j \tag{4.10}$$

The vector potential describing the total magnetic field at the position of electron j, $\boldsymbol{A}(\boldsymbol{r}_j)$ is given by

$$\boldsymbol{A}(\boldsymbol{r}_j) = \tfrac{1}{2}\boldsymbol{H} \wedge \boldsymbol{r}_j \tag{4.11}$$

where $\boldsymbol{r}_j$ is the distance from electron j to some arbitrary origin. In equation (4.9), $\boldsymbol{r}_{j\mathrm{B}}$ and $\boldsymbol{r}_{ij}$ are $\boldsymbol{r}_j - \boldsymbol{R}_{\mathrm{B}}$ and $\boldsymbol{r}_i - \boldsymbol{r}_j$ respectively, $\boldsymbol{R}_{\mathrm{B}}$ being the distance from nucleus B (charge Z_{B}) to the origin.

Substituting equations (4.10) and (4.11) into equation (4.9) and working in the Coulomb gauge[14] ($\nabla \boldsymbol{A} = 0$) enables us to rewrite equation (4.9) as

$$H(\boldsymbol{H}) = H^{(0)} + \sum_{\alpha} H_{\alpha} H_{\alpha}^{(1,0)} + \tfrac{1}{2} \sum_{\alpha} \sum_{\beta} H_{\alpha} H_{\alpha\beta}^{(2,0)} H_{\beta} \tag{4.12}$$

where H_{α} and H_{β} are components of the external magnetic field and

$$H^{(0)} = \sum_j \left(-\tfrac{1}{2}\nabla_j^2 - \sum_{\mathrm{B}} Z_{\mathrm{B}} r_{j\mathrm{B}}^{-1} \right) + \sum_{j \neq l} \sum r_{jl}^{-1} \tag{4.13}$$

$$H_{\alpha}^{(1,0)} = -(i/2c) \sum_j L_{j\alpha} \tag{4.14}$$

$$H_{\alpha\beta}^{(2,0)} = (1/4c^2) \sum_j (r_j^2 \delta_{\alpha\beta} - r_{j\alpha} r_{j\beta}) \tag{4.15}$$

$$L_{j\alpha} = (\boldsymbol{r}_j \wedge \boldsymbol{\nabla}_j)_{\alpha} \tag{4.16}$$

In equation (4.15) $\delta_{\alpha\beta}$ is the Kronecker delta.

The Hamiltonian in the absence of the external magnetic field is $H^{(0)}$ and solving the Schrödinger equation based on this Hamiltonian gives a set of unperturbed wave functions Ψ_0, Ψ_1, Ψ_2 ... Ψ_n (energies $E_0 < E_1 < E_2 < \ldots < E_n$) where the lowest energy wavefunction describes the ground state. When the molecule is placed in a magnetic field, the ground-state wave function changes from Ψ_0 to $\Psi(\boldsymbol{H})$ and the energy changes from E_0 to $E(\boldsymbol{H})$, these changes being clearly related to the diamagnetic polarisation of the molecule. For small values of the magnetic field strength, $E(\boldsymbol{H})$ can be written as

$$E(\boldsymbol{H}) = E_0 - \sum_{\alpha} H_{\alpha} \gamma_{\alpha} - \tfrac{1}{2} \sum_{\alpha} \sum_{\beta} H_{\alpha} \chi_{\alpha\beta} H_{\beta} \tag{4.17}$$

where γ_{α} is a component of the permanent magnetic moment of the molecule. For singlet-state molecules with zero orbital angular momentum, γ is zero.

For small values of H_α and H_β, $E(\boldsymbol{H})$ and $\Psi(\boldsymbol{H})$ can be expanded as a Taylor series about their zero-field values

$$\Psi(\boldsymbol{H}) = \Psi_0+\sum_\alpha(\partial\Psi(H)/\partial H_\alpha)_0 H_\alpha+\ldots = \Psi_0+\sum_\alpha H_\alpha\Psi_\alpha^{(1,0)}\ldots \tag{4.18}$$

$$E(\boldsymbol{H}) = E_0+\sum_\alpha(\partial E(\boldsymbol{H})/\partial H_\alpha)_0 H_\alpha+\tfrac{1}{2}\sum_\alpha\sum_\beta(\partial^2 E(\boldsymbol{H})/\partial H_\alpha\partial H_\beta)_0 H_\alpha H_\beta+\ldots$$

$$= E_0+\sum_\alpha H_\alpha E_\alpha^{(1,0)}+\tfrac{1}{2}\sum_\alpha\sum_\beta H_\alpha E_{\alpha\beta}^{(2,0)}H_\beta+\ldots \tag{4.19}$$

and a comparison of equations (4.17) and (4.19) shows that the calculation of $\chi_{\alpha\beta}$ requires a knowledge of $E_{\alpha\beta}^{(2,0)}$. Using the Rayleigh–Schrödinger perturbation theory[15], this is given by

$$E_{\alpha\beta}^{(2,0)} = \langle\Psi_0\,|\,H_{\alpha\beta}^{(2,0)}\,|\,\Psi_0\rangle+\langle\Psi_0\,|\,H_\alpha^{(1,0)}\,|\,\Psi_\beta^{(1,0)}\rangle. \tag{4.20}$$

We can expand $\Psi_\beta^{(1,0)}$ in terms of the complete set of unperturbed states $\{\Psi_k\}$

$$\Psi_\beta^{(1,0)} = \sum_{k\neq 0}^{\infty} c_{k\beta}^{(1,0)}\Psi_k \tag{4.21}$$

where

$$c_{k\beta}^{(1,0)} = -(E_k-E_0)^{-1}\langle\Psi_k\,|\,H_\beta^{(1,0)}\,|\,\Psi_0\rangle \tag{4.22}$$

Therefore, the zz component of the magnetic susceptibility tensor $\boldsymbol{\chi}$ is given by

$$\chi_{zz} = \chi_{zz}^{\mathrm{d}}+\chi_{zz}^{\mathrm{p}} \tag{4.23}$$

$$= -(1/4c^2)\langle\Psi_0\,|\sum_j(x_j^2+y_j^2)\,|\,\Psi_0\rangle+(1/2c^2)\sum_{k\neq 0}^{\infty}\Big\{(E_k-E_0)^{-1}$$
$$\times\langle\Psi_0\,|\sum_j L_{jz}\,|\,\Psi_k\rangle\langle\Psi_k\,|\sum_l L_{lz}\,|\,\Psi_0\rangle\Big\} \tag{4.24}$$

This general type of theory was first developed by Van Vleck in the early years of quantum mechanics[12]. The first term in equation (4.24), usually called the diamagnetic term, $\boldsymbol{\chi}^{\mathrm{d}}$, is proportional to the mean-square distance $\overline{x^2+y^2}$ of the electrons from the z axis and corresponds to the uniform circulation of electron currents in spherical atoms which is given by the Langevin formula[16]

$$\chi_{xx}^{\mathrm{d}} = \chi_{yy}^{\mathrm{d}} = \chi_{zz}^{\mathrm{d}} = -(1/6c^2)\sum_j\overline{r_j^2} \tag{4.25}$$

For an isolated atom this is the only contribution. However, for an atom in a molecule, this circulation cannot occur freely because of the hindering effect of the electric fields due to the other atoms and so a second paramagnetic (positive) contribution, $\boldsymbol{\chi}^{\mathrm{p}}$, has to be added. The second part of equation (4.24) describes this temperature-independent paramagnetic term and gives the reduction of the diamagnetism which arises from the mixing of the ground state (Ψ_0) with the excited states (Ψ_1, Ψ_2 ... Ψ_n) by the magnetic field.

If we also consider the dipole field arising from a nuclear magnetic moment μ_B in the molecule, then the total magnetic field at the position of electron j is described by the vector potential

$$A'(r_j) = \tfrac{1}{2}H \wedge r_j + (\mu_B \wedge r_{jB})r_{jB}^{-3} \tag{4.26}$$

Inserting this vector potential into equation (4.9) and rewriting as before, we obtain

$$H(H, \mu_B) = H^{(0)} + \sum_\alpha H_\alpha H_\alpha^{(1,0)} + \sum_\alpha \mu_{B\alpha} H_{B\alpha}^{(0,1)} + \tfrac{1}{2} \sum_\alpha \sum_\beta H_\alpha H_{\alpha\beta}^{(2,0)} H_\beta$$

$$+ \sum_\alpha \sum_\beta H_\alpha H_{B\alpha\beta}^{(1,1)} \mu_{B\beta} \tag{4.27}$$

where the additional terms $H_{B\alpha}^{(0,1)}$ and $H_{B\alpha\beta}^{(1,1)}$ are given by

$$H_{B\alpha}^{(0,1)} = -(i/c) \sum_j L_{jB\alpha} r_{jB}^{-3} \tag{4.28}$$

$$H_{B\alpha\beta}^{(1,1)} = (1/2c^2) \sum_j (r_j r_{jB} \delta_{\alpha\beta} - r_{j\alpha} r_{jB\beta}) r_{jB}^{-3} \tag{4.29}$$

$$L_{jB\alpha} = (r_{jB} \wedge \nabla_j)_\alpha \tag{4.30}$$

Proceeding in a way similar to that described above for magnetic susceptibility and noting that

$$(\sigma_B)_{\alpha\beta} = (\partial^2 E(H, \mu)/\partial H_\alpha \partial \mu_{B\beta})_{H = \mu = 0} \tag{4.31}$$

we obtain an equation for the magnetic shielding constant, σ_{Bzz}, similar to that derived by Ramsey[17]

$$\sigma_{Bzz} = \sigma_{Bzz}^{d} + \sigma_{Bzz}^{p}$$

$$= (1/2c^2) \langle \Psi_0 | \sum_j (x_j x_{jB} + y_j y_{jB}) r_{jB}^{-3} | \Psi_0 \rangle - (1/2c^2) \sum_{k \neq 0}^{\infty} (E_k - E_0)^{-1}$$

$$\Big\{ \langle \Psi_0 | \sum_j L_{jz} | \Psi_k \rangle \langle \Psi_k | \sum_l L_{lBz} r_{lB}^{-3} | \Psi_0 \rangle + \langle \Psi_0 | \sum_j L_{jBz} r_{jB}^{-3} | \Psi_k \rangle$$

$$\cdot \langle \Psi_k | \sum_l L_{lz} | \Psi_0 \rangle \Big\} \tag{4.32}$$

The diamagnetic term σ_{Bzz}^{d} in equation (4.32) is the contribution to the secondary magnetic field at nucleus B due to the diamagnetic Langevin-type currents. For an isolated spherical atom it is the only contribution and is given by the Lamb formula[18]

$$\sigma_{Bzz}^{d} = (1/3c^2) \sum_j r_{jB}^{-1} \tag{4.33}$$

The second paramagnetic term (σ_{Bzz}^{p}) in equation (4.32) gives the reduction in shielding due to the mixing of the ground state Ψ_0 with the excited states Ψ_1, Ψ_2, ... Ψ_n by the magnetic field.

The difficulty in calculating χ and σ by this perturbation method is that not only does one have little knowledge of the wave functions describing the high-energy discrete states but the summations in equations (4.24) and (4.32) include integration over continuum states and little is known about these also for most molecules. The few detailed calculations that are available for simple systems suggest that continuum contributions may be at least as important as those from the discrete states[19, 20].

For these reasons early theoretical studies avoided the sum-over-states approach and made calculations on the basis of the closure approximation.

The matrix sum rule[21] in quantum mechanics is represented by the identity

$$\sum_{k}^{\infty} \langle \Psi_0 | P | \Psi_k \rangle \langle \Psi_k | Q | \Psi_0 \rangle = \langle \Psi_0 | PQ | \Psi_0 \rangle \tag{4.34}$$

Equation (4.34) is valid if the functions $P \,|\, \Psi_k\rangle$ and $Q \,|\, \Psi_k\rangle$ can be expanded in terms of the complete set of functions $\{\Psi_k\}$. The closure approximation amounts to choosing an average excitation energy for all states such that one can write

$$\begin{aligned} &\sum_{k \neq 0}^{\infty} \langle \Psi_0 | P | \Psi_k \rangle \langle \Psi_k | Q | \Psi_0 \rangle (E_k - E_0)^{-1} \\ &\approx \Delta E^{-1} \sum_{k \neq 0}^{\infty} \langle \Psi_0 | P | \Psi_k \rangle \langle \Psi_k | Q | \Psi_0 \rangle = \Delta E^{-1} \{ \langle \Psi_0 | PQ | \Psi_0 \rangle \\ &\qquad \langle \Psi_0 | P | \Psi_0 \rangle \langle \Psi_0 | Q | \Psi_0 \rangle \} \end{aligned} \tag{4.35}$$

Applying this to the Van Vleck formula (equation (4.24)) gives

$$\chi_{zz}^{\mathrm{p}} = (1/2c^2) \Delta E^{-1} \langle \Psi_0 | \sum_j L_{jz} \sum_l L_{lz} | \Psi_0 \rangle \tag{4.36}$$

This equation was first used by Van Vleck and Frank[22] together with a value of $\Delta E = 16.6$ eV to calculate the magnetic susceptibility of H_2. The most elaborate calculation by this method however was performed by Witmer[23] for H_2. Using the James–Coolidge wave function[24] for Ψ_0 and the value of ΔE suggested by Van Vleck and Frank he calculated a susceptibility of -3.8 e.m.u.* which may be compared with the experimental value[25] of -3.94 e.m.u. Some theoretical calculations have also been made for N_2 and for CH_4 [26–28] but the results are in poor agreement with experimental values largely because of the difficulties encountered in evaluating χ^{p}. More details of these calculations are given in the review by Pacault, Hoarau and Marchand[29] and the problems associated with the evaluation of χ^{p} have been discussed by Weltner[30].

4.2.2 Variational methods

An alternative method which avoids these difficulties is a variational approach in which the wave function in the presence of the magnetic field is taken in the form

$$\Psi(\boldsymbol{H}) = \Psi_0 + \boldsymbol{H}.\Psi_1 \tag{4.37}$$

*All susceptibility values in this article are in units of 10^{-6} erg G^{-2} mol^{-1}.

It is then possible to obtain approximate values for χ by taking various trial functions for Ψ_0 and Ψ_1, each containing a certain number of parameters, and by minimising the energy with respect to these parameters. A very complete calculation of χ for H_2 was performed by Ishiguro and Koide[31] who substituted James–Coolidge type functions for Ψ_0 and Ψ_1 in equation (4.37). This method gave a good value of χ for H_2 but extension to larger molecules does not seem feasible.

To make this method more applicable, Tillieu and Guy[32] proposed a special type of variational function given by

$$\Psi(\boldsymbol{H}) = \Psi_0(1+i\boldsymbol{H}.\boldsymbol{g}) \tag{4.38}$$

where $\boldsymbol{g}$ is a vector function of electronic coordinates to be determined by variational methods. For example, taking the external magnetic field along the z direction

$$\Psi(\boldsymbol{H}) = \Psi_0(1+iH_z g_z) \tag{4.39}$$

The energy corresponding to this function is

$$E(\boldsymbol{H}) = \langle \Psi(\boldsymbol{H}) \,|\, H(\boldsymbol{H}) \,|\, \Psi(\boldsymbol{H})\rangle \tag{4.40}$$

$$\begin{aligned} &= E_0 + H_z^2\langle \Psi_0 \,|\, H_{zz}^{(2,0)} \,|\, \Psi_0\rangle + iH_z^2\langle \Psi_0 \,|\, H_z^{(1,0)} \,|\, g_z\Psi_0\rangle \\ &\quad - iH_z^2\langle g_z\Psi_0 \,|\, H_z^{(1,0)} \,|\, \Psi_0\rangle + H_z^2\langle g_z\Psi_0 \,|\, H^{(0)} \,|\, g_z\Psi_0\rangle + \ldots \end{aligned} \tag{4.41}$$

To calculate the susceptibility for H_2, Tillieu and Guy replaced g_z by

$$g_z = ax + bxy \tag{4.42}$$

where the parameters a and b are determined by the variational conditions $\partial E/\partial a = \partial E/\partial b = 0$. If the origin is taken at the centre of the electronic charge distribution, χ_{zz} for a two-electron system is given by

$$\chi_{zz} = (-1/2c^2)\left\{\overline{(x^2+y^2)} + \overline{(x^2-y^2)}^2\,\overline{(x^2+y^2)}^{-1}\right\} \tag{4.43}$$

The two parts of this equation correspond to the diamagnetic and paramagnetic terms of the Van Vleck formula. Equation (4.43) was also derived by Gans and Mrowka[33] who further expressed the mean values of x^2, y^2 and z^2 in terms of the xx, yy and zz components of the electrical polarisability tensor $\boldsymbol{\alpha}$ using additional approximations.

Values of χ for H_2 calculated using various forms of variational function

Table 4.1 Diamagnetic susceptibilities for H_2^*

Method	χ^d	χ^p	χ	*Reference*
James–Coolidge + full variational method	−4.2196	0.1507	−4.0306	31
Heitler–London†	−4.22	0.11	−4.11	32
Weinbaum†	−5.55	0.26	−5.29	32
Molecular orbital†	−4.03	0.09	−3.94	32
James–Coolidge†	−3.92	0.07	−3.85	32

*Throughout this article the unit of susceptibility will be 10^{-6} erg G^{-2} mol^{-1}.

†All these methods used the Tillieu–Guy type of variational function.

$\Psi(\boldsymbol{H})$ are given in Table 4.1. It appears that the simplified variational function of Tillieu and Guy gives results for H_2 which compare quite well with the more accurate study of Ishiguro and Koide.

The main disadvantage of variational functions of the type given in equation (4.38) is that $\Psi(\boldsymbol{H})$ is restricted to have nodal surfaces in the same position as Ψ_0 thereby limiting the flexibility of the variation[34]. Thus, although Tillieu and Guy's method gives good results for some simple saturated molecules[35,36] (H_2, CH_4) in which the localised wave functions have no nodes, molecules which are described by wave functions with nodes will be poorly treated. Because of this deficiency, Guy, Tillieu and Baudet[37] were led to the false conclusion that χ will be negative for all closed-shell molecules with zero orbital angular momentum.

So far it has been assumed that one can solve the Schrödinger equation

$$H^{(0)}\Psi_k = E_k\Psi_k \quad (k = 0, 1, 2 \ldots n) \tag{4.44}$$

to obtain exact eigenfunctions Ψ_k and energies E_k. For molecules other than the hydrogen molecular ion this is not possible and approximate forms of Ψ_k have to be used.

4.2.3 Molecular-orbital theory

The approximation usually made for closed-shell ground states of molecules is to write Ψ_0 as a normalised Slater determinant of doubly-occupied molecular orbitals ψ_1. Thus, for a molecule with $2n$ electrons, Ψ_0 is given by

$$\Psi_0 = |\psi_1\bar{\psi}_1\psi_2\bar{\psi}_2 \ldots \psi_n\bar{\psi}_n| \tag{4.45}$$

where the bar indicates that ψ_1 is associated with a β spin function. In most cases a further approximation is made in which the molecular orbitals ψ_1 are written as linear combinations of atomic orbitals ϕ_ν(LCAO)

$$\psi_1 = \sum_\nu c_{\nu 1}\phi_\nu \tag{4.46}$$

It is clear that more accurate molecular orbitals can be obtained from large basis sets of ϕ-functions which permit increased flexibility in the representation. However, this increases the complexity of the calculations and frequently limits applications to small molecules. As a general rule, molecular orbitals are simplest to apply and interpret if the basis set is *minimal*, that is, consists only of the least number of atomic orbitals required to describe the atomic ground state. For example, minimal basis sets for H and for C to F are 1s and 1s, 2s and 2p atomic orbitals, respectively. If a larger number of ϕ-functions is used the basis set is usually described as *extended*.

Once the molecular orbitals are determined the charge density can be analysed in terms of the basis functions ϕ_ν. If there are two electrons per molecular orbital, the total charge density is

$$\text{Charge density} = 2\sum_j^{\text{occ}} \psi_j^2 = \sum_\nu \sum_\lambda P_{\nu\lambda}\phi_\nu\phi_\lambda \tag{4.47}$$

where

$$P_{\nu\lambda} = 2\sum_{j}^{\text{occ}} c_{\nu j}^{*} c_{\lambda j} \tag{4.48}$$

$P_{\nu\lambda}$ is the density matrix and contains the detailed information about charge distribution that is implicit in the ground state molecular-orbital wave function. The diagonal element $P_{\nu\nu}$ is a measure of an electron population for orbital ϕ_{ν}.

By minimising the energy E_0, based on the single-determinant wave function Ψ_0, subject to orthonormality constraints between molecular orbitals, Roothaan-derived equation (4.49) for the LCAO coefficients $c_{\lambda j}$

$$\sum_{\lambda} (F_{\nu\lambda} - \varepsilon_j S_{\nu\lambda}) c_{\lambda j} = 0 \tag{4.49}$$

where $S_{\nu\lambda}$ is the overlap matrix given by

$$S_{\nu\lambda} = \int \phi_{\nu}^{*} \phi_{\lambda} \mathrm{d}\tau \tag{4.50}$$

ε_j are one-electron energies and F is the Fock matrix whose elements are given by

$$F_{\nu\lambda} = H_{\nu\lambda}^{\text{core}} + \sum_{\rho\sigma} P_{\rho\sigma} \left\{ (\nu\lambda \mid \rho\sigma) - \tfrac{1}{2} (\nu\sigma \mid \rho\lambda) \right\} \tag{4.51}$$

Here $H_{\nu\lambda}^{\text{core}}$ is the matrix of the one-electron Hamiltonian for motion in the field of the bare nuclei

$$H_{\nu\lambda}^{\text{core}} = \int \phi_{\nu}^{*}(1) \left\{ -\tfrac{1}{2}\nabla_1^2 - \sum_{\mathrm{B}} Z_{\mathrm{B}} r_{1\mathrm{B}}^{-1} \right\} \phi_{\lambda}(1) \mathrm{d}\tau_1 \tag{4.52}$$

and $(\nu\lambda \mid \rho\sigma)$ is the two-electron integral given by

$$(\nu\lambda \mid \rho\sigma) = \iint \phi_{\nu}^{*}(1)\phi_{\lambda}(1)\,(r_{12})^{-1}\phi_{\rho}^{*}(2)\phi_{\sigma}(2)\,\mathrm{d}\tau_1 \mathrm{d}\tau_2 \tag{4.53}$$

Since the density matrix $P_{\nu\lambda}$ depends on the LCAO coefficients $c_{\lambda j}$, equation (4.49) is not linear and has to be solved by an iterative procedure. It is frequently described as a self-consistent (LCAO–SCF) equation.

The most difficult part of LCAO–SCF theory is usually the evaluation of the large number of two-electron integrals (equation (4.53)). Semi-empirical methods treat these in a simplified manner and avoid the difficulties. *Ab initio* methods evaluate the integrals but have to use ϕ-functions for which such integration is possible.

If the number of basis functions ϕ_{ν} is greater than n, the self-consistent procedure also leads to unoccupied molecular orbitals $\psi_{n+1}, \psi_{n+2}, \ldots$ The simplest treatment of excited states is one in which electrons are reassigned to the self-consistent ground-state molecular orbitals in a different configuration. Because $\sum_j \boldsymbol{L}_j$ is a sum of one-electron operators, the only excited states, Ψ_k, that can give non-zero matrix elements in equation (4.24) are those described by configurations in which an electron is promoted from an occupied orbital ψ_l to an unoccupied one ψ_m. Such states are either singlets or triplets and are described by the wave functions

$$
\begin{aligned}
{}^1\Psi_l^m &= \{|\psi_1\bar{\psi}_1\cdots\psi_l\bar{\psi}_m\cdots| - |\psi_1\bar{\psi}_1\cdots\bar{\psi}_l\psi_m\cdots|\}/\sqrt{2} \\
{}^3\Psi_{l(1)}^m &= |\psi_1\bar{\psi}_1\cdots\psi_l\psi_m\cdots| \\
{}^3\Psi_{l(0)}^m &= \{|\psi_1\bar{\psi}_1\cdots\psi_l\bar{\psi}_m\cdots| + |\psi_1\bar{\psi}_1\cdots\bar{\psi}_l\psi_m\cdots|\}/\sqrt{2} \\
{}^3\Psi_{l(-1)}^m &= |\psi_1\bar{\psi}_1\cdots\bar{\psi}_l\bar{\psi}_m\cdots|
\end{aligned}
\tag{4.54}
$$

Using the usual rules[38] for reducing the many-electron matrix elements to one-electron matrix elements, it can be shown that only singly-excited singlet states give non-zero matrix elements of $\boldsymbol{L}_j$. These are given by

$$\langle\Psi_0|\sum_j \boldsymbol{L}_j|{}^1\Psi_l^m\rangle = \sqrt{2}\{\langle\psi_l(1)|\boldsymbol{L}_1|\psi_m(1)\rangle\} \tag{4.55}$$

Therefore, within the molecular-orbital framework, equation (4.24) reduces to

$$\chi_{zz} = -(1/2c^2)\sum_l^{\text{occ}}\langle\psi_l|x^2+y^2|\psi_l\rangle + (1/c^2)\sum_l^{\text{occ}}\sum_m^{\text{unocc}}({}^1E_l^m - E_0)^{-1}\times \langle\psi_l|L_z|\psi_m\rangle\langle\psi_m|L_z|\psi_l\rangle \tag{4.56}$$

This approximation removes the problem associated with the infinite sum of equation (4.24). However, high-energy excited states are poorly described at this level of theory and contributions from such states, which are often as important as those from low-lying excited states, are probably in error. There is an additional problem. As was noted in equation (4.11), all distances are referred to an arbitrary origin. Clearly, the calculated susceptibility should not depend on the choice of this origin[12], that is, should not depend on the gauge of the vector potential $\boldsymbol{A}(\boldsymbol{r}_i)$[39]. The sum-over-states formula (equation (4.56)), however, will only give gauge-invariant results if the set of molecular orbitals $\{\psi_l\}$ is complete. This limit is not usually attainable and thus equation (4.56) produces calculated results which depend on the position chosen for the origin. A full discussion of the gauge-invariance of self-consistent wave functions has been given by Epstein[40].

Single-determinant wave functions have also been used in variational methods for calculating χ. Karplus and Kolker[41] generalised the approach of Tillieu and Guy and calculated χ for some diatomic molecules. They used single-determinant wave functions, writing the molecular orbitals in the presence of the magnetic field in the form

$$\psi_l(\boldsymbol{H}) = \psi_l^{(0)} + i\boldsymbol{H}.\boldsymbol{g}_l\psi_l^{(0)} \tag{4.57}$$

where g_{lx}, for example, is the polynomial

$$g_{lx} = \sum_{a,b,c,d}\lambda(abcd)r^a x^{2b}y^{2c+1}z^d \tag{4.58}$$

and $\psi_l^{(0)}$ are the molecular orbitals in the absence of the magnetic field. The coefficients $\lambda(abcd)$ are determined by minimising χ_{xx}^{p} as a function of the $\lambda(abcd)$[42]. The odd powers of y and even powers of x in equation (4.58) have the form required to yield non-zero interaction terms with the magnetic perturbation Hamiltonian $H_x H_x^{(1,0)}$. A summary of the results obtained by this method is given in Table 4.2.

A comparison of calculated results for χ with the limited experimental data suggests that this method can give satisfactory values for average susceptibilities. However, magnetic anisotropies are poorly calculated. A further point worth noting is the dependence of the calculated values of χ on the choice of origin. This is particularly evident for CO. This reflects the

Table 4.2 Diamagnetic susceptibilities calculated by the Karplus–Kolker variational method†

	χ (*theory*)	χ (*exp*)	$\chi_\perp$ (*theory*)	$\chi_\parallel$ (*theory*)	$\Delta\chi$ (*theory*)	$\Delta\chi$ (*exp*)
H_2	−4.02	−3.94	−4.19	−3.69	0.50	0.553
Li_2	−32.97	—	−34.06	−30.81	3.25	—
N_2	−17.3	−13.3±3	−17.85	−16.28	1.57	−8.56
F_2	−18.6	—	−21.13	−13.69	7.44	—
LiH*	−9.13	—	−8.99	−9.44	−0.45	−2.14
Li*H	−9.15	—	—	—	—	—
H*F	−8.48	−8.6	−8.79	−8.15	0.64	2.00
HF*	−8.65	−8.6	—	—	—	—
C*O	−15.65	—	−15.40	−16.14	−0.74	−5.23
CO*	−18.11	—	—	—	—	—
N^*H_3	−18.03	−16 to −18	−17.37	−19.34	−1.97	+0.37±0.04

†Units are 10^{-6} erg G^{-2} mol^{-1}.
All diatomic values are taken from Karplus, M. and Kolker, H. J. (1963). *J. Chem. Phys.*, **38**, 1263. The ammonia results are from Sadlej, A. J. (1971). *Molec. Phys.*, **20**, 593. The asterisk indicates that all distances are referred to this nucleus as origin.

use of a restricted form of g_{lx} since if g_{lx} were sufficiently flexible, gauge invariant results would be obtained[43].

Recently, this method has been revived by Sadlej[44] within a semi-empirical framework. His results for diatomics are similar to those calculated by Karplus and Kolker. Sadlej also calculated the susceptibility tensor for NH_3 obtaining good agreement with the experimental value for χ, but rather poorer agreement for $\Delta\chi$.

4.2.4 Methods based on self-consistent perturbation theory

Although one or two other variational methods have been used to calculate magnetic susceptibilities[45–47], in the remainder of this section we will concentrate on recent approaches which have been based on the SCF equations (4.49)–(4.53). These self-consistent perturbation theories also have the advantage that they avoid the difficulties associated with the sum-over-states theory.

In this method one seeks a single determinant solution

$$\Psi(\boldsymbol{H}) = |\psi_1(\boldsymbol{H})\bar{\psi}_1(\boldsymbol{H})\psi_2(\boldsymbol{H})\bar{\psi}_2(\boldsymbol{H})\ldots\psi_n(\boldsymbol{H})\bar{\psi}_n(\boldsymbol{H})| \tag{4.59}$$

to the Schrödinger equation based on the Hamiltonian of equation (4.9). The corresponding energy is

$$E(\boldsymbol{H}) = \langle\Psi(\boldsymbol{H})|H(\boldsymbol{H})|\Psi(\boldsymbol{H})\rangle \tag{4.60}$$

Differentiating $E(\boldsymbol{H})$ with respect to H_β gives

$$(\partial E(\boldsymbol{H})/\partial H_\beta)_0 = \langle(\partial\Psi(H)/\partial H_\beta)_0 \mid H^{(0)} \mid \Psi_0\rangle + \langle\Psi_0 \mid (\partial H(\boldsymbol{H})/\partial H_\beta)_0 \mid \Psi_0\rangle + \langle\Psi_0 \mid H^{(0)} \mid (\partial\Psi(\boldsymbol{H})/\partial H_\beta)_0\rangle \tag{4.61}$$

From equation (4.59) we obtain

$$(\partial\Psi/\partial H_\beta)_0 = \sum_i^{\text{occ}} \{ \mid \psi_1^{(0)}\overline{\psi_1^{(0)}} \dots (\partial\psi_i(\boldsymbol{H})/\partial H_\beta)_0\overline{\psi_i^{(0)}} \dots \psi_n^{(0)}\overline{\psi_n^{(0)}} \mid - \mid \psi_1^{(0)}\overline{\psi_1^{(0)}} \dots (\overline{\partial\psi_i(\boldsymbol{H})/\partial H_\beta)_0}\psi_i^{(0)} \dots \psi_n^{(0)}\overline{\psi_n^{(0)}} \mid \} \tag{4.62}$$

where $\psi_i^{(0)}$ are the zero-field molecular orbitals. If the change of the molecular orbitals with the magnetic field is restricted to occur only through a variation of the LCAO coefficient, that is

$$\psi_j(\boldsymbol{H}) = \sum_\nu c_{\nu j}(\boldsymbol{H})\phi_\nu \tag{4.63}$$

then $\{(\partial\psi_i(\boldsymbol{H})/\partial H_\beta)_0\}$ is fully contained in the space spanned by $\{\psi_l^{(0)}\}$ and we may write

$$(\partial\psi_i(\boldsymbol{H})/\partial H_\beta)_0 = \sum_{l \neq i}^{\text{all MOs}} b_{il}\psi_l^{(0)} \tag{4.64}$$

Substituting this into equation (4.62) and noting equation (4.54) gives

$$(\partial\Psi/\partial H_\beta)_0 = \sqrt{2}\sum_i^{\text{occ}}\sum_{l \neq i}^{\text{all}} b_{il}^1\Psi_i^l$$

$$= \sqrt{2}\sum_i^{\text{occ}}\sum_l^{\text{unocc}} b_{il}^1\Psi_i^l \tag{4.65}$$

If the ground-state wave function is truly self-consistent, then equation (4.61) assumes the simplified form

$$(\partial E(\boldsymbol{H})/\partial H_\beta)_0 = \langle\Psi_0 \mid (\partial H(H)/\partial H_\beta)_0 \mid \Psi_0\rangle \tag{4.66}$$

since $\langle{}^1\Psi_i^l \mid H^{(0)} \mid \Psi_0\rangle$ vanishes for SCF wave functions[48]. This approach, which follows approximately that given by Pople, McIver and Ostlund[49], leads to equation (4.67) for $\chi_{\alpha\beta}$

$$\chi_{\alpha\beta} = -\langle\Psi_0 \mid H_{\alpha\beta}^{(2,0)} \mid \Psi_0\rangle - [\partial/\partial H_\alpha(\langle\Psi(H_\alpha) \mid H_\beta^{(1,0)} \mid \Psi(H_\alpha)\rangle)]_{H=0} \tag{4.67}$$

The two terms represent the diamagnetic and paramagnetic contributions to the susceptibility. It can be seen from equation (4.67) that $\Psi(\boldsymbol{H})$ correct to first-order in H_α is all that is required to evaluate $\chi_{\alpha\beta}$. Thus the Hamiltonian to be used in the perturbed wave function calculation is

$$H(H_\alpha) = H^{(0)} + H_\alpha H_\alpha^{(1,0)} \tag{4.68}$$

Because of the imaginary nature of the perturbation $H_\alpha^{(1,0)}$ it is necessary to allow the molecular orbitals to become complex in the presence of the magnetic field. For the restricted type of expansion given in equation (4.63) this is easily accomplished by introducing complex LCAO coefficients $c_{\nu l}(H_\alpha)$

$$c_{\nu l}(H_\alpha) = c_{\nu l}^{(0)} + iH_\alpha(c_{\nu l}^{(1)})_\alpha \tag{4.69}$$

Consequently, the density matrix also becomes complex

$$P_{\nu\lambda}(H_\alpha) = P_{\nu\lambda}^{(0)} + iH_\alpha(P_{\nu\lambda}^{(1)})_\alpha + \dots \tag{4.70}$$

where

$$P_{\nu\lambda}^{(0)} = 2 \sum_j^{\text{occ}} c_{\nu j}^{(0)} c_{\lambda j}^{(0)}$$

and

$$(P_{\nu\lambda}^{(1)})_\alpha = 2 \sum_j^{\text{occ}} \{c_{\nu j}^{(0)}(c_{\lambda j}^{(1)})_\alpha - (c_{\nu j}^{(1)})_\alpha c_{\lambda j}^{(0)}\} \tag{4.71}$$

The LCAO coefficients $c_{\nu j}(H_\alpha)$ are determined by solving the perturbed Roothaan equations

$$\sum_\lambda (F_{\nu\lambda}(H_\alpha) - \varepsilon_j(H_\alpha) S_{\nu\lambda}^{(0)}) c_{\lambda j}(H_\alpha) = 0 \tag{4.72}$$

where

$$F_{\nu\lambda}(H_\alpha) = (H_{\nu\lambda}^{\text{core}})^{(0)} + iH_\alpha(H_{\nu\lambda}^{(1,0)})_\alpha + \sum_\rho \sum_\sigma P_{\rho\sigma}(H_\alpha) \times \{(\nu\lambda \mid \rho\sigma)^{(0)} - \tfrac{1}{2}(\nu\sigma \mid \rho\lambda)^{(0)}\} \tag{4.73}$$

In this equation the zero superscripts on the overlap, core and two-electron integrals indicate that these are independent of the field and

$$(H_{\nu\lambda}^{(1,0)})_\alpha = -(1/2c)\langle \phi_\nu \mid L_\alpha \mid \phi_\lambda \rangle \tag{4.74}$$

Equation (4.67) for $\chi_{\alpha\beta}$ can now be further simplified to

$$\chi_{\alpha\beta} = -\sum_\nu \sum_\lambda \{P_{\nu\lambda}^{(0)}(H_{\nu\lambda}^{(2,0)})_{\alpha\beta} + (P_{\nu\lambda}^{(1)})_\alpha (H_{\nu\lambda}^{(1,0)})_\beta\} \tag{4.75}$$

where

$$(H_{\nu\lambda}^{(2,0)})_{\alpha\beta} = (1/4c^2)\langle \phi_\nu \mid (r^2\delta_{\alpha\beta} - r_\alpha r_\beta) \mid \phi_\lambda \rangle \tag{4.76}$$

The perturbed wave function $\Psi(H_\alpha)$ may be evaluated numerically by solving the Roothaan equations for various values of the magnetic field strength and using finite difference methods[49,50] to extract $P_{\nu\lambda}^{(1)}$. Alternatively, the perturbed Roothaan equations may be formally differentiated, terms of order H_α separated and the first-order (in H_α) equations solved either directly[51] or iteratively[52,53].

It is worth mentioning at this point that the same perturbed wave function ($\Psi(H_\alpha)$) can be used for calculating the magnetic shielding tensors, $\sigma_{\alpha\beta}$, for all the nuclei in the molecule. These are given by

$$(\sigma_B)_{\alpha\beta} = \sum_\nu \sum_\lambda \{P_{\nu\lambda}^{(0)}(H_{B\nu\lambda}^{(1,1)})_{\alpha\beta} + (P_{\nu\lambda}^{(1)})_\alpha (H_{B\nu\lambda}^{(0,1)})_\beta\} \tag{4.77}$$

Equations (4.75) and (4.77) can be used with LCAO SCF wave functions constructed from any field-independent basis set and at any level of approximation. This type of theory was first used by Stevens, Pitzer and Lipscomb[51,54–61] to calculate the magnetic properties of diatomic molecules. They used an extended basis set of Slater-type atomic orbitals (STOs) to describe Ψ_0.

This set of ϕ-functions was then further extended in order to have increased flexibility for the calculation of $\Psi(\boldsymbol{H})$. If all the additional ϕ-functions are of different symmetry to all those used for Ψ_0, then this procedure is straightforward. However, if some of the additional functions are of the same symmetry as those used in Ψ_0 this can lead to problems.

This method has also been used for calculations on polyatomic molecules by Arrighini, Maestro and Moccia[62–68] and by Ditchfield, Miller and

Table 4.3 Ab initio susceptibility results calculated by SCF perturbation theory†

Molecule	χ^d *(calc)*	χ^p *(calc)*	χ^p *(exp)*	χ^{Total} *(calc)*	*Reference*
Li*H	−20.47	12.83	12.71	−7.63	54
LiH*	−23.81	15.85	16.05	−7.96	54
F*H	−10.98	0.58	—	−10.40	55
FH*	−32.08	21.18	—	−10.90	55
F_2^*	−67.31	56.33	—	−10.98	56
F_2‡	−41.70	31.08	—	−10.62	56
Li_2‡	—	—	—	−28.94	58
B*H	−18.02	36.77	—	18.75	59
BH*	−37.78	56.30	—	18.52	59
B*F	−60.78	46.47	—	−14.32	59
BF*	−44.07	29.86	—	−14.21	59
Co‡	—	—	—	−13.26	60
Al*H	−33.93	32.56	—	−1.37	61
OH_2‡	−15.59	1.21	1.504	−14.38	65
NH_3‡	−20.47	3.83	3.89	−16.64	62
CH_4‡	−27.97	9.02	9.31	−18.95	63
H_2O_2‡	−51.95	19.52	—	−32.43	62
CH_3F‡	−60.02	32.64	38.05	−27.38	64
H_2O‡	−15.0	0.9	1.504	−14.1	§
NH_3‡	−20.7	1.8	3.89	−18.9	§
CH_4‡	−28.3	3.9	9.31	−24.4	§
C_2H_6‡	−86.8	28.2	—	−58.6	§
C_2H_4‡	−65.2	24.8	—	−40.4	§
C_2H_2‡	−47.6	12.9	—	−34.7	§
CH_3F‡	−60.3	16.6	38.05	−43.7	§
CH_3OH‡	−75.2	22.6	—	−52.6	§
HCHO‡	−47.8	24.9	—	−22.9	§
HCN‡	−39.4	14.7	—	−24.7	§

†Units are 10^{-6} erg G^{-2} mol^{-1}.
‡The origin of the vector potential was taken at the centre of mass. For all other molecules the origin was taken at the nucleus marked*.
§These unpublished results were calculated using the method of Reference 50 together with the ϕ-functions specified in Reference 72. See also References 53 and 69.

Pople[50, 53, 69]. The Italian group used the same fairly large extended set of STOs to describe both Ψ_0 and $\Psi(\boldsymbol{H})$. Minimal[70, 71] and slightly extended sets[72] of contracted Gaussian functions were used in our work.

Table 4.3 presents some susceptibility results obtained by this type of method. To obtain some indication of the accuracy of their calculated susceptibility values for diatomic molecules, Stevens *et al.*[51, 54–61] compared rotational magnetic moments derived from calculated χ^p values with experimental moments from molecular-beam studies. The agreement for LiH and HF was good, but somewhat worse for F_2. Since errors in calculated values of χ^d are expected to be less than those in χ^p, this suggests that χ values are also calculated well. There is, however, a discrepancy between the calculated and experimental (-8.6) χ values for HF. Stevens and Lipscomb[55] have attributed this to the fact that the experimental value is a liquid-phase measurement. It should be noted, however, that other theories[41,84] calculate values of χ for HF which are fairly close to the experimental value.

An interesting result of Table 4.3 is the prediction that BH is paramagnetic. The large value of χ^p arises from the presence of a low-lying $\sigma \rightarrow \pi^*$ excited state which is easily mixed into the ground state by the magnetic field. The diamagnetic contribution is fairly small because of the small number of electrons. This result, however, has been regarded with some reservations since other calculations[84] have found BH to be diamagnetic. To obtain more information on this problem, Laws *et al.*[61] have calculated χ for AlH and find a small negative value. AlH has a similar ground-state electronic structure to BH and also possesses a similar low-lying $\sigma \rightarrow \pi^*$ excited state. The calculated value of χ^p is quite close to that in BH, but the value of χ^d is more negative due to the larger number of electrons in AlH. Laws *et al.*[61] thus attribute the difference between χ values for BH and AlH to differences in χ^d and suggest that experimental verification of the small negative value of χ in AlH would provide additional evidence that BH is paramagnetic.

It should be noted that this type of theory will only lead to susceptibility values which are independent of the gauge of the vector potential when the set of ϕ-functions is complete[40]. Even the diatomic values which are obtained with sets of ϕ-functions about three to four times the size of a minimal basis set show a slight gauge dependence. Thus it is unlikely that gauge-invariant susceptibility values for large molecules will be obtained by such methods.

Arrighini *et al.*[62–68] have calculated χ values for small polyatomic molecules using basis sets in which first-row atoms are described by s, p and d functions and hydrogen atoms by s and p functions. The theoretical values of χ^p for water, ammonia and methane are in good agreement with the experimental data. Calculated χ values for these three hydrides are also in satisfactory agreement with the experimental values of -13.0 (H_2O), -16.3 (NH_3) and -17.4 (CH_4). For methyl fluoride the theoretical results for both χ and χ^p are considerably worse.

To investigate the performance of a *slightly* extended basis set we calculated values of χ using the 4–31 G set[72] of ϕ-functions. Each first-row atom 1s orbital is described by a single ϕ-function, but valence orbitals (2s and 2p for first-row atoms, 1s for hydrogen) are each described by two ϕ-functions. It is apparent from Table 4.3 that the use of SCF perturbation theory with such basis sets is inadequate for calculating absolute and relative susceptibility values. A comparison of our results with those of Arrighini *et al.* shows that the main deficiency is in the calculated χ^p contribution; our calculated χ^d values are quite close to those obtained by the Italian group. The calculated values of χ^p are markedly improved by adding d functions on first-row atoms and p-functions on hydrogen.

Magnetic shielding results calculated using the same basis sets are presented in Table 4.4. For this property even the large basis-set calculations of Stevens *et al.* show a marked gauge dependence. Calculated values of spin–rotation constants (derived from σ^p), however, are in good agreement with the experimental values for LiH and HF. Also the theoretical value of σ for F_2 is in moderate agreement with the antishielding found experimentally.

In our magnetic-shielding work[53,69] we were interested in changes in shielding rather than in absolute values. It was hoped that errors in the description of individual molecules would largely cancel and that the effects responsible for changes in magnetic shielding would be handled satisfactorily by the perturbed SCF method using the 4–31 G basis set. The calculated results are in moderate overall agreement with experiment, a

Table 4.4 Ab initio magnetic-shielding results calculated by SCF perturbation theory†

Molecule	*σ(calc)*	*σ(exp)*	*Reference*
Li*H	90.2	—	54
F*H	414	—	55
FH*	437	—	55
F_2^*	−276	−210	56
F_2‡	−200	−210	56
Li_2‡	98.6	—	58
B*H	−262	—	59
H_2O	331.06	—	67
NH_3	272.34	—	67
CH_4	193.89	0.011	67
CH_3F§	161.66	−77.5	67
CH_4‡§	221.1	0.0	69
C_2H_6‡§	230.1	−8.0	69
CH_3NH_2‡§	213.2	−14.8	69
CH_3OH‡§	195.7	−50.1	69
CH_3F‡§	179.8	−77.5	69
C_2H_4‡§	94.8	−126	69
H_2CO‡§	14.1	−196.8	69
C_2H_2‡§	142.9	−76	69
CH_3CHO‡§	213.2	−31.7	69
CH_3CHO‡¶	45.2	−202.6	69
$HCONH_2$‡§	87.9	−167.9	69

*This indicates the nucleus under study. The origin of the vector potential is also taken at this nucleus.
†Units are parts per million.
‡Origin of the vector potential taken at the centre of mass.
§σ refers to the ^{13}C magnetic shielding constant.
¶σ refers to the carbonyl ^{13}C magnetic shielding constant.
||All experimental ^{13}C values are given as chemical shifts relative to CH_4. A negative value indicates a down-field shift from CH_4.

number of important effects being well reproduced. The observed low-field ^{13}C shifts of CH_3X compounds relative to ethane are well described, as are the magnetic-shielding effects associated with multiple bonds. In contrast, the chemical shifts of ethane relative to methane, and acetaldehyde relative to formaldehyde, are poorly calculated. In these cases (unlike the CH_3X compounds), the positions of the carbon atoms relative to the origin of the vector potential are very different and it seems likely that these poor theoretical values are due, at least in part, to the gauge dependence of this method.

The results of Table 4.4, however, suggest that this method will successfully calculate the chemical shift of a particular nucleus in a group of molecules of comparable size (e.g. X—CH_2^*—Y, X, Y=CH_3, NH_2, OH and F) provided that the position of that nucleus relative to the origin of the vector potential does not markedly change.

Therefore this type of method can give a good description of the magnetic properties of diatomic molecules (if very large basis sets are used) and of the relative shielding in certain polyatomic systems. However, it is unlikely that this approach will be useful for a general study of the diamagnetism of large molecules.

4.2.5 Self-consistent perturbation theory with gauge-invariant atomic orbitals

All the theoretical methods discussed so far are attempts to evaluate the diamagnetic and paramagnetic parts of the susceptibility separately. For large molecules, these two parts become big and largely cancel so that calculations become subject to considerable error. In addition, the SCF theory outlined above cannot give gauge-invariant results using sets of ϕ-functions that are practical for calculations on large molecules.

A way of avoiding these difficulties is to use gauge-invariant atomic orbitals* (GIAO), where the atomic basis functions ϕ_ν are modified by a multiplicative complex factor which depends on the gauge of the vector potential. The use of GIAO with SCF perturbation theory not only avoids the problem of the summation over excited states, but with this method even minimal sets of ϕ-functions will give gauge-invariant results for χ[74]. Use of GIAO in problems of molecular diamagnetism was first made by London[75] in connection with ring currents in aromatic hydrocarbons, and they have since been employed in more general studies of magnetic properties by Hameka[76–86], Pople[11, 87–93], McWeeny[94], Hall and Hardisson[95], Davies[96], Amos and Roberts[74] and others[97–100].

Self-consistent perturbation theory of magnetic susceptibility and magnetic shielding was first developed by Davies[96] and by Hall and Hardisson[95]. These authors, however, were interested in the ring-current contributions to $\boldsymbol{\chi}$ and $\boldsymbol{\sigma}$ and consequently their developments reflected the approximations of the π-electron method[101, 102].

Here we develop the theory in a general form and then show how the introduction of various types of approximation leads to the theories proposed by Bley[97, 98], Pople and others. This presents all the theories in a unified framework and allows a realistic appraisal of the various approximate methods.

GIAO are easily introduced into the SCF theory by writing

$$\psi_j(\boldsymbol{H}) = \sum_{\nu} c_{\nu j}(\boldsymbol{H})\chi_\nu(\boldsymbol{H}) \tag{4.78}$$

*This is an unfortunate phrase since these orbitals include a factor which depend on the gauge. A better description would be that proposed by Pople of 'gauge-dependent atomic orbitals'.

where the GIAO, $\chi_\nu(\boldsymbol{H})$, is given by

$$\chi_\nu(\boldsymbol{H}) = \exp\left((-i/c)\boldsymbol{A}_\nu \cdot \boldsymbol{r}\right)\phi_\nu$$
$$= f_\nu \phi_\nu \tag{4.79}$$

Here $\boldsymbol{A}_\nu = \frac{1}{2}\boldsymbol{H} \wedge \boldsymbol{R}_\nu$, where $\boldsymbol{R}_\nu$ is the distance vector from the arbitrary origin to the nucleus on which ϕ_ν is located. The LCAO coefficients $c_{\nu j}(\boldsymbol{H})$, which are again allowed to be complex, can be found by solving the perturbed Roothaan equations

$$\sum_\lambda F_{\nu\lambda}(\boldsymbol{H}) - \varepsilon_j(\boldsymbol{H}) S_{\nu\lambda}(\boldsymbol{H})\, c_{\lambda j}(\boldsymbol{H}) = 0 \tag{4.80}$$

However, since the atomic orbitals χ_ν depend upon the magnetic field, there are some important differences in the self-consistent perturbation theory. The Fock matrix elements $F_{\nu\lambda}(\boldsymbol{H})$ are given by

$$F_{\nu\lambda}(\boldsymbol{H}) = H_{\nu\lambda}(\boldsymbol{H}) + G_{\nu\lambda}(\boldsymbol{H}) \tag{4.81}$$

where

$$H_{\nu\lambda}(\boldsymbol{H}) = \langle \chi_\nu(\boldsymbol{H}) | \{ \tfrac{1}{2}(-i\nabla + (1/c)\boldsymbol{A}(\boldsymbol{r}))^2 - \sum_{\mathrm{B}} Z_{\mathrm{B}} r_{\mathrm{B}}^{-1} \} | \chi_\lambda(\boldsymbol{H}) \rangle \tag{4.82}$$

$$G_{\nu\lambda}(\boldsymbol{H}) = \sum_{\rho\sigma} P_{\rho\sigma}(\boldsymbol{H}) \{ (\chi_\nu^* \chi_\lambda | \chi_\rho^* \chi_\sigma) - \tfrac{1}{2}(\chi_\nu^* \chi_\sigma | \chi_\rho^* \chi_\lambda) \}$$
$$= \sum_{\rho\sigma} P_{\rho\sigma}(\boldsymbol{H}) G_{\nu\lambda\rho\sigma}(\boldsymbol{H}) \tag{4.83}$$

We note that use of GIAO leads to overlap, core and two-electron integrals with an explicit field dependence. The energy, $E(\boldsymbol{H})$ corresponding to $\Psi(\boldsymbol{H})$ is given by

$$E(\boldsymbol{H}) = \sum_{\nu\lambda} P_{\nu\lambda}(\boldsymbol{H}) \{ H_{\nu\lambda}(\boldsymbol{H}) + \tfrac{1}{2} G_{\nu\lambda}(\boldsymbol{H}) \} \tag{4.84}$$

Recalling that $E_\alpha^{(1,0)} = (\partial E(\boldsymbol{H})/\partial H_\alpha)_{H_\alpha = 0}$, we obtain

$$E_\alpha^{(1,0)} = i \sum_{\nu\lambda} \left\{ (P_{\nu\lambda}^{(1)})_\alpha (H_{\nu\lambda}^{(0)} + G_{\nu\lambda}^{(0)}) + P_{\nu\lambda}^{(0)} \left[(H_{\nu\lambda}^{(1,0)})_\alpha + \tfrac{1}{2} \sum_{\rho\sigma} P_{\rho\sigma}^{(0)} (G_{\nu\lambda\rho\sigma}^{(1)})_\alpha \right] \right\} \tag{4.85}$$

Since the solutions of equation (4.80) are constrained to be orthonormal for each value of the magnetic field strength we may write

$$\sum_{\nu\lambda} c_{\nu j}^*(\boldsymbol{H}) c_{\lambda l}(\boldsymbol{H}) S_{\nu\lambda}(\boldsymbol{H}) = \delta_{jl} \tag{4.86}$$

and

$$\sum_{\nu\lambda} \left\{ c_{\nu j}^{(0)} c_{\lambda l}^{(1)} S_{\nu\lambda}^{(0)} - c_{\nu j}^{(1)} c_{\lambda l}^{(0)} S_{\nu\lambda}^{(0)} + c_{\nu j}^{(0)} c_{\lambda l}^{(0)} S_{\nu\lambda}^{(1)} \right\} = 0 \tag{4.87}$$

Using equations (4.87) and (4.71) allows equation (4.85) to be simplified to

$$E_\alpha^{(1,0)} = i \sum_{j\nu\lambda} c_{\nu j}^{(0)} \left\{ 2(H_{\nu\lambda}^{(1,0)})_\alpha + \sum_{\rho\sigma} P_{\rho\sigma}^{(0)} (G_{\nu\lambda\rho\sigma}^{(1)})_\alpha - 2(S_{\nu\lambda}^{(1)})_\alpha \varepsilon_j^{(0)} \right\} c_{\lambda j}^{(0)} \tag{4.88}$$

An important point to note about equation (4.88) is that the evaluation of $E_\alpha^{(1,0)}$ does not require the first-order correction to the coefficients $(c_{\nu j}^{(1)})_\alpha$

or the orbital energies $(\varepsilon_j^{(1)})_\alpha$. Thus the calculation of the magnetic susceptibility tensor and the magnetic shielding constants require only the first-order corrections $c_v^{(1)}$ and $\varepsilon_j^{(1)}$.

The equation for the zz component of the magnetic susceptibility tensor is

$$\chi_{zz} = -\sum_{j\nu\lambda}\Bigg\{-(c_{\nu j}^{(1)})_z\Big[2(H_{\nu\lambda}^{(1,0)})_z+\sum_{\rho\sigma}P_{\rho\sigma}^{(0)}(G_{\nu\lambda\rho\sigma}^{(1)})_z-2(S_{\nu\lambda}^{(1)})_z\varepsilon_j^{(0)}\Big]c_{\lambda j}^{(0)}$$
$$+c_{\nu j}^{(0)}\Big[2(H_{\nu\lambda}^{(2,0)})_{zz}+\sum_{\rho\sigma}\Big((P_{\rho\sigma}^{(1)})_z(G_{\nu\lambda\rho\sigma}^{(1)})_z+P_{\rho\sigma}^{(0)}(G_{\nu\lambda\rho\sigma}^{(2)})_{zz}\Big)$$
$$-2(S_{\nu\lambda}^{(2)})_{zz}\varepsilon_j^{(0)}-2(S_{\nu\lambda}^{(1)})_z(\varepsilon_j^{(1)})_z\Big]c_{\lambda j}^{(0)}+c_{\nu j}^{(0)}\Big[2(H_{\nu\lambda}^{(1,0)})_z+$$
$$\sum_{\rho\sigma}P_{\rho\sigma}^{(0)}(G_{\nu\lambda\rho\sigma}^{(1)})_z-2(S_{\nu\lambda}^{(1)})_z\varepsilon_j^{(0)}\Big](c_{\lambda j}^{(1)})_z \tag{4.89}$$

where

$$H_{\nu\lambda}^{(0)} = H_{\nu\lambda}^{\text{core}} \tag{4.90}$$

$$(H_{\nu\lambda}^{(1,0)})_z = -(1/2c)\{\langle\phi_\nu\,|\,L_z^\lambda\,|\,\phi_\lambda\rangle-Q_{\nu\lambda}^z H_{\nu\lambda}^{(0)}-\langle T_{\nu\lambda}^z\phi_\nu\,|\,H^{\text{core}}\,|\,\phi_\lambda\rangle\} \tag{4.91}$$

$$(H_{\nu\lambda}^{(2,0)})_{zz} = (1/4c^2)\{\langle\phi_\nu\,|\,x_\lambda^2+y_\lambda^2\,|\,\phi_\lambda\rangle-Q_{\nu\lambda}^z\langle\phi_\nu\,|\,L_z^\lambda\,|\,\phi_\lambda\rangle-\langle T_{\nu\lambda}^z\phi_\nu|L_z^\lambda|\phi_\lambda\rangle$$
$$-(Q_{\nu\lambda}^z)^2H_{\nu\lambda}^{(0)}-2Q_{\nu\lambda}^z(T_{\nu\lambda}^z\phi_\nu\,|\,H^{\text{core}}\,|\,\phi_\lambda\rangle-\langle(T_{\nu\lambda}^z)^2\phi_\nu\,|\,H^{\text{core}}\,|\,\phi_\lambda\rangle\} \tag{4.92}$$

with similar equations for $(G_{\nu\lambda\rho\sigma}^{(1)})_z$ and $(G_{\nu\lambda\rho\sigma}^{(2)})_{zz}$. In equations (4.91) and (4.92)

$$L_z^\lambda = (\boldsymbol{r}_\lambda\wedge\boldsymbol{\nabla})_z \tag{4.93}$$

$$\boldsymbol{r}_\lambda = \boldsymbol{r}-\boldsymbol{R}_\lambda \tag{4.94}$$

and in deriving these equations we have made use of the relationship

$$\langle\chi_\nu\,|\,\tfrac{1}{2}(-i\boldsymbol{\nabla}+(1/c)\boldsymbol{A}(\boldsymbol{r}))^2\,|\,\chi_\lambda\rangle = \langle f_\nu^* f_\lambda\phi_\nu\,|\,\tfrac{1}{2}(-i\boldsymbol{\nabla}+(1/c)\boldsymbol{A}(\boldsymbol{r}_\lambda))^2\,|\,\phi_\lambda\rangle \tag{4.95}$$

The gauge factor product in equation (4.95) is given by

$$f_\nu^* f_\lambda = \exp\{(i/c)(\boldsymbol{A}_\nu-\boldsymbol{A}_\lambda).\boldsymbol{r}\} \tag{4.96}$$

and we have written equations (4.91) and (4.92) in terms of the vectors $\boldsymbol{Q}_{\nu\lambda}$ and $\boldsymbol{T}_{\nu\lambda}$ which are defined by

$$\boldsymbol{Q}_{\nu\lambda} = \boldsymbol{R}_\nu\wedge\boldsymbol{R}_\lambda \tag{4.97}$$

$$\boldsymbol{T}_{\nu\lambda} = \boldsymbol{R}_{\nu\lambda}\wedge\boldsymbol{r}_\nu \tag{4.98}$$

Similarly, the zz component of the magnetic shielding tensor at nucleus B is given by

$$\sigma_{\text{B}zz} = \sum_{\nu\lambda}\Big\{P_{\nu\lambda}^{(0)}(H_{\text{B}\nu\lambda}^{(1,1)})_{zz}+(P_{\nu\lambda}^{(1)})_z(H_{\text{B}\nu\lambda}^{(0,1)})_z\Big\} \tag{4.99}$$

where

$$(H^{(0,1)}_{\mathrm{B}\nu\lambda})_z = -(1/c)\langle\phi_\nu|L_{\mathrm{B}z}r_{\mathrm{B}}^{-3}|\phi_\lambda\rangle \tag{4.100}$$

$$(H^{(1,1)}_{\mathrm{B}\nu\lambda})_{zz} = (1/2c^2)\{\langle\phi_\nu|(x_{\mathrm{B}}x_\lambda+y_{\mathrm{B}}y_\lambda)r_{\mathrm{B}}^{-3}|\phi_\lambda\rangle - Q^z_{\nu\lambda}H^{(0,1)}_{\mathrm{B}\nu\lambda} - \langle T^z_{\nu\lambda}\phi_\nu|L_{\mathrm{B}z}r_{\mathrm{B}}^{-3}|\phi_\lambda\rangle\} \tag{4.101}$$

The first-order coefficients $c^{(1)}_{\nu j}$ are determined by expanding the perturbed Roothaan equation (4.80), separating orders of $\boldsymbol{H}$ and solving the first-order equation given by

$$\sum_\lambda \left\{(F^{(0)}_{\nu\lambda} - \varepsilon^{(0)}_j S^{(0)}_{\nu\lambda})c^{(1)}_{\lambda j} + (F^{(1)}_{\nu\lambda} - \varepsilon^{(0)}_j S^{(1)}_{\nu\lambda})c^{(0)}_{\lambda j}\right\} = 0 \tag{4.102}$$

where the elements of $\boldsymbol{F}^{(1)}$ are

$$F^{(1)}_{\nu\lambda} = H^{(1,0)}_{\nu\lambda} + \sum_{\rho\sigma}(P^{(0)}_{\rho\sigma}G^{(1)}_{\nu\lambda\rho\sigma} + P^{(1)}_{\rho\sigma}G^{(0)}_{\nu\lambda\rho\sigma}) \tag{4.103}$$

Using standard perturbation theory gives the projection of $c^{(1)}_{\lambda j}$ on the empty orbitals. Since the atomic orbitals depend on the magnetic field, there will also be a contribution to $c^{(1)}_{\lambda j}$ from the occupied orbitals. This contribution can be determined from the orthonormality conditions (equation (4.86)) so that the total is given by (in matrix notation)

$$\boldsymbol{c}^{(1)}_j = \sum_k^{\mathrm{occ}}(-\tfrac{1}{2}\boldsymbol{c}^{(0)\dagger}_k\boldsymbol{S}^{(1)}\boldsymbol{c}^{(0)}_j)\boldsymbol{c}^{(0)}_k - \sum_l^{\mathrm{unocc}}(\varepsilon^{(0)}_l - \varepsilon^{(0)}_j)^{-1}[\boldsymbol{c}^{(0)\dagger}_l\{\boldsymbol{F}^{(1)} - \varepsilon^{(0)}_j\boldsymbol{S}^{(1)}\}\boldsymbol{c}^{(0)}_j]\boldsymbol{c}^{(0)}_l \tag{4.104}$$

It is clear that these formulae reduce to the fixed basis set results if the gauge factors are set equal to unity. Similar equations have been derived by Moccia[103] for the general first, second and third derivatives of the SCF energy.

There have been few applications of this theory at the *ab initio* level. The calculations that are available[100, 104] have so far been confined to calculations of $\boldsymbol{\sigma}$ since this is a somewhat easier problem. Although calculations of $\boldsymbol{\chi}$ and $\boldsymbol{\sigma}$ require the same perturbed wave function, there are many more terms in the equation for $\boldsymbol{\chi}$, some of which are difficult to evaluate (e.g. $G^{(2)}_{\nu\lambda\rho\sigma}$). However, it is likely that this theory will be applied to a study of $\boldsymbol{\chi}$ in the near future.

To illustrate the promise of this theory we summarise some preliminary results for magnetic shielding in Table 4.5. These results were obtained using a minimal basis set of atomic Hartree–Fock ϕ-functions (LEMAO-5G)[71] appropriately modified by gauge factors. The calculated chemical shifts for both carbon and hydrogen are in good overall agreement with experiment. The ^{13}C chemical shift for methyl fluoride relative to methane shows a marked improvement over the values of Table 4.4. The ^{13}C chemical shifts for multiply-bonded molecules, however, are somewhat overestimated. This may be due to a lack of flexibility afforded by the isotropic LEMAO-5G minimal basis set. The calculated proton chemical shifts show all the observed trends and the overall accuracy of these results suggests that a meaningful analysis of the various contributions to proton shielding could be made.

We conclude this section by briefly mentioning the GIAO work of Hameka which has already been fully reviewed[80]. In Hameka's approach, a single determinant wave function is first obtained for the ground state in the absence of the field. The usual atomic orbitals are then replaced by GIAO and χ and σ are determined by perturbation theory rather than by resolving self-consistently to find $\Psi(\boldsymbol{H})$. Thus, although Hameka's method gives gauge-invariant results for χ and σ, the wave function $\Psi(\boldsymbol{H})$ is not self-consistent and the use of perturbation theory introduces unknown energy

Table 4.5 Ab initio magnetic-shielding results using SCF perturbation theory and GIAO†

Molecule	σ(*calc*)	$\Delta\sigma$(*calc*)	$\Delta\sigma$(*exp*)
C^*H_4	233.1	0.0	0.0
C^*H_3F	158.6	−74.5	−77.5‡
$C_2^*H_2$	142.6	−90.5	−76.0‡
$C_2^*H_4$	80.3	−152.8	−126.0‡
H_2C^*O	−17.8	−250.9	−197‡
CH_4^*	32.9	0.0	0.0
CH_3^*F	29.7	−3.2	−4.0
$C_2H_2^*$	30.6	−2.3	−2.65§
$C_2H_4^*$	26.7	−6.2	−5.61§
H_2^*CO	22.8	−10.1	−9.4§

*This indicates the nucleus under study.
†Units are parts per million.
‡See Reference 69.
§These experimental values are taken from Reference 233, p. 25 and p. 63.

denominators into the equations for χ and σ similar to those found in the sum-over-states theory. The results obtained for χ are usually similar to those obtained by variational procedures[41]. However, Hameka's approach often gives quite different values for $\Delta\chi$. Although the original method has recently been extended to triatomic molecules[86], it is likely that the SCF perturbation method with GIAO will be preferred in future calculations of χ and σ for polyatomic molecules.

4.2.6 Semi-empirical molecular-orbital methods

In contrast to the few *ab initio* calculations, several groups have calculated χ and σ using semi-empirical approaches, many of which can conveniently be derived by introducing approximations into the theory developed in Section 4.2.5.

The first approximate theory of diamagnetism which considered all the valence electrons was that given by Pople[11,89] several years ago for non-cyclic molecules. His method was developed within the independent-electron framework and thus all explicit two-electron terms (e.g. $G_{\nu\lambda\rho\sigma}$) vanish from the theory. In this approximation the LCAO coefficients are determined by solving the well-known secular equations

$$\sum_{\lambda} \{H_{\nu\lambda}(\boldsymbol{H}) - \varepsilon_j(\boldsymbol{H}) S_{\nu\lambda}(\boldsymbol{H})\} c_{\lambda j}(\boldsymbol{H}) = 0 \tag{4.105}$$

It should be noted that this equation is linear in the coefficients $c_{\lambda j}$.

In this type of theory the zero-field core matrix elements, $H_{\nu\lambda}^{\text{core}(0)}$, are usually approximated in some simple way without ever specifying the nature of $H^{(0)}$. Pople also systematically neglected all two-centre overlap integrals, $S_{\nu\lambda}$, thus setting $S_{\nu\lambda}^{(1)}$ and $S_{\nu\lambda}^{(2)}$ of equation (4.89) to zero. With these simplifications the first-order matrix elements corresponding to $F_{\nu\lambda}^{(1)}$ of equation (4.81) are given by

$$F_{\nu\lambda}^{(1)} = H_{\nu\lambda}^{(1,0)} \tag{4.106}$$

and the first-order LCAO coefficients by

$$\boldsymbol{c}_j^{(1)} = -\sum_l^{\text{unocc}} (\varepsilon_l^{(0)} - \varepsilon_j^{(0)})^{-1} [\boldsymbol{c}_l^{(0)\dagger} \boldsymbol{H}^{(1,0)} \boldsymbol{c}_j^{(0)}] \boldsymbol{c}_l^{(0)} \tag{4.107}$$

Thus it only remains to specify $H_{\nu\lambda}^{(1,0)}$ and $H_{\nu\lambda}^{(2,0)}$. Before defining these it is usual to introduce an approximation first suggested by London[75]. This involves replacing $\boldsymbol{r}$ in $f_\nu{}^* f_\lambda$ of equation (4.96) by $(\boldsymbol{R}_\nu + \boldsymbol{R}_\lambda)/2$. With this approximation $T_{\nu\lambda}^z$ vanishes and $(H_{\nu\lambda}^{(1,0)})_z$ and $(H_{\nu\lambda}^{(2,0)})_{zz}$ are given by

$$(H_{\nu\lambda}^{(1,0)})_z = -(1/2c)\langle \phi_\nu | L_z^\lambda | \phi_\lambda \rangle;\ \phi_\nu \text{ and } \phi_\lambda \text{ on the same atom} \tag{4.108}$$

$$(H_{\nu\lambda}^{(1,0)})_z = (1/2c) Q_{\nu\lambda}^z H_{\nu\lambda}^{\text{core}(0)};\ \phi_\nu \text{ on atom A, } \phi_\lambda \text{ on atom B} \tag{4.109}$$

$$(H_{\nu\lambda}^{(2,0)})_{zz} = (1/4c^2)\langle \phi_\nu | (x_\lambda^2 + y_\lambda^2) | \phi_\lambda \rangle;\ \phi_\nu \text{ and } \phi_\lambda \text{ on the same atom} \tag{4.110}$$

$$(H_{\nu\lambda}^{(2,0)})_{zz} = -(1/4c^2)(Q_{\nu\lambda}^z)^2 H_{\nu\lambda}^{\text{core}(0)};\ \phi_\nu \text{ on atom A, } \phi_\lambda \text{ on atom B} \tag{4.111}$$

The two-centre $\langle \phi_\nu | L^\lambda | \phi_\lambda \rangle$ and $\langle \phi_\nu | x_\lambda^2 + y_\lambda^2 | \phi_\lambda \rangle$ integrals are neglected since they may be reduced to two-centre overlap integrals which are systematically set to zero. Pople also neglected the above two-centre $H_{\nu\lambda}^{(1,0)}$ and $H_{\nu\lambda}^{(2,0)}$ terms showing that these would make no contribution to the susceptibility of non-cyclic molecules within the approximations of the independent-electron method. He also introduced a further approximation to the one-centre elements $H_{\nu\lambda}^{(2,0)}$ setting

$$H_{\nu\lambda}^{(2,0)} = 0 \quad \text{unless } \phi_\nu = \phi_\lambda \tag{4.112}$$

Thus the only remaining non-zero matrix elements are $H_{\nu\lambda}^{(1,0)}$ of equation (4.108) and

$$(H_{\nu\nu}^{(2,0)})_{zz} = (1/4c^2)\langle \phi_\nu | (x_\nu^2 + y_\nu^2) | \phi_\nu \rangle \tag{4.113}$$

The equations for χ_{zz}^{d} and χ_{zz}^{p} then become

$$\chi_{zz}^{\text{d}} = -(1/4c^2) \sum_\nu P_{\nu\nu} (x_\nu^2 + y_\nu^2)_{\nu\nu} \tag{4.114}$$

and

$$\chi_{zz}^{\text{p}} = (1/c^2) \sum_j^{\text{occ}} \sum_l^{\text{unocc}} (\varepsilon_l - \varepsilon_j)^{-1} \sum_{\nu<\lambda} \sum_{\rho<\sigma} \{(c_{\nu j} c_{\lambda l} - c_{\lambda j} c_{\nu l}) \times (c_{\rho j} c_{\sigma l} - c_{\sigma j} c_{\rho l}) H_{\nu\lambda}^{(1,0)} H_{\rho\sigma}^{(1,0)}\} \tag{4.115}$$

with similar expressions for the other elements. These diamagnetic and paramagnetic contributions can be split into atomic parts to give

$$\boldsymbol{\chi}^{\text{d}} = \sum_{\text{A}} \boldsymbol{\chi}_{\text{A}}^{\text{d}} \tag{4.116}$$

where

$$(\chi_{\mathrm{A}}^{\mathrm{d}})_{zz} = -(1/4c^2)\sum_{\nu}^{\mathrm{A}} P_{\nu\nu}(x_{\mathrm{A}}^2+y_{\mathrm{A}}^2)_{\nu\nu} \tag{4.117}$$

Here the sum is restricted to atomic orbitals on centre A. Equations (4.116) and (4.117) give the diamagnetism that would be calculated for a molecule if the Langevin atomic formula were applied to each atomic orbital with the appropriate electron population $P_{\nu\nu}$.

Separating the paramagnetic contribution into atomic terms we obtain

$$\chi^{\mathrm{p}} = \sum_{\mathrm{A}} \chi_{\mathrm{A}}^{\mathrm{p}} \tag{4.118}$$

where

$$(\chi_{\mathrm{A}}^{\mathrm{p}})_{zz} = (1/c^2)\sum_{j}^{\mathrm{occ}}\sum_{l}^{\mathrm{unocc}} (\varepsilon_l-\varepsilon_j)^{-1} \sum_{\nu<\lambda}^{\mathrm{A}}\sum_{\mathrm{B}}\sum_{\rho<\sigma}^{\mathrm{B}} \{(c_{\nu j}c_{\lambda l}-c_{\lambda j}c_{\nu l}) \times(c_{\rho j}c_{\sigma l}-c_{\sigma j}c_{\rho l})(L_z^{\lambda})_{\nu\lambda}(L_z^{\sigma})_{\rho\sigma}\} \tag{4.119}$$

This can be interpreted as the paramagnetic moment per unit field induced on atom A. Unlike $\chi_{\mathrm{A}}^{\mathrm{d}}$, this involves a sum over all atoms in the molecule since a magnetic field acting on atom B will induce currents on atom A. If we consider only hydrogen and the first-row atoms carbon to fluorine and work with a minimal set of ϕ-functions, then equation (4.119) may be further simplified by noting that

$$\begin{aligned}(L_z^{\lambda})_{\nu\lambda} &= 1 \text{ if } \phi_\nu = 2\mathrm{p}_x,\ \phi_\lambda = 2\mathrm{p}_y;\ \phi_\nu \text{ and } \phi_\lambda \text{ on the same centre}\\ &= 0 \text{ otherwise}\end{aligned} \tag{4.120}$$

Therefore $(\chi_{\mathrm{A}}^{\mathrm{p}})_{zz}$ may be written in the form

$$(\chi_{\mathrm{A}}^{\mathrm{p}})_{zz} = (1/c^2)\sum_{j}^{\mathrm{occ}}\sum_{l}^{\mathrm{unocc}} \{(\varepsilon_l-\varepsilon_j)^{-1}(c_{x\mathrm{A}j}c_{y\mathrm{A}l}-c_{y\mathrm{A}j}c_{x\mathrm{A}l}) \times\sum_{\mathrm{B}}(c_{x\mathrm{B}j}c_{y\mathrm{B}l}-c_{y\mathrm{B}j}c_{x\mathrm{B}l})\} \tag{4.121}$$

where $c_{x\mathrm{A}j}$ is written for the LCAO coefficient of the p_x orbital on atom A, etc. (Since all the quantities in equation (4.121) are zero-order in H we have omitted the superscripts.)

Equation (4.121) leads to the following important conclusions:

(a) The local contribution to $\chi_{\mathrm{A}}^{\mathrm{p}}$ (i.e. when B = A in $\sum_{\mathrm{B}}$) which is due to the currents induced on centre A is zero unless atom A has 2p electrons. Thus this contribution is zero for hydrogen and has finite values for the atoms carbon to fluorine.

(b) The local contribution to $\chi_{\mathrm{A}}^{\mathrm{p}}$ may be anisotropic.

(c) Because of the $(\varepsilon_l-\varepsilon_j)^{-1}$ factor, the local contribution to $\chi_{\mathrm{A}}^{\mathrm{p}}$ is likely to be larger for systems which have low-lying excited states (of the correct symmetry) localised on centre A.

(d) The contributions to $\chi_{\mathrm{A}}^{\mathrm{p}}$ due to the magnetic field acting on other atoms (i.e. B $\neq$ A in $\sum_{\mathrm{B}}$) are zero unless *both* atoms A and B have 2p electrons. Thus this contribution is zero for both atoms in X—H bonds.

(e) The other-atom contributions are likely to be larger if the molecule possesses low-lying excited states.

An additional simplification to equation (4.121) can be made by replacing $(\varepsilon_l - \varepsilon_j)^{-1}$ by an average value $\langle \Delta E^{-1} \rangle$, and noting that equation (4.122)

$$\sum_j^{\text{occ}} c_{\nu j} c_{\lambda j} + \sum_l^{\text{unocc}} c_{\nu l} c_{\lambda l} = \delta_{\nu\lambda} \tag{4.122}$$

is valid for the neglect of differential overlap case. The final equation obtained by Pople[11] is

$$(\chi_{\text{A}}^{\text{p}})_{zz} = (1/4c^2) \langle (\Delta E)^{-1} \rangle Q_{\text{A}} \tag{4.123}$$

where

$$\boldsymbol{Q}_{\text{A}} = \sum_{\text{B}} \boldsymbol{Q}_{\text{AB}} \tag{4.124}$$

$$\boldsymbol{Q}_{\text{AB}} = P_{x_{\text{A}} x_{\text{B}}}(2\delta_{\text{AB}} - P_{y_{\text{A}} y_{\text{B}}}) + P_{y_{\text{A}} y_{\text{B}}}(2\delta_{\text{AB}} - P_{x_{\text{A}} x_{\text{B}}}) + 2P_{x_{\text{A}} y_{\text{B}}} P_{y_{\text{A}} x_{\text{B}}} \tag{4.125}$$

Pople also developed a theory of n.m.r. chemical shifts at the same level of approximation using the complete vector potential of equation (4.26). The expressions for the diamagnetic and paramagnetic contributions to the shielding at nucleus A are

$$(\sigma_{\text{A}}^{\text{d}})_{zz} = (1/2c^2) \sum_{\nu} P_{\nu\nu}^{(0)} [(x_{\text{A}} x_\nu + y_{\text{A}} y_\nu) r_{\text{A}}^{-3}]_{\nu\nu} \tag{4.126}$$

and

$$(\sigma_{\text{A}}^{\text{p}})_{zz} = -(2/c^2) \sum_j^{\text{occ}} \sum_l^{\text{unocc}} (\varepsilon_l - \varepsilon_j)^{-1} \sum_{\nu < \lambda} \sum_{\rho < \sigma} \{(c_{\nu j} c_{\lambda l} - c_{\lambda j} c_{\nu l}) \times (c_{\rho j} c_{\sigma l} - c_{\sigma j} c_{\rho l}) (L_{\text{A}z} r_{\text{A}}^{-3})_{\nu\lambda} (L_z^{\sigma})_{\rho\sigma}\} \tag{4.127}$$

These contributions can also be broken down into various parts. The local (to atom A) diamagnetic and paramagnetic terms are given by

$$(\sigma_{\text{A}}^{\text{d}})_{zz}^{\text{AA}} = (1/2c^2) \sum_{\nu}^{\text{on A}} P_{\nu\nu}^{(0)} [(x_{\text{A}}^2 + y_{\text{A}}^2) r_{\text{A}}^{-3}]_{\nu\nu} \tag{4.128}$$

and

$$(\sigma_{\text{A}}^{\text{p}})_{zz}^{\text{AA}} = -(2/c^2) \sum_j^{\text{occ}} \sum_l^{\text{unocc}} \{(\varepsilon_l - \varepsilon_j)^{-1} (c_{x_{\text{A}} j} c_{y_{\text{A}} l} - c_{y_{\text{A}} j} c_{x_{\text{A}} l}) \times \sum_{\text{B}} (c_{x_{\text{B}} j} c_{y_{\text{B}} l} - c_{y_{\text{B}} j} c_{x_{\text{B}} l}) \langle r^{-3} \rangle_{2\text{p}_{\text{A}}} \} \tag{4.129}$$

$$= -2 \langle r^{-3} \rangle_{2\text{p}_{\text{A}}} (\chi_{\text{A}}^{\text{p}})_{zz} \tag{4.130}$$

where $\langle r^{-3} \rangle_{2\text{p}_\text{A}}$ is the mean value of r^{-3} for the 2p atomic orbitals on centre A. For minimal sets of ϕ-functions in which all 2p atomic orbitals on a given centre are assumed to have the same size, there is no approximation implied by replacing the matrix element $H_{\text{A}\nu\lambda}^{(0,1)}$ by this average. It should be noted, however, that this simplification is only possible if $H_{\text{A}\nu\lambda}^{(0,1)}$ terms are neglected when ϕ_ν and ϕ_λ are on different centres.

The local diamagnetic term is similar to the Lamb formula. In so far as we can neglect p functions, χ_A^p is zero for hydrogen and consequently there is no local paramagnetic contribution to proton shielding. However, for atoms with 2p electrons, equation (4.130) suggests a correlation of the chemical shift with diamagnetic susceptibility data. The $\langle r^{-3} \rangle$ proportionality factor makes the local paramagnetic contribution large and as Saika and Slichter[105] first showed for fluorine, this contribution dominates the shielding at the first-row atoms carbon to fluorine.

The only other non-zero terms in equation (4.127) are those in which ϕ_ν and ϕ_λ are both on some other centre B. To simplify these terms, Pople introduced a further approximation in which the secondary field at A due to the currents on centre B is calculated on the basis that the induced moment in the electron group at B may be replaced by a point dipole. This contribution to the shielding at A is then given by the following simple expression which was first derived by McConnell[10] and by Pople[87]

$$(\sigma_A)_{zz}^{AB} = \sum_\gamma \chi_{Bz\gamma} R_{AB}^{-5} (R_{AB}^2 \delta_{\gamma z} - 3(R_{AB})_\gamma (R_{AB})_z) \tag{4.131}$$

The average value of the $\boldsymbol{\sigma}_A^{AB}$ tensor is given by

$$\boldsymbol{\sigma}_A^{AB} = \tfrac{1}{3} \sum_{\alpha\beta} \chi_{B\alpha\beta} R_{AB}^{-5} (R_{AB}^2 \delta_{\alpha\beta} - 3(R_{AB})_\alpha (R_{AB})_\beta) \tag{4.132}$$

If the susceptibility on atom B is isotropic, this formula gives zero, so the contribution is frequently referred to as the *neighbour anisotropy effect.* If the local susceptibility tensor on atom B is axially symmetric, equation (4.132) reduces to

$$\boldsymbol{\sigma}_A^{AB} = \tfrac{1}{3}(\Delta\chi_B)(1 - 3\cos^2\theta_B) R_{AB}^{-3} \tag{4.133}$$

where θ_B is the angle between the principal axis of $\boldsymbol{\chi}_B$ and the internuclear vector $\boldsymbol{R}_{AB}$. It should be emphasised that this simple expression is only valid if R_{AB} is very much greater than the dimensions of the charge distribution on centre B. Didry and Guy[106] have shown that the dipolar approximation is only valid for values of R_{AB} of the order of 3–4 Å.

Thus within the approximations of this theory, the susceptibility of non-cyclic molecules is given by the expression

$$\boldsymbol{\chi} = \sum_A^{\text{atoms}} (\boldsymbol{\chi}_A^d + \boldsymbol{\chi}_A^p) \tag{4.134}$$

where the anisotropy can be associated with the bonds formed by each atom. Similarly, the total shielding constant for atom A can be written as

$$\boldsymbol{\sigma}^A = (\boldsymbol{\sigma}^d)^{AA} + (\boldsymbol{\sigma}^p)^{AA} + \sum_{B \neq A} \boldsymbol{\sigma}^{AB} \tag{4.135}$$

Pople used these formulae to make approximate estimates of atomic contributions to the diamagnetic susceptibilities of non-cyclic compounds containing hydrogen and the first-row elements carbon to fluorine. These calculations were based on localised orbitals for bonds and an average excitation energy (ΔE) of 10 eV with some corrections for states of lower energy. One of the main aims of this work was to provide a theoretical framework within which one could interpret Pascal's rules.

For the first-row hydrides, XH_n, non-polar localised bonding orbitals were used and lone pairs were represented by pure sp^3 hybrids directed in those tetrahedral directions complementary to the bonds. Within the approximations of this theory all Q_{XH} terms are zero and the total susceptibilities for these hydrides may be written as a sum of the atomic contributions given in Table 4.6.

Table 4.6 Comparison of atomic contributions calculated by Pople's method and Pascal constants*
(From Pople, J. A. (1962). *J. Chem. Phys.*, **37**, 62, reproduced by courtesy of the American Institute of Physics)

Atom A	χ_A^d*(calc)*	χ_A^p*(calc)*	χ_A*(calc)*	χ_A*(Pascal)*
H	−2.37	—	−2.37	−2.93
He	−1.64	—	−1.64	−2.02
C	−9.14	6.46	−2.68	−6.00
N	−7.91	6.46	−1.45	−5.55 (open chain)
O	−6.96	5.39	−1.58	−4.61 (alcohols)
F	−6.21	3.23	−2.98	−6.3
Ne	−5.60	—	−5.60	−6.96

*Units are 10^{-6} erg G^{-2} mol^{-1}.

It is clear that the calculated values are too small in all cases. However, relative values are predicted much more satisfactorily and the theory succeeds in reproducing the fall in the constants after carbon and the subsequent rise to higher values for fluorine and neon. This is due to a competition between a falling paramagnetic term and a decreasing diamagnetic term as the atomic number increases. In this scheme (non-polar localised orbitals) the charges on the atoms in a molecule are the same as those in the free atoms. Thus the diamagnetic terms are determined by the average value of r^2. If Slater-type orbitals of the type

$$\phi_{2s} = N_{2s} r \exp(-\zeta r) \tag{4.136}$$

$$\phi_{2p_x} = N_{2p_x} x \exp(-\zeta r) \tag{4.137}$$

are used, then $\langle r^2 \rangle$ is simply proportional to $(1/\zeta^2)$ where the scale factor ζ is a measure of the size of the atomic orbital. As an orbital contracts, ζ increases and consequently $\langle r^2 \rangle$ decreases. As we go from carbon to fluorine, the valence electrons become more tightly bound due to the increased nuclear charge and, since $\langle r^2 \rangle$ decreases more quickly than the number of valence electrons increases, the diamagnetic term becomes more positive. Qualitatively the lower paramagnetic term for fluorine is largely due to it being zero along the H—F direction.

Pople attributed the disagreement between the calculated contributions and the Pascal constants in Table 4.6 to the poor value of $\langle r^2 \rangle$ afforded by Slater-type atomic orbitals. Therefore, before attempting to calculate constitutive corrections, Pople empirically adjusted the calculated χ_A^d values to give total χ_A values in agreement with Pascal's constants. These values were then used with the calculated results for χ_A^p to give the principal susceptibility values of Table 4.7.

The theory predicts that all carbon atoms in saturated compounds have the same χ_A^p and are isotropic since all Q_{AB} terms are zero. Accordingly, all carbon atoms in unstrained paraffins should have the same chemical shift. In fact, such resonances are spread over a range of *c.* 30 p.p.m. to low field from methane. Since carbon atoms are predicted to be isotropic, calculated

Table 4.7 Calculated principal components of atomic contributions in saturated molecules*

(From Pople, J. A. (1962). *J. Chem. Phys.*, **37**, 63 reproduced by courtesy of the American Chemical Society)

Atom and coordinate system	*Principal components of* χ_A
−H	−2.93 (isotropic)
>C<	−6.00 (isotropic)
>N −	−5.55 (isotropic)
O (coordinate axes x, y)	$\chi_{xx} = \chi_{zz} = -3.54$; $\chi_{yy} = -6.77$
−F → z	$\chi_{xx} = \chi_{yy} = -4.68$; $\chi_{zz} = -9.53$

*Units are 10^{-6} erg G^{-2} mol^{-1}.

chemical shifts for all paraffinic hydrogens are the same. Again there is a spread of 1.5 p.p.m. to low field from methane in the experimental data, but this may be due to changing electron density.

Some authors[107] have postulated an intrinsic diamagnetic anisotropy for the C—C single bond. Recent experimental evidence[7] suggests that both C—C and C—H bonds may be magnetically anisotropic.

Similarly, χ_A^p for nitrogen atoms in saturated compounds is also predicted to be isotropic. Kukolich[108] has recently found a small anisotropy of +1.3 e.m.u. The anistropic χ_A value for oxygen atoms in saturated molecules arises from the anisotropy in χ_O^p. A number of workers[109–111] have measured the anisotropy of the susceptibility tensor for water and find a small anisotropy with two of the principal components approximately equal and different from the third.

In the Pascal scheme, positive constitutive corrections are used for C—C double and triple bonds. Pople[89] determined such constitutive terms by comparing calculated values of χ_C^p for ethylene and acetylene with χ_C^p obtained for saturated hydrocarbons.

For ethylene, it was assumed that the molecule may be described by non-polar localised C—H and C—C bonds formed from trigonal hybrids together with a π-type bond. If the z axis is chosen along the C—C bond and the molecule taken to lie in the zx plane, the appropriate elements of the density matrix are

$$P_{x_A x_A} = P_{y_A y_A} = P_{z_A z_A} = 1 \tag{4.138}$$

$$P_{z_A z_B} = -\tfrac{2}{3} P_{x_A x_B} = 0; \; P_{y_A y_B} = 1 \tag{4.139}$$

Therefore unlike saturated hydrocarbons, olefins have a non-zero $\boldsymbol{Q}_{AB}$ tensor between different atoms. The total $\boldsymbol{Q}$ tensor on each carbon atom is

$$Q_{zz} = Q_{yy} = 2; \quad Q_{xx} = 10/3; \quad (Q_C)_{av} = 22/9 \tag{4.140}$$

Thus for ethylene and other olefins the theory predicts larger χ_A^p terms on the carbon atoms ($Q_{\alpha\alpha} = 2$ for saturated hydrocarbons) but only if the magnetic field is perpendicular to the C—C double bond and in the molecular plane. This additional contribution arises from the mixing of $\sigma \rightarrow \pi^*$ and $\pi \rightarrow \sigma^*$ excited states with the ground state by the magnetic field. The calculated constitutive correction for the C—C double bond is 2.9 which is somewhat lower than the experimental value of 5.5.

Using the value of χ_C^p in equation (4.130) gives a calculated value for the carbon chemical shift of ethylene of 46 p.p.m. downfield from ethane. Although in the correct direction this shift is much smaller than the experimental value of −120 p.p.m. Also since χ_C^p is anisotropic the proton shift in ethylene (relative to ethane) due to the neighbour anisotropy effect may be evaluated from equation (4.132). The calculated value of −2.1 p.p.m. is smaller than the experimental result of −4.88 p.p.m.

The predicted diamagnetic anisotropy should have other consequences for proton shifts. Protons on or close to the axis of low diamagnetism such as the protons furthest from the double bond in cyclopentene will be shifted to low-field. This is confirmed by the measurement of Wiberg and Nist[112] who found a shift of −0.39 p.p.m. for these protons relative to cyclopentane. Also protons which approach a C—C double bond from above or below should show small high-field shifts.

The C—C triple bond is predicted to have a high diamagnetic anisotropy with high diamagnetism along the molecular axis. This implies a high-field shift for the protons in acetylene due to the neighbour anisotropy effect. The

Table 4.8 Principal components of the susceptibility tensors for some simple hydrocarbons calculated using Pople's theory*

Molecule†	χ_{xx}^d‡	χ_{xx}^p	χ_{yy}^p	χ_{zz}^p	χ^p	χ(*calc*)	χ(*exp*)	$\Delta\chi$(*calc*)§
CH_4	−24.18	6.48	6.48	6.48	6.48	−17.72	−17.4	0.0
C_2H_6	−42.50	12.96	12.96	12.96	12.96	−29.54	−26.8	0.0
C_2H_4	−36.64	21.56	12.96	12.96	15.83	−20.81	−18.8	+8.66
C_2H_2	−30.78	19.44	19.44	0.0	12.96	−17.82	−20.8	−19.44
$H_2C{=}C{=}CH_2$	−49.10	26.88	26.88	19.44	24.40	−24.70	−25.5	−7.44
H_2CO	−28.32	20.60	10.77	17.29	16.22	−12.10	—	—

*Units are 10^{-6} erg G^{-2} mol^{-1}.
†The z axis is taken along the C—C (C=O) bonds and planar molecules lie in the xz plane.
‡The diamagnetic contribution is evaluated from empirically adjusted atomic diamagnetic contributions. This contribution is assumed isotropic.
§$\Delta\chi$ is taken as $\chi_{zz} - \frac{1}{2}(\chi_{xx} + \chi_{yy})$ for acetylene and allene and as $\chi_{xx} - \frac{1}{2}(\chi_{yy} + \chi_{zz})$ for ethylene.

observed position of the proton resonance in acetylene (−1.92 p.p.m. relative to ethane) can be interpreted as a cancellation of this with a low-field shift due to the reduced charge density on the hydrogen atoms.

In Table 4.8 we summarise the results for χ and $\Delta\chi$ for some simple systems. All the anisotropies can be explained using the simple ideas presented above.

This theoretical attempt to explain Pascal's rules has been criticised by Hameka[113] who argued that the approximations introduced by Pople[114] may change the calculated results by as much as 50%. Hameka and his co-workers[78, 115] prefer to consider the molecular susceptibility as a sum of bond contributions, with small corrections for the interactions between adjacent bonds and for the atomic inner-shell electrons. In many cases such a description is equivalent or almost equivalent to Pascal's rules but is more successful in describing the susceptibilities of isomeric molecules. Nevertheless, Pople's theory, although clearly very approximate, successfully explains some of the principal features of observed susceptibilities.

In an attempt to remove some of the deficiencies associated with the independent-electron method, the average-energy approximation and the localised orbital model, we developed a similar type of theory[116] based on the complete neglect of differential overlap (CNDO) approximations[117, 118]. This simple treatment is still semi-empirical and yet attempts to take a broad account of the main features of electron–electron interaction. The CNDO method neglects the smaller two-electron integrals $(\nu\lambda \mid \rho\sigma)$ which involve the overlap between different ϕ-functions and treats all other integrals in a simplified manner. In this approach the Fock matrix elements are given by

$$F_{\nu\nu} = -\tfrac{1}{2}(I_\nu + A_\nu) + [(P_{AA} - Z_A) - \tfrac{1}{2}(P_{\nu\nu} - 1)]\gamma_{AA} + \sum_{B \neq A} (P_{BB} - Z_B)\gamma_{AB} \quad (\phi_\nu \text{ on atom A})$$

$$F_{\nu\lambda} = \beta^0_{AB} S_{\nu\lambda} - \tfrac{1}{2} P_{\nu\lambda}\gamma_{AB} \quad (\phi_\nu \text{ on atom A, } \phi_\lambda \text{ on atom B}) \tag{4.141}$$

Here $-\frac{1}{2}(I_\nu + A_\nu)$ is the Mulliken electronegativity of the orbital ϕ_ν, obtained as the average of the ionisation potential and the electron affinity. γ_{AB} is a spherically averaged interaction between an electron on A and one on B, and P_{AA} is the total electronic charge on atom A. In the expression for $F_{\nu\lambda}$, β^0_{AB} is a semi-empirical bonding parameter. Within the CNDO framework the first-order Fock matrix elements are given by

$$F^{(1)}_{\nu\lambda} = H^{(1,0)}_{\nu\lambda} - \tfrac{1}{2} P^{(1)}_{\nu\lambda}\gamma_{AB} \tag{4.142}$$

Since only two-electron integrals of the type $(\nu\nu \mid \lambda\lambda)$ are retained in CNDO theory, the $G^{(1)}_{\nu\lambda\rho\sigma}$ and $G^{(2)}_{\nu\lambda\rho\sigma}$ terms are zero and the field dependence of the two-electron integrals disappears. Also the systematic neglect of all two-centre overlap integrals sets $S^{(1)}_{\nu\lambda}$ and $S^{(2)}_{\nu\lambda}$ to zero. $H^{(1,0)}_{\nu\lambda}$ and $H^{(2,0)}_{\nu\lambda}$ are given by equations (4.108)–(4.111) where

$$H^{\text{core}(0)}_{\nu\lambda} = \beta^0_{AB} S^{(0)}_{\nu\lambda} \tag{4.143}$$

We also introduced a further approximation neglecting all two-centre $H^{(1,0)}_{\nu\lambda}$ and $H^{(2,0)}_{\nu\lambda}$ terms. Since we were only interested in non-cyclic molecules, we felt that this would not introduce serious error. In Table 4.9 we present some of our results using the intermediate neglect of differential overlap (INDO)[119] version of the theory. This differs from CNDO only in the inclusion of one-centre integrals of the type $(\nu\lambda \mid \nu\lambda)$ and experience has shown[120] that these two methods give very similar results for diamagnetic molecules. It can be seen from Table 4.9 that although absolute INDO results are in rather poor agreement with experiment, relative values are moderately reproduced. The INDO values are consistent with many of the predictions of Pople's theory.

In particular, the low diamagnetism of ethylene in the direction perpendicular to the C—C double bond and in the molecular plane is reproduced as is the high diamagnetism along the C—C triple bond in acetylene. The INDO result for $\Delta\chi$ in acetylene, however, is considerably smaller than that calculated by Pople. The large value of χ_{xx} in formaldehyde may be explained

Table 4.9 INDO magnetic susceptibilities (two-centre terms neglected)*

Molecule	χ^{d}_{xx}	χ^{d}_{yy}	χ^{d}_{zz}	χ^{p}_{xx}	χ^{p}_{yy}	χ^{p}_{zz}	χ	$\chi(exp)$
CH_4	−15.6	−15.6	−15.6	3.8	3.8	3.8	−11.8	−17.4
C_2H_6	−27.9	−27.9	−27.7	7.5	7.5	7.2	−20.4	−26.8
C_2H_4	−24.7	−24.5	−24.5	15.5	6.8	7.8	−14.6	−18.8
C_2H_2	−21.3	−21.3	−21.5	9.9	9.9	0.0	−14.8	−20.0
H_2CO	−18.5	−19.2	−18.9	21.4	7.6	13.9	−4.6	—
CH_3F	−19.6	−19.6	−20.1	6.6	6.6	3.9	−14.0	−17.8
CH_3CHO	−31.3	−31.5	−30.8	18.4	11.4	24.7	−13.0	−22.7
$(CH_3)_2O$	−34.4	−34.0	−34.7	12.2	11.9	11.4	−22.5	−26.3
$CH_2{=}C{=}CH_2$	−33.6	−33.6	−33.6	16.5	16.5	11.1	−18.9	−25.5
$CH_2{=}C{=}O$	−28.4	−27.4	−28.2	20.4	19.1	12.3	−10.7	—
$H_2C{=}CF_2$	−33.6	−32.6	−33.0	16.8	12.2	12.9	−19.1	—
$FC{\equiv}CH$	−25.4	−25.4	−26.1	12.9	12.9	0.0	−17.0	—
HCN	−18.5	−18.5	−18.3	15.3	15.3	0.0	−8.2	—
CH_3CN	−30.7	−30.7	−30.5	19.4	19.4	3.9	−16.4	−27.6
H_2O	−10.1	−9.7	−10.3	6.0	5.0	4.0	−5.0	−13.0
NH_3	−13.0	−12.0	−13.0	6.6	4.0	6.6	−7.1	−16.3

*Units are 10^{-6} erg G^{-2} mol^{-1}

Table 4.10 INDO ^{13}C magnetic-shielding constants (two-centre terms neglected)†

Molecule	$\sigma(calc)$	$\Delta\sigma(calc)$‡	$\Delta\sigma(exp)$‡
CH_4	−63.3	0.0	0.0
C_2H_6	−62.0	+1.3	−8.0
CH_3F	−60.9	+2.4	−77.5
C_2H_4	−103.4	−40.1	−126
H_2CO	−124.0	−60.7	−197
C_2H_2	−50.2	+10.1	−76
$H_2C{=}C^*{=}CH_2$	−113.7	−50.4	−216
$CH_2{=}C{=}C^*H_2$	−94.7	−31.4	−76

*Indicates nucleus under study.
†Units are p.p.m.
‡All $\Delta\sigma$ values are given as chemical shifts relative to CH_4.

in the same way as the χ_{xx} value in ethylene. The fairly large χ_{zz} value, on the other hand, arises from the mixing of the low-lying $n \rightarrow \pi^*$ state with the ground state by the magnetic field.

The INDO method predicts a very small anisotropy for χ^{p} in ethane and a somewhat larger value for $\Delta\chi^{p}$ in ammonia. However, the total values of $\Delta\chi$ for ethane and ammonia are small, being −0.1 and −1.9 e.m.u., respectively.

This method was also used to calculate chemical shifts, and to illustrate the performance of the theory we present some typical results for ^{13}C magnetic-shielding constants in Table 4.10. The theory appears quite insensitive to the role which substituents play in determining ^{13}C chemical shifts in CH_3X compounds and although the low-field ^{13}C shifts of ethylene and formaldehyde are qualitatively reproduced the low-field shift of acetylene is not. Clearly this level of theory does not give an adequate description of magnetic shielding.

Recently, Bley has also developed a similar theory of magnetic susceptibility for diatomic[97] and small polyatomic[98] molecules. However, since he was interested in non-cyclic and cyclic molecules, Bley included both the one- and the two-centre $H_{\nu\lambda}^{(1,0)}$ and $H_{\nu\lambda}^{(2,0)}$ terms of equations (4.108)–(4.111). Some of his results are presented in Table 4.11. The absolute values reported in entry C are in good agreement with experiment except for acetylene.

Table 4.11 INDO magnetic susceptibilities (two-centre terms included)*
(From W. R. Bley[98], by courtesy of Taylor and Francis, Ltd.)

Molecule		χ_{xx}	χ_{yy}	χ_{zz}	χ	$\chi(exp)$
CH_4	A§	−14.0	−14.0	−14.0	−14.0	
	C‖	−17.3	−17.3	−17.3	−17.3	−17.4
C_2H_6	A	−24.3†	−24.9	−24.9	−24.7	
	B¶	−24.2	−24.4	−24.4	−24.4	
	C	−28.3	−29.5	−29.5	−29.1	−26.8
C_2H_4	A	−16.6†	−8.4	−20.9‡	−15.3	
	B	−17.0	−9.6	−20.5	−15.7	
	C	−19.1	−9.9	−24.2	−17.7	−18.8
C_2H_2	A	−21.7†	−11.2	−11.2	−14.7	
	B	−21.8	−11.7	−11.7	−15.1	
	C	−23.0	−12.0	−12.0	−15.7	−20.8
Cyclo-C_3H_6	A	−29.9	−29.9	−43.6‡	−34.5	
	C	−33.9	−33.9	−47.5	−38.4	−39.2

*Units are 10^{-6} erg G^{-2} mol^{-1}.
†The x-direction is along the C—C bond.
‡The z-direction is perpendicular to the molecular plane.
§Level A; CNDO, $\zeta_H = 1.2$.
¶Level B; INDO, $\zeta_H = 1.2$.
‖Level C; CNDO, $\zeta_H = 1.0$.

Comparing his results with those of Table 4.9 indicates that although two-centre $H_{\nu\lambda}^{(1,0)}$ and $H_{\nu\lambda}^{(2,0)}$ terms can make contributions to the susceptibilities of non-cyclic molecules, such contributions are found to be considerably larger for cyclopropane. Except for acetylene, the inclusion of such terms improves the agreement between calculated and experimental values. The results for acetylene are not changed by these terms since for any linear molecule it is easily shown that $\boldsymbol{Q}_{\nu\lambda}$ terms are zero. However, if the London approximation was not made, $\boldsymbol{T}_{\nu\lambda}$ terms would give non-zero two-centre contributions even for linear molecules. As far as we are aware, no semi-empirical studies have yet been performed which have included these $\boldsymbol{T}_{\nu\lambda}$ terms.

Because two-centre terms are included, the total susceptibility can no longer be written as a sum of atomic contributions. Within the INDO framework, equation (4.89) becomes

$$\chi_{\alpha\beta} = \sum_{A}^{\text{atoms}} (\chi_A)_{\alpha\beta}^{\text{one-centre}} + \chi_{\alpha\beta}^{\text{two-centre}} \tag{4.144}$$

where

$$(\chi_A)_{\alpha\beta}^{\text{one-centre}} = -\sum_{\nu\lambda}^{A} \{P_{\nu\lambda}^{(0)}(H_{\nu\lambda}^{(2,0)})_{\alpha\beta} + (P_{\nu\lambda}^{(1)})_{\alpha}(H_{\nu\lambda}^{(1,0)})_{\beta}\} \tag{4.145}$$

and

$$\chi_{\alpha\beta}^{\text{two-centre}} = -\sum_{C \neq D}\sum\sum_{\rho}^{C}\sum_{\sigma}^{D} \{P_{\rho\sigma}^{(0)}(H_{\rho\sigma}^{(2,0)})_{\alpha\beta} + (P_{\rho\sigma}^{(1)})_{\alpha}(H_{\rho\sigma}^{(1,0)})_{\beta}\} \tag{4.146}$$

However, χ_A of equation (4.145) is not the same as that defined by Pople[11]. The difference is illustrated by considering the first-order coefficients $c_{\lambda j}^{(1)}$ which in Bley's approach[98] can be written as

$$(c_{\lambda j}^{(1)})_{\alpha} = (c_{\lambda j}^{(1)})_{\alpha}^{L} + (c_{\lambda j}^{(1)})_{\alpha}^{R} \quad (\phi_{\lambda} \text{ on atom A}) \tag{4.147}$$

where

$$(c_{\lambda j}^{(1)})_{\alpha}^{L} = -\sum_{l}^{\text{unocc}} (\varepsilon_l^{(0)} - \varepsilon_j^{(0)})^{-1} \sum_{A}\sum_{\nu\sigma}^{A} [c_{\nu l}^{(0)}(F_{\nu\sigma}^{(1)})_{\alpha} c_{\sigma j}^{(0)}] c_{\lambda l}^{(0)} \tag{4.148}$$

and

$$(c_{\lambda j}^{(1)})_{\alpha}^{R} = -\sum_{l}^{\text{unocc}} (\varepsilon_l^{(0)} - \varepsilon_j^{(0)})^{-1} \sum_{C \neq D}\sum\sum_{\rho}^{C}\sum_{\sigma}^{D} \{[c_{\rho l}^{(0)}(F_{\rho\sigma}^{(1)})_{\alpha} c_{\sigma j}^{(0)}] c_{\lambda l}^{(0)}\} \tag{4.149}$$

Consequently, the first-order density matrix $(P_{\nu\lambda}^{(1)})_{\alpha}$ becomes

$$(P_{\nu\lambda}^{(1)})_{\alpha} = (P_{\nu\lambda}^{(1)})_{\alpha}^{L} + (P_{\nu\lambda}^{(1)})_{\alpha}^{R} \quad \phi_{\nu} \text{ and } \phi_{\lambda} \text{ both on atom A} \tag{4.150}$$

and substituting this into equation (4.145) we obtain

$$(\chi_A)_{\alpha\beta}^{\text{one-centre}} = -\sum_{\nu\lambda}^{A} \{P_{\nu\lambda}^{(0)}(H_{\nu\lambda}^{(2,0)})_{\alpha\beta} + (P_{\nu\lambda}^{(1)})_{\alpha}^{L}(H_{\nu\lambda}^{(1,0)})_{\beta} + (P_{\nu\lambda}^{(1)})_{\alpha}^{R}(H_{\nu\lambda}^{(1,0)})_{\beta}\} \tag{4.151}$$

$$= (\chi_A^{d})_{\alpha\beta}^{\text{intra}} + (\chi_A^{p})_{\alpha\beta}^{\text{intra}} + (\chi_A)_{\alpha\beta}^{\text{cross}} \tag{4.152}$$

In equation (4.152), the sum of the first two terms, $(\chi_A)_{\alpha\beta}^{\text{intra}}$, corresponds to the atomic contributions defined by Pople. Bley, however, defined new local terms which include the $(\chi_A)_{\alpha\beta}^{\text{cross}}$ terms, setting

$$\boldsymbol{\chi}_A^{\text{local}} = \boldsymbol{\chi}_A^{\text{one-centre}} \tag{4.153}$$

and

$$\boldsymbol{\chi}_A^{\text{non-local}} = \boldsymbol{\chi}_A^{\text{two-centre}} \tag{4.154}$$

Table 4.12 presents local values for carbon and hydrogen atoms in some simple hydrocarbons. For saturated hydrocarbons these results can be compared directly with Pascal's atomic constants. The average theoretical

χ_C^{local} and χ_H^{local} values for the saturated hydrocarbons in Table 4.12 are -7.0 and -2.0 which are quite close to the Pascal values of -6 and -2.93, respectively. Bley finds that the $(\chi_A)^{cross}$ terms make significant contributions to these values.

Comparing χ_C^{local} values for unsaturated hydrocarbons with those from saturated systems allows theoretical estimates of the constitutive corrections

Table 4.12 Carbon and hydrogen local susceptibility contributions*
(From Bley, W. R.[98], reproduced by courtesy of Taylor and Francis Ltd.)

Molecule		C		H
		χ_{intra}	χ_{local}	χ_{local}
CH_4	A	-5.37	-6.71	-1.63
	C	-4.91	-7.08	-2.41
C_2H_6	A	-5.33	-7.12	-1.65
	C	-4.93	-7.98	-2.40
C_2H_4	A	-3.94	-4.52	-1.63
	C	-3.53	-4.56	-2.34
C_2H_2	A	-5.82	-5.82	-1.55
	C	-5.63	-5.63	-2.20
Cyclo-C_3H_6	A	-4.95	-7.19	-1.64
	C	-4.70	-7.63	-2.36

*Units are 10^{-6} erg G^{-2} mol^{-1}.

to be made. The calculated correction for the C—C double bond is $+5.0$ which is in good agreement with Pascal's value of $+5.5$. The result for the triple C—C bond is rather poorer, the calculated value of $+2.4$ being much larger than the Pascal constant of $+0.8$.

From Table 4.11 it is clear that the calculated results are sensitive to the size of the hydrogen atomic orbitals. Changing the orbital scale factor ζ_H

Table 4.13 Local and non-local contributions to the principal components of the susceptibility tensor*†
(From data in Bley[98])

Molecule	χ_{xx} *local*	χ_{xx} *non-local*	χ_{yy} *local*	χ_{yy} *non-local*	χ_{zz} *local*	χ_{zz} *non-local*
C_2H_6	-23.1‡	-1.2	-24.7	-0.2	-24.7	-0.2
C_2H_4	-16.7‡	$+0.1$	-8.5	$+0.1$	-21.6§	$+0.7$
C_2H_2	-21.7‡	0.0	-11.2	0.0	-11.2	0.0
Cyclo-C_3H_6	-30.7	$+0.8$	-30.7	$+0.8$	-32.8§	-10.8

*Units are 10^{-6} erg G^{-2} mol^{-1}.
†These results were evaluated using Level A (see footnote § of Table 4.11).
‡The x-direction is along the C—C bond.
§The z-direction is perpendicular to the molecular plane.

from 1.2 to 1.0 makes the hydrogen orbitals more diffuse and increases χ_H^{local} as shown in Table 4.12. The only contribution to χ_H^{local} is $(\chi_H^d)^{intra}$ which depends on the electronic charge on the hydrogen atom and the size of the hydrogen orbitals. It appears that the change of orbital size is probably more important since the pairs of values for χ_H^{local} in Table 4.12 have ratios close to 1.44 which is the value expected if P_{HH} remains unchanged. In all cases, the differences between the total susceptibilities calculated at levels A and C are very close to those calculated by only considering the changes in χ_H^{local}.

Although $\chi^{non\text{-}local}$ consists of several terms, the results of Table 4.13 indicate that these are either individually small or largely cancel to give a total susceptibility for non-cyclic molecules which is essentially local in character[121]. For cyclopropane, however, the non-local term is significant and arises almost totally from a large non-local contribution when the magnetic field is perpendicular to the plane of the three-membered ring (z-direction). This large interatomic contribution results in an increased diamagnetism in the z-direction and is consistent with the interpretation of the long-range shielding effects of three-membered rings[122].

Ellis, Maciel and McIver[99] have also used this method to calculate ^{13}C shielding constants in hydrocarbons. Their results for chemical shifts of hydrocarbons (Table 4.14) are in good agreement with experimental values;

Table 4.14 INDO ^{13}C magnetic shielding constants (two-centre terms included)†

(From Ellis, P. D., Maciel, G. E. and McIver, J. W., Jr., *J. Amer. Chem. Soc.*, in the press, reproduced by courtesy of the American Chemical Society)

Molecule	σ	$\Delta\sigma(calc)$‡	$\Delta\sigma(exp)$‡
CH_4	6.06	120.62	130.8
Cyclo-C_3H_6	−6.31	108.25	130.7
C_2H_6	3.06	117.62	122.8
$C^*H_3CH_2CH_3$	−0.42	114.14	113.1
$CH_3C^*H_2CH_3$	3.31	117.87	112.6
C_2H_4	−108.98	5.58	5.4
$H_2C^*{=}C{=}CH_2$	−82.64	31.92	54.0
$H_2C{=}C^*{=}CH_2$	−176.56	−62.07	−84.0
C_2H_2	−74.49	40.07	54.8
C_6H_6	−114.56	0.0	0.0

*Indicates nucleus under study.
†Units are parts per million.
‡All $\Delta\sigma$ values are given relative to C_6H_6.

the only deficiency is the relatively poor description of the β-methyl effect (e.g. the shift between $H_3C^*{-}CH_3$ and C^*H_4). These workers used INDO parameters which differ from the original values suggested by Pople and co-workers[119] and it is not clear if the improvement in calculated ^{13}C shielding constants shown by comparing Tables 4.10 and 4.14 can be attributed to the re-parameterisation or to the inclusion of two-centre $H_{\nu\lambda}$ terms. The results for acetylene, however, suggest that re-parameterisation has produced major changes in calculated ^{13}C shielding constants.

Thus SCF perturbation theory with INDO wave functions and inclusion of *both* one and two-centre $H_{\nu\lambda}$ terms appears to provide a unified theory capable of interpreting both magnetic susceptibilities and ^{13}C magnetic shielding constants for hydrocarbons. Applications to larger, particularly condensed ring hydrocarbons and to systems containing atoms other than carbon and hydrogen are awaited with interest.

4.2.7 Semi-empirical methods for planar condensed cyclic hydrocarbons

In the previous section we concentrated on the magnetic effects of current circulations which can be regarded as localised on individual atoms since such considerations are the main ones required when dealing with non-cyclic molecules. However, when an external magnetic field is perpendicular to the plane of a flat cyclic conjugated molecule, there are additional contributions to the susceptibility which can be interpreted physically as being due to the π electrons flowing around the rings in the molecule. Consequently, equation (4.134) has to be modified and the total susceptibility of such cyclic systems is often written in the form

$$\boldsymbol{\chi} = \sum_{\mathrm{A}}^{\mathrm{atoms}} \chi_{\mathrm{A}} + \boldsymbol{\chi}^{\mathrm{ring}} \tag{4.155}$$

where it is assumed that the molecular or 'ring current' part of $\boldsymbol{\chi}$ can be treated separately from the atomic contributions. Similarly, the chemical shift of an atom A in such a molecule may be written as[105]

$$\boldsymbol{\sigma}_{\mathrm{A}} = (\boldsymbol{\sigma}_{\mathrm{A}}^{\mathrm{d}})^{\mathrm{AA}} + (\boldsymbol{\sigma}_{\mathrm{A}}^{\mathrm{p}})^{\mathrm{AA}} + \sum_{\mathrm{A} \neq \mathrm{B}} \boldsymbol{\sigma}^{\mathrm{AB}} + \boldsymbol{\sigma}^{\mathrm{ring}} \tag{4.156}$$

These molecular contributions have aroused a great deal of interest since the early attempt by Pauling[123] to give a quantitative explanation of the large diamagnetic susceptibilities found in aromatic hydrocarbons. These large average values have been shown by crystal studies[124] to arise from a large susceptibility perpendicular to the plane of the ring.

Most quantum-mechanical studies have assumed that $\boldsymbol{\chi}^{\mathrm{ring}}$ may be equated to the non-local π-electron contribution ($\boldsymbol{\chi}^{\pi}$) to the total susceptibility. Consequently, theoretical approaches have treated the π-electrons separately from the rest of the electrons, calculated $\boldsymbol{\chi}^{\pi}$ and then determined a total susceptibility by estimating local contributions using Pascal's rules or some related empirical procedure.

The first quantum mechanical treatment of the susceptibilities of aromatic hydrocarbons was given by London[125] who introduced GIAO into Hückel[126] molecular-orbital theory. In this approach, the matrix elements of the electronic Hamiltonian are simply

$$H_{\nu\nu} = \langle \phi_\nu | \tilde{H} | \phi_\nu \rangle = \alpha_\nu \tag{4.157}$$

$$H_{\nu\lambda}(\boldsymbol{H}) = \langle \chi_\nu | \tilde{H} | \chi_\lambda \rangle = \langle f_\nu^* f_\lambda \phi_\nu | \tilde{H} | \phi_\lambda \rangle \tag{4.158}$$

where $\tilde{H}$ indicates some unspecified total electronic Hamiltonian and α_ν is the usual Hückel Coulomb integral. Using the London approximation in the gauge factor product $f_\nu^* f_\lambda$ allows equation (4.158) to be simplified to

$$H_{\nu\lambda}(\boldsymbol{H}) = \beta_{\nu\lambda}^0 \exp((i/2c)\boldsymbol{Q}_{\nu\lambda}.\boldsymbol{H}) \tag{4.159}$$

where $\beta_{\nu\lambda}^0$ is the Hückel carbon–carbon resonance integral and is equal to β^0 for bonded carbon atoms and zero otherwise. If we consider the z direction to be perpendicular to the plane of the ring, then π-type molecular orbitals are formed from $2p_z$ atomic orbitals. A calculation of χ^{ring} (χ_{zz}) thus requires the $(H^{(1,0)}_{\nu\lambda})_z$ and $(H^{(2,0)}_{\nu\lambda})_{zz}$ terms of equations (4.108) to (4.111). All one-centre $(H^{(1,0)}_{\nu\lambda})_z$ terms are zero since ϕ_ν and ϕ_λ are restricted to be $2p_z$ atomic orbitals and $L_z \mid \phi_{2p_z}\rangle$ vanishes. One-centre $(H^{(2,0)}_{\nu\lambda})_{zz}$ terms are not zero but since these terms are local in character they are neglected in evaluating the interatomic term $\boldsymbol{\chi}^{\text{ring}}$. Thus only two-centre terms remain and in the Hückel scheme these are given by

$$(H^{(1,0)}_{\nu\lambda})_z = (1/2c)Q^z_{\nu\lambda}\beta^0 \tag{4.160}$$

$$(H^{(2,0)}_{\nu\lambda})_{zz} = -(1/4c^2)(Q^z_{\nu\lambda})^2\beta^0 \tag{4.161}$$

The equation for $\boldsymbol{\chi}^{\text{ring}}$ then becomes

$$\boldsymbol{\chi}^{\text{ring}} = -\sum_{\nu\to\lambda}\left\{P^{(0)}_{\nu\lambda}(H^{(2,0)}_{\nu\lambda})_{zz} + (P^{(1)}_{\nu\lambda})_z(H^{(1,0)}_{\nu\lambda})_z\right\} \tag{4.162}$$

where $\nu \to \lambda$ indicates that the sum is restricted to p_z atomic orbitals on adjacent atoms. If we consider benzene as an example, and take the arbitrary origin in the vector potential at the centre of the benzene ring, then it is easily shown that $(\boldsymbol{Q}_{\nu\lambda})_z$ is equal to twice the area ($W_{\nu\lambda}$) of the triangle with vertices, the origin, the position of ϕ_ν and the position of ϕ_λ. Also with this particular choice of origin, $(P^{(1)}_{\nu\lambda})_z$ is zero and $\boldsymbol{\chi}^{\text{ring}}$ assumes the simple form

$$\boldsymbol{\chi}^{\text{ring}} = (1/4c^2)\sum_{\nu\to\lambda} P^{(0)}_{\nu\lambda}(Q^z_{\nu\lambda})^2\beta^0 \tag{4.163}$$

$$= (1/4c^2)\sum_{\nu\to\lambda} P^{(0)}_{\nu\lambda}4W^2_{\nu\lambda}\beta^0 \tag{4.164}$$

Thus for benzene, where $P^{(0)}_{\nu\lambda} = \frac{2}{3}$ for bonded atoms, equation (4.164) gives

$$\boldsymbol{\chi}^{\text{ring}} = (1/4c^2)\tfrac{2}{3}(2)4W^2_{\nu\lambda}\beta^0\,(6) \tag{4.165}$$

and if we define W_{B} to be the area of the benzene ring, this becomes

$$\boldsymbol{\chi}^{\text{ring}} = -(2/9c^2)W^2_{\text{B}} \mid \beta^0 \mid \tag{4.166}$$

which is the equation obtained by London[75]. Thus $\boldsymbol{\chi}^{\text{ring}}$ is proportional to the square of the area of the ring and to the resonance integral β^0. It should be noted that for systems other than benzene $(P^{(1)}_{\nu\lambda})_z$ will not in general be zero and the full form of equation (4.162) must be used.

This method has been applied with some success to a wide variety of hydrocarbons, notably by McWeeny[127]. Other extensions of the original London treatment include calculations on porphyrins and heterocyclic systems by Berthier *et al.*[128] and others[129,130]. In general, the London theory appears to work well for relative values of diamagnetism. To obtain values for diamagnetic anisotropy London assumed that $\Delta\chi$ could be equated to $\boldsymbol{\chi}^{\text{ring}}$ and absolute values of $\Delta\chi$ calculated in this way are too small even if more refined methods of calculation are used[131,132]. This assumption, however, has been questioned by a number of workers[92,133–135] and is now generally accepted to be incorrect. For example, using the independent-electron theory of Section 4.2.6, Pople[92] derived equation (4.167) for a planar conjugated cyclic molecule

$$\Delta\chi = \boldsymbol{\chi}^{\text{ring}} - \{\mathscr{E}_\pi/(6c^2\Delta E\beta^0)\} \tag{4.167}$$

where $\mathscr{E}_\pi$ is the total Hückel π-electron energy, ΔE is a mean excitation energy and β^0 is the carbon–carbon resonance integral. The second term on the right is the contribution from the local terms and since this and χ^{ring} are both negative for $(4n+2)$ π-electron systems the theory predicts that the overall anisotropy is increased (over the London value) by these localised atomic contributions. The second term is -17.2 e.m.u. for benzene suggesting that local terms can contribute about 30% to the observed anisotropy of -59.7 e.m.u.; the remainder is due to the ring contribution.

Hoarau[136] obtained a similar formula to equation (4.167). He proposed that one could account for the local contributions to χ_{zz} by defining anisotropic Pascal constants for aromatic hydrocarbons and suggested the values $\chi_{\text{C}\perp} = -7.4$ and $\chi_{\text{H}\perp} = -2.0$. Davies[134], on the other hand, suggested that the local π-electron contribution is proportional to the number of atoms, n and obtained equation (4.168) for the total π-electron contribution to χ

$$\chi^\pi = \chi^{\text{ring}} + \chi^{\text{local}}$$

$$\chi^{\text{local}} = -1.35\, n \tag{4.168}$$

Here χ^{local} is effectively describing the one-centre $H^{(2,0)}_{2\text{p}_z 2\text{p}_z}$ terms which are omitted when calculating χ^{ring}. Other support for the view that local terms make significant contributions to $\Delta\chi$ has been provided by Itoh *et al.*[131], Craig *et al.*[130] and by Pople and Ferguson[93].

Although London's theory calculates χ^{ring} to be negative for $(4n+2)$ π-electron systems, Pople and Untch[137] have shown that this theory also predicts paramagnetic ring contributions for conjugated monocyclic polyenes with $4n$ electrons. This is in agreement with the earlier results of Berthier *et al.*[138] who calculated positive ring contributions for pentalene and heptalene. However, Pople and Untch also found that these paramagnetic currents will be partially quenched by alternation of bond lengths and molecular non-planarity.

The effects of interatomic currents on the shielding of nuclei in the vicinity of planar conjugated rings have received a great deal of attention. The first attempt to calculate the shielding effects of interatomic currents in aromatic molecules was made by Pople[139]. In this classical treatment, the π-electron system of benzene, for example, is regarded as a circular loop in the plane of the ring of radius a, and the current induced by the external field is calculated by Larmor's theorem[140]. The secondary magnetic field associated with this current loop is then calculated by replacing the current loop by an equivalent point magnetic dipole situated at the centre of the ring and using equation (4.132). This approach was extended to polycyclic systems, where the shielding at a particular nucleus A is given by [141]

$$\sigma_{\text{A}}^{\text{ring}} = -(a^2/2c^2)\sum_i R_i^{-3} \tag{4.169}$$

where R_i is the distance from the ith ring centre to atom A. This clearly approximate approach has been refined by Waugh and Fessenden[142] and by Johnson and Bovey[143] who substituted for the dipole a pair of current loops situated above and below the plane of the ring approximately at the point of maximum electron density in the π orbitals of benzene. All these classical methods are attempts to calculate the current density in the molecule when it is placed in an external magnetic field, and, then to evaluate the secondary field due to this density at a particular nucleus. For a $(4n+2)$ π-electron systems all the methods predict deshielding of protons outside of and in the plane of the ring. However, unlike the dipole approximation, the Johnson–Bovey model predicts a positive shielding effect at all points inside the ring, a result verified by experimental studies of [14]- and [18]-annulenes[144]. For $4n$ π-electron systems which exhibit positive χ^{ring} contributions, these shielding rules must be reversed, protons outside the ring being shielded while those inside the ring are deshielded. Examples of this type of shielding are provided by the [16]- and [24]-annulenes[144].

The first quantum mechanical treatments of this type of shielding were presented by Pople[88] and by McWeeny[94], who both extended the London theory of susceptibility to magnetic-shielding calculations. Pople's approach[88] involves calculating the currents induced by the external magnetic field using London's theory. Thus it determines the secondary fields due to these current loops and represents an attempt to incorporate the earlier classical ideas within a quantum mechanical framework. Jonathon *et al.*[145] have used this method to interpret the trends in proton chemical shifts in polycyclic hydrocarbons.

McWeeny[94], on the other hand, re-developed the London procedure starting with the complete vector potential of equation (4.26). This method effectively introduces a test magnetic dipole at the nucleus of interest to probe the secondary magnetic field and has the advantage of giving a direct calculation of magnetic shielding. McWeeny also introduced different gauge factors f_ν' setting

$$f_\nu' = \exp\{(-i/c)\boldsymbol{A}_\nu' \cdot \boldsymbol{r}\} \tag{4.170}$$

where

$$\boldsymbol{A}_\nu' = \tfrac{1}{2}\boldsymbol{H} \wedge \boldsymbol{R}_\nu + (\boldsymbol{\mu} \wedge \boldsymbol{R}_\nu)R_\nu^{-3} \tag{4.171}$$

The matrix elements $H_{\nu\lambda}(\boldsymbol{H}, \boldsymbol{\mu})$ are then given by equation (4.172).

$$\begin{aligned} H_{\nu\lambda}(\boldsymbol{H}, \boldsymbol{\mu}) &= \langle f_\nu'\phi_\nu \,|\, \tfrac{1}{2}(-i\nabla + (1/c)\boldsymbol{A}'(\boldsymbol{r}))^2 + V \,|\, f_\lambda\phi_\lambda\rangle \\ &= \langle f_\nu' f_\lambda'\phi_\nu \,|\, -\tfrac{1}{2}\nabla^2 + V \,|\, \phi_\lambda\rangle - (i/c)\langle f_\nu' f_\lambda'\phi_\nu \,|\, (\boldsymbol{A}' - \boldsymbol{A}_\lambda')\cdot\nabla \,|\, \phi_\lambda\rangle \\ &\quad + (1/2c^2)\langle f_\nu' f_\lambda'\phi_\nu \,|\, (\boldsymbol{A}' - \boldsymbol{A}_\lambda')\cdot(\boldsymbol{A}' - \boldsymbol{A}_\lambda') \,|\, \phi_\lambda\rangle \end{aligned} \tag{4.172}$$

where V represents the total potential energy terms. Using the London approximation and neglecting the last two terms, McWeeny obtained equation (4.173)

$$H_{\nu\lambda}(\boldsymbol{H}, \boldsymbol{\mu}) = \beta_{\nu\lambda}^{\circ} \exp\{(i/2c)\boldsymbol{Q}_{\nu\lambda} \cdot (\boldsymbol{H} + \boldsymbol{\mu}R_\nu^{-3} + \boldsymbol{\mu}R_\lambda^{-3})\} \tag{4.173}$$

Re-deriving the London procedure based on this Hamiltonian matrix and picking out energy terms bilinear in $\boldsymbol{H}$ and $\boldsymbol{\mu}$ leads to equation (4.174) for the isotropic shielding constant at nucleus A

$$\sigma_{\text{A}}^{\text{ring}} = (2\beta^{\circ}/3c^2)W_{\text{B}}^2 a^{-3}\left(\sum_i J_i[-K(\boldsymbol{r}_i)]\right) \tag{4.174}$$

Here J_i is identified with the contribution from the ith ring of the molecule, $K(\boldsymbol{r}_i)$ is a geometric factor which depends merely on the physical position of nucleus A relative to the ith ring and a is the length of a C—C bond. This theory has been widely used to obtain the ring contributions to magnetic shielding in polycyclic hydrocarbons[146–152]. A recent critical evaluation of the numerical performance of this type of theory has been given by Haigh *et al.*[152] who found that the McWeeny theory gives a good account of the magnetic shielding of non-sterically hindered protons in planar cyclic hydrocarbons.

All the methods presented in this section so far have been based on generalisations of the Hückel theory of conjugated molecules and so suffer from the uncertainties which are known to surround that theory. Attempts to improve on this model by using SCF wave functions were made by Davies[96], by Hall and Hardisson[95] and by Black *et al.*[153]. Hall and Hardisson, for example, introduced magnetic-field dependent terms into Pariser–Parr–Pople (PPP)[101, 102] SCF π-electron theory, leading to equation (4.175) for the Fock matrix elements

$$
\begin{aligned}
F^{(0)}_{\nu\nu} &= \omega_\nu + \tfrac{1}{2}P^{(0)}_{\nu\nu}G^{(0)}_{\nu\nu} + \sum_{\lambda \neq \nu} (P^{(0)}_{\lambda\lambda} - Z_\lambda)G^{(0)}_{\nu\lambda} \\
F^{(0)}_{\nu\lambda} &= \beta^{\circ}_{\nu\lambda} - \tfrac{1}{2}P^{(0)}_{\nu\lambda}G^{(0)}_{\nu\lambda} \\
F^{(1)}_{\nu\nu} &= (i/2)P^{(1)}_{\nu\nu}G^{(0)}_{\nu\nu} + i\sum_{\lambda \neq \nu} P^{(1)}_{\lambda\lambda}G^{(0)}_{\nu\lambda} \\
F^{(1)}_{\nu\lambda} &= (i/2c)\beta^{\circ}_{\nu\lambda}\boldsymbol{Q}_{\nu\lambda} - (i/2)P^{(1)}_{\nu\lambda}G^{(0)}_{\nu\lambda}
\end{aligned}
\tag{4.175}
$$

where ω_ν is an approximate p-electron ionisation potential. Their early work[95, 154] and the recent study by Roberts[155] have indicated that although absolute SCF values of susceptibilities and proton magnetic shielding constants differ markedly from the corresponding Hückel results, the SCF ratios (relative to benzene) are similar to the Hückel ratios (see Tables 4.15 and 4.16). Amos and Roberts[74] have also used PPP theory to calculate magnetic susceptibilities but they neglected the changes induced in the two-electron terms by the external field. This so-called uncoupled procedure[156] can then be perturbatively improved to allow for the changes in the two-electron terms. It is worth noting that the uncoupled method, although no longer giving SCF wave functions in the presence of the magnetic field represents a considerable simplification over the full theory at the PPP level. However, this is only obtained at the expense of gauge invariance. For *ab initio* theories, in which evaluation of the two-electron integrals is the time-consuming step, such simplifications result in a relatively minor saving of effort.

The introduction of the gauge factors f_ν' is unnecessary and, as McWeeny has recently commented[157], is of doubtful validity. A similar equation to equation (4.174) was obtained by Parker and Memory[158] who used London gauge factors and a particular form of the London approximation for the r^{-3} terms. The r^{-3} terms have been approximated in several ways

$$\text{LAI}\quad r^{-3} \approx \left(\tfrac{1}{2}(R_s + R_t)\right)^{-3} \tag{4.176}$$

$$\text{LAII}\quad r^{-3} \approx \tfrac{1}{2}(R_s^{-3} + R_t^{-3}) \tag{4.177}$$

$$\text{LAIII}\quad r^{-3} \approx c(R_s^{-3} + R_t^{-3}) \tag{4.178}$$

where c is usually chosen to give agreement with experiment. Use of LAII in Parker and Memory's work[158] leads to a similar equation to that obtained by McWeeny.

Recently, the validity of LAI to LAIII has been fully examined in a series of papers by Amos and Roberts[159–161] and some of their results are presented in Table 4.17. For a few systems they calculated chemical shifts treating the r^{-3} terms exactly, and, using the approximations LAI to LAIII. Their results show that none of the LA for the r^{-3} term are particularly satisfactory and that

Table 4.15 Calculated diamagnetic anisotropies due to ring currents (ratios relative to benzene)

(From Hall, G. G. and Hardisson, A. (1962). *Proc. Roy. Soc.* (*London*), A268, 337, reproduced by courtesy of the Royal Society)

Molecule	*Hückel –London*	*PPP –London*	*Exp.*
Benzene	(1)	(1)	(1)
Naphthalene	2.18	2.14	2.06
Styrene	0.91	0.98	–
Azulene	2.26	2.28	2.38

Table 4.16 Calculated ring-current contributions to proton chemical shifts

(From Roberts[155], reproduced by courtesy of Springer-Verlag)

Molecule	*Proton*	*Absolute Values**		*Ratios relative to benzene*		
		Hückel† –McWeeny	*SCF*	*Hückel –McWeeny*	*SCF*	*Exp‡*
Benzene		−4.39	−2.50	1	1	1
Naphthalene	1	−5.75	−3.22	1.31	1.28	1.30
	2	−5.14	−2.87	1.16	1.15	1.07
Styrene	1	−0.79	−0.48	0.18	0.19	–
	2	−0.26	−0.17	0.06	0.07	—
	3	−0.61	−0.37	0.14	0.15	—
Azulene	1	−5.00	−3.10	1.14	1.23	1.35
	2	−5.58	−3.43	1.27	1.37	1.02
	3	−5.88	−3.32	1.34	1.32	1.62
	4	−5.36	−2.99	1.22	1.18	0.86
	5	−5.27	−2.92	1.20	1.16	1.14

*Units are p.p.m.
†Hückel $\beta^0 = -4.15$ eV.
‡Based on an experimental estimate of 1.55 p.p.m. for benzene (Spiesecke, H. and Schneider, W. G. (1961). *J. Chem. Phys.*, 35, 731.

any agreement between exact and LA results are due to a fortuitous cancellation of errors. They also examined the McWeeny theory and find that the neglected second term of equation (4.172), if included, would cancel many of the terms which are retained. Their overall conclusions are that considerable errors in calculated values can arise if any LA is used for the r^{-3} terms and that any quantitative agreement between experiment and values calculated using McWeeny's theory must be regarded with caution.

In conclusion of this section it should be mentioned that the idea of separating the magnetic effects due to the σ- and the π-electrons has been strongly criticised by Musher[162–164]. In fact, he suggests that the concept of a ring current is an unsatisfactory one and that the diamagnetic anisotropy of benzene and particularly polycyclic aromatic hydrocarbons is better treated as the sum of local contributions. He claims that ring currents have no physical reality and arise because of the approximations inherent in the London theory. Hopefully, *ab initio* calculations of the susceptibility and shielding in aromatic molecules will determine the validity of these statements. In the meantime, the simple model of a ring current and its associated shielding effects provides the organic chemist with an excellent understanding of the contribution which aromatic systems make to the shielding of neighbouring protons.

Table 4.17 Ring-current contributions to proton chemical shifts calculated at various levels of approximation

(From Roberts[161], reproduced by courtesy of the North-Holland Publishing Company)

Molecule	*Proton*	'*Exact*'* *calculation*	LAI†	LAII†	LAIII†	*Exp.*
Benzene‡		−0.72	−1.06	−2.40	−0.96	—
Naphthalene§	1	−0.47	−0.60	−0.74	−0.30	−0.46
	2	−0.21	−0.25	−0.35	−0.14	−0.11
Anthracene§	1	−0.70	−0.85	−1.09	−0.43	−0.66
	2	−0.29	−0.32	−0.44	−0.18	−0.12
	9	−0.92	−1.17	−1.47	−0.59	−1.09

*These results are obtained evaluating the r^{-3} terms 'exactly' using numerical integration.
†These results are obtained by using various forms of the London approximation for the r^{-3} terms (see text).
‡Absolute values p.p.m.
§ Values relative to benzene.

4.3 EXPERIMENTAL METHODS

Experimental values of the average magnetic susceptibility of liquids and solids are usually determined by the conventional Gouy technique[1]. Magnetic susceptibilities of gases, on the other hand, have been measured by Bitter[165] and by Barter *et al.*[166] who used a Pauling-type oxygen meter. To obtain reliable results by any of these techniques it is essential to ensure that all oxygen has been removed from the sample and it is likely that failure to do this has led to the widely differing values of χ_m reported for the same molecule by independent groups of workers. Bulk magnetic susceptibilities and the various experimental techniques used to obtain these have recently been reviewed by Mulay and Mulay[167].

Since we are mainly interested here in those aspects of susceptibility which are directly related to magnetic shielding, we will concentrate on experimental methods which have been used to measure $\Delta\chi$.

4.3.1 Magnetic anisotropies from crystal data

Until a few years ago most direct experimental data on diamagnetic anisotropies had been derived from measurements on molecular crystals with lower than cubic symmetry. Such crystals possess bulk diamagnetic anisotropy, and, if it is assumed that the susceptibility tensor for the crystal can be obtained by superimposing those for the individual molecules (taking proper account of their orientation relative to crystal axes), then information about molecular components can be derived if the crystal structure is known. The amount of information that can be obtained in this way depends on the type of crystal. The connection between crystal and molecular tensors for the principal crystal systems has been discussed in detail by Lonsdale and Krishnan[4] who consider the orientations of the principal axes of the molecular susceptibility to the crystal axes. An alternative procedure is to transform the complete tensor including off-diagonal terms[93]. In some cases, thermal molecular motion leads to temperature-dependent crystal susceptibilities, but corrections can usually be applied[124].

Table 4.18 Experimental magnetic-susceptibility anisotropies from crystal studies*

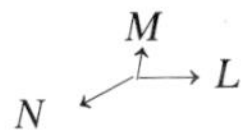 M, L, N (axes)	χ_L	χ_M	χ_N	*Reference*
	−34.9	−34.9	−94.6	124
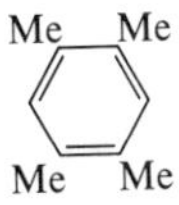 Me Me / Me Me	−77.3	−82.4	−143.9	4
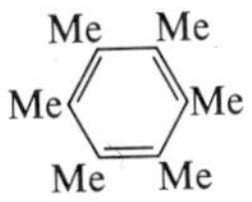 Me Me / Me Me / Me Me	−101.1	−102.7	−163.8	4†
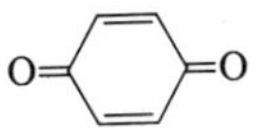 O= =O	−23.0	−27.0	−65.2	169
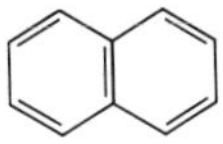	−54.7	−52.6	−173.5	4
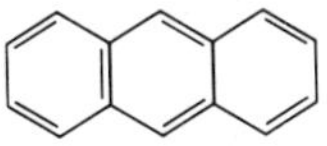	−75.8	−62.6	−251.6	4

Table 4.18 *continued*

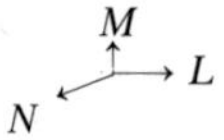 M ↑ → L, N ↙	χ_L	χ_M	χ_N	*Reference*
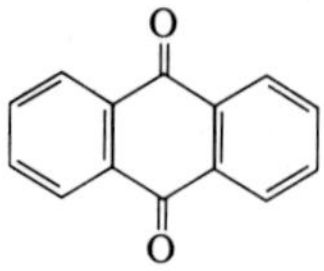 O, O	−76.1	−64.5	−217.6	169
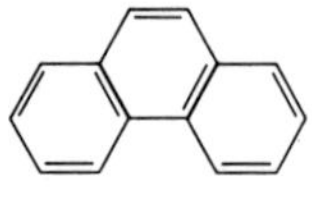	−74.0	−74.0	−240.0	170‡
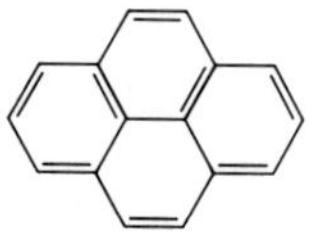	−80.6	−80.6	−303.0	4‡
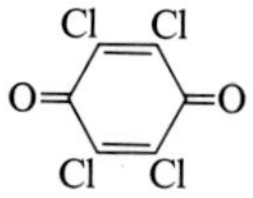 Cl Cl, O= =O, Cl Cl	−84.4	−98.5	−138.5	169
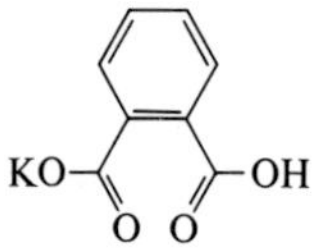 KO– –OH, O O	−73.92	−79.12	−144.79	172
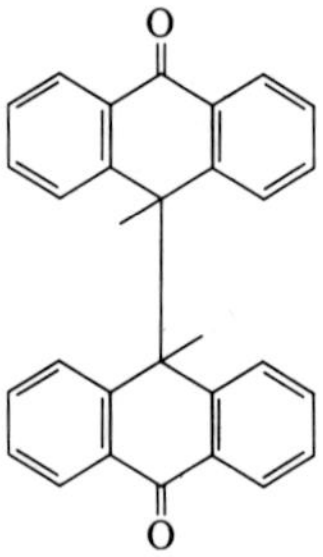 O, O	−134.2	−141.4	−414.8	171
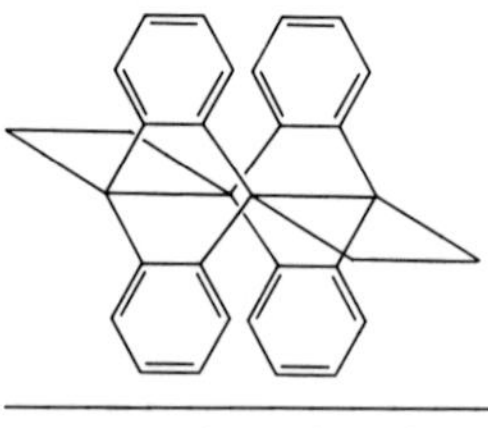	−181.9	−216.7	−430.4	171

*Units are 10^{-6} erg G^{-2} mol^{-1}.

†*L*-direction displaced by *c.* 10 degrees from line joining opposite methyl groups.

‡In-plane components assumed equal.

Bothner-By and Pople[168] have summarised experimental anisotropies derived from crystal measurements up to 1965. We include a few of their results and values obtained since their review in Table 4.18. Many of the crystal measurements are on aromatic molecules where there is a marked anisotropy with high diamagnetism perpendicular to the molecular plane. These effects are usually interpreted in terms of the interatomic ring contributions discussed in Section 4.2.7.

4.3.2 Magnetic anisotropies from Cotton–Mouton studies

Other estimates of magnetic-susceptibility anisotropies have been based on experimental measurements of magnetic birefringence in liquids[173–176] and compressed gases[7]. Provided the temperature-independent terms[177] can be neglected and the molecule has effectively axial symmetry, then the molar Cotton–Mouton constant is given by[177]

$$ {}_{\mathrm{m}}C = (4\pi N/405kT)(\alpha_{||}-\alpha_{\perp})(\chi_{||}-\chi_{\perp}) \tag{4.179}$$

Deduction of accurate magnetic anisotropies from Cotton–Mouton measurements has been hindered by a lack of reliable values of $\boldsymbol{\alpha}$. However, electric

Table 4.19 Experimental magnetic-susceptibility anisotropies from magnetic-birefringence studies of compressed gases*†

Molecule	$\chi_{\|\|}-\chi_{\perp}$	*Molecule*	$\chi_{\|\|}-\chi_{\perp}$
N_2	−8.43	C_6H_6	−53.6¶
CO	−10.24	$C_6H_3F_3$	−39.2¶
CO_2	−6.02	C_6F_6	−31.9¶
N_2O	−10.8		
C_2H_6	−3.6‡		
cyclo-C_3H_6	−19.3		

*Units are 10^{-6} erg G^{-2} mol^{-1}.
†These values are taken from Buckingham, A. D., Prichard, W. H. and Whiffen, D. H. (1967). *Trans. Faraday Soc.*, **63**, 1057.
‡More recent value quoted in Raynes, W. T. (1971). *Molec. Phys.*, **20**, 321.
¶Recent values taken from Bogaard, M. P., Buckingham, A. D., Corfield, M. G., Dunmur, D. A. and White, A. H. (1972). *Chem. Phys. Lett.*, **12**, 558.

polarisability components are now becoming more available[229] and susceptibility anisotropies derived from magnetic-birefringence studies of compressed gases are given in Table 4.19.

If the anisotropies of bond susceptibilities are additive properties and regular tetrahedral geometry is assumed,

$$\Delta\chi(C_2H_6) = (\chi_{||}-\chi_{\perp})_{C-C}-2(\chi_{||}-\chi_{\perp})_{C-H} \tag{4.180}$$

where parallel and perpendicular refer to bond directions. Accurate values for C—C and C—H bond anisotropies would be a valuable aid in understanding the shielding effects of C—C and C—H bonds. Previous estimates of $\Delta\chi(C_2H_6)$ were −2.6 e.m.u.[178] based on crystal studies[179] of long chain acids and −2.0 e.m.u. based on a wide survey[180] of the experimental data, both values being smaller than that found in Table 4.19. Values of $\Delta\chi$ for nitrogen are in good agreement with those calculated by Karplus and Kolker[41] (Table 4.2) and by Hameka[82]. The value for carbon monoxide,

Table 4.20 Experimental magnetic-susceptibility anisotropies from magnetic-birefringence studies of liquids and solutions*

	χ†	χ	*Reference*
Benzene	−35.5	−93.5	174
Toluene	−46.4	−104.0	173
p-Xylene	−56.4	−118.0	173
Mesitylene	−57.8	−130.2	173
Durene	−78.6	−146.5	173
Hexamethylbenzene	−97.9	−171.7	173
Fluorobenzene	−38.6	−97.7	173
Chlorobenzene	−53.9	−102.0	173
Bromobenzene.	−63.4	−110.0	173
Iodobenzene	−79.5	−115.1	173
Nitrobenzene	−32.7	−120.2	173
Benzonitrile	−44.7	−106.2	173
Naphthalene	−51.7	−175.6	174
Acenaphthene	−74.7	−186.0	173
Phenanthrene	−74.1	−239.1	173
Pyrene	−69.3	−305.1	174
Anthracene	−70.3	−249.4	174
Chrysene	−88.2	−305.8	174
Triphenylene	−77.8	−314.1	174
p-Benzoquinone	−26.4	−68.9	175
Chloranil	−84.8	−136.7	175
1,4-Naphthaquinone	−38.3	−144.8	175
9,10-Anthraquinone	−66.8	−217.1	175
9,10-Phenanthraquinone	−59.8	−194.0	175

*Units are 10^{-6} erg G^{-2} mol^{-1}.
†The in-plane principal components χ_L and χ_M are assumed equal.

Table 4.21 Values of $\chi_\perp$ for C—X bonds determined from Cotton–Mouton studies*

(From Le Fevre, R. J. W. and Murthy, D. S. N. (1966). *Aust. J. Chem.*, **19**, 185, reproduced by courtesy of CSIRO)

Molecule	X	$\chi_\perp$(C—X)
Toluene	CH_3	−10.36
p-Xylene	CH_3	−12.14
Mesitylene	CH_3	−12.17
Durene	CH_3	−13.19
Hexamethylbenzene	CH_3	−12.95
Fluorobenzene	F	−4.04
Chlorobenzene	Cl	−8.43
Bromobenzene	Br	−16.32
Iodobenzene	I	−21.38
Nitrobenzene	NO_2	−26.50
Cyanobenzene	CN	−12.53

*Units are 10^{-6} erg G^{-2} mol^{-1}.

however, is larger than the calculated value[41]. Similarly, the value for cyclopropane is larger than the semi-empirical value calculated by Bley[98]. It should be noted, however, that lack of information on the temperature-independent contributions to ${}_{\mathrm{m}}C$ makes the anisotropies derived from equation (4.179) somewhat uncertain.

Cotton–Mouton constants of liquids and solutions have been measured by Le Fevre and co-workers using photometric techniques[173–176]. They have studied monocyclic and polycyclic aromatic hydrocarbons and several quinones and a survey of their results is presented in Table 4.20. A comparison of these results with those of Table 4.18 indicates that magnetic anisotropies derived from Cotton–Mouton studies are close to those obtained from crystal measurements. However, the crystal studies do suggest that the assumption made by the Le Fevre group that $\chi_L = \chi_M$ is often incorrect.

From these results Le Fevre *et al.*[173,174] derived $\chi_\perp$ values of -15.7 and $+0.06$ e.m.u. for aromatic C—C and C—H bonds, respectively. Assuming that these values may be used as additive 'constants' allowed them to estimate $\chi_\perp$ for C—X bonds and these results are collected in Table 4.21. The results on quinones indicate that the diamagnetic anisotropy is evidently diminished when C=O replaces C—H. Using assumed values for $\chi_\perp$ (C—C) and $\chi_\perp$(C—H), Le Fevre and Murthy[175] estimated $\chi_\perp$ (C=O) to be 8.4 e.m.u. This value and those of Table 4.21, however, are subject to several uncertainties and should be regarded with caution.

4.3.3 Magnetic anisotropies from microwave and beam studies

The most important source of diamagnetic anisotropies in the last few years has been the Zeeman microwave studies of Flygare and his group[181,182]. They obtained the first measurement of magnetic susceptibility by microwave spectroscopy in 1968[9] and applications to a wide range of molecules have rapidly followed.

The Hamiltonian of a rotating rigid rotor in the presence of a static magnetic field along the laboratory-fixed z axis, H_z, is given by (retaining up to quadratic terms in H_z)

$$H = \sum_{\alpha} \{J_\alpha^2/2I_{\alpha\alpha}\} - (H_z\mu_0/2\hbar) \sum_{\alpha\beta} J_\beta(g_{\beta\alpha} + g^*_{\alpha\beta})\phi_{\alpha z} - (H_z^2/2) \sum_{\alpha\beta} \phi_{z\alpha}\chi_{\alpha\beta}\phi_{\beta z} \quad (4.181)$$

where J_α is the rotational angular momentum and $I_{\alpha\alpha}$ is the moment of inertia along the α principal inertial axis. The sums over α and β are over the three principal inertial axes a, b and c, $\phi_{\alpha z}$ is the direction cosine between the space-fixed z axis and the molecule-fixed α axis and μ_0 is the nuclear magneton. To a high approximation, the molecular Zeeman parameters $g_{\alpha\beta}$ and $\chi_{\alpha\beta}$ are defined by equations (4.182) and (4.183).

$$g_{aa} = (M_p/I_{aa}) \left\{ \sum_n Z_n M_n^{-1}(b_n^2 + c_n^2) + 2 \sum_{k>0} (E_0 - E_k)^{-1} \times |\langle \Psi_0 | L_a | \Psi_k \rangle|^2 \right\} \quad (4.182)$$

$$\chi_{aa} = -(1/4c^2)\sum_n Z_n^2 M_n^{-1}(b_n^2+c_n^2)-(1/4c^2)\langle\Psi_0|\sum_i(b_i^2+c_i^2)|\Psi_0\rangle$$

$$-(1/4c^2)\sum_{k>0}(E_0-E_k)^{-1}|\langle\Psi_0|L_a|\Psi_k\rangle|^2 \qquad (4.183)$$

Here M_p is the proton mass (in atomic units), M_n is the mass of the nth nucleus, Z_n is the atomic number of the nth nucleus, b_n is the projection on the b axis of the distance from the molecular centre of mass (c.m.) to the nth nucleus and L_a is the component of the electronic angular momentum along the a axis. b_i is the projection on the b axis of the distance from the c.m. to the ith electron.

The first term in equation (4.183) is the nuclear contribution to the magnetic susceptibility and is considerably smaller than the other two terms due to the inverse M_n dependence. The second and third terms in equation (4.183) are the conventional susceptibility terms, χ_{aa}^{d} and χ_{aa}^{p}, of the previous sections. It should be noted, however, that in equation (4.183) these terms are defined with respect to the c.m. as origin.

Huttner and Flygare[183] have discussed the rotational dependence of the Hamiltonian of equation (4.181) and obtained equation (4.184) for the rotational energy of a general molecule in the uncoupled basis first order in J

$$E(J\tau M) = E^0(J\tau M)-\{(\mu_0 MH)/[J(J+1)]\}\sum_\alpha g_{\alpha\alpha}\langle J_\tau||J_\alpha^2||J_\tau\rangle$$

$$-\{H^2/[J(J+1)]\}C(JM)\sum_\alpha(\chi_{\alpha\alpha}-\chi)\langle J_\tau||J_\alpha^2||J_\tau\rangle \qquad (4.184)$$

where

$$C(JM) = [3M^2-J(J+1)]/[(2J-1)(2J+3)] \qquad (4.185)$$

In equation (4.184) $E^0(J\tau M)$ is the zero-field energy and $\langle J_\tau||J_\alpha^2||J_\tau\rangle$ is the reduced matrix element of J_α^2 in the general asymmetric top basis indicated by quantum number τ. Thus, we note that, in general, the three independent diagonal elements of the molecular g-value tensor can be measured. However, the sign of M cannot be determined experimentally with the normal plane-polarised electromagnetic field and so only the relative signs of the three g-values can be determined. Also only two independent magnetic susceptibility anisotropies can be determined because

$$\sum_\alpha(\chi_{\alpha\alpha}-\chi) = 0 \qquad (4.186)$$

Flygare takes these two independent anisotropies to be

$$\chi_{aa}-\chi = \tfrac{1}{3}(2\chi_{aa}-\chi_{bb}-\chi_{cc})$$
$$\chi_{bb}-\chi = \tfrac{1}{3}(2\chi_{bb}-\chi_{aa}-\chi_{cc}) \qquad (4.187)$$

Thus, in general, the Zeeman perturbation provides five unique parameters. Symmetry, however, often reduces this number and permits simplification of equation (4.184). For example, for a symmetric top molecule there are only two independent g values and a single independent susceptibility anisotropy and

$$\langle J_a^2 \rangle = \tfrac{1}{2}[J(J+1)-K^2] = \langle J_b^2 \rangle \qquad (4.188)$$

$$\langle J_c^2 \rangle = K^2 \qquad (4.189)$$

For a linear molecule there is a further reduction to a single independent g-value and a single anisotropy and since K also vanishes, equation (4.184) becomes (c is the internuclear axis)

$$E(JM) = E^0(JM) - H\mu_0 M g_{bb} - (H^2/3)C(JM)(\chi_{bb} - \chi_{cc}) \qquad (4.190)$$

Therefore, the g-values and the susceptibility anisotropies can be obtained to fairly high accuracy by using equation (4.184) to interpret the rotational spectrum as a function of the magnetic field.

Comparing equations (4.182) and (4.183) we find that χ_{aa}^{p} can be related to g_{aa} by

$$\chi_{aa}^{p} = -(1/2c^2)\left[g_{aa}/(8\pi A M_p) - \tfrac{1}{2}\sum_n Z_n(b_n^2 + c_n^2)\right] \qquad (4.191)$$

where A is the rotational constant in Hz. If the molecular structure is known, values of $\sum_n Z_n b_n^2$ etc. can be computed leading directly to values of the diagonal elements of the paramagnetic susceptibility tensor. Finally, if the average magnetic susceptibility, χ, is also known the individual elements χ_{aa}, χ_{bb}, χ_{cc} and χ_{aa}^{d}, χ_{bb}^{d}, χ_{cc}^{d} can be extracted as well. However, accurate molecular structures are sometimes unavailable and experimental values of average susceptibilities are often unreliable. These two factors place unfortunate limitations on the accuracy with which the individual elements of the $\boldsymbol{\chi}^{d}$, $\boldsymbol{\chi}^{p}$ and $\boldsymbol{\chi}$ tensors can be obtained.

Molecular-beam techniques can also be used to determine molecular g-values and magnetic susceptibility anisotropies. These methods include the nuclear and molecular magnetic resonance methods, the electric-field molecular-beam methods and the recent beam-maser method. In contrast to the microwave techniques which usually observe the $\Delta J = 0, \pm 1, \Delta M_J = 0, \pm 1$ transitions, beam methods observe changes in nuclear ($\Delta M_I = \pm 1$) or molecular ($\Delta M_J = \pm 1$) orientations where $\Delta J = 0$. Although the molecular beam methods are more accurate, microwave spectroscopy is presently a more powerful tool for the measurement of magnetic susceptibility anisotropies because of its wider applicability. A limitation of this technique, however, is the requirement that the molecules possess a permanent electric dipole moment.

In Tables 4.22 and 4.23 we summarise the susceptibility anisotropies obtained by molecular beam (MB) and microwave spectroscopy (MS). The values of $\Delta\chi$ for nitrogen and carbon dioxide obtained by MS are in close agreement with the results obtained from magnetic birefringence studies[7]. In contrast, the MS values for carbon monoxide and nitrous oxide are somewhat smaller than those in Table 4.19. Although cyclopropane has not been measured by the MS technique, Benson and Flygare[228] have estimated a value for $\Delta\chi$ of -10 from considerations of $\Delta\chi$ in similar ring systems. This value is again much smaller than that found by Buckingham *et al.*[7] and is fairly close to the value calculated by Bley[98]. Using the Cotton–Mouton constants of Buckingham *et al.*[7], polarisability anisotropies[229] and the MS results, Flygare *et al.*[190] have estimated the temperature-independent contributions to ${}_{m}C$ for nitrous oxide, carbon dioxide and nitrogen. It appears that neglect of these terms in magnetic birefringence studies can lead to uncertainties in the derived values of $\Delta\chi$.

The value of $\Delta\chi$ in methyl acetylene is only slightly larger than the result

Table 4.22 Experimental magnetic-susceptibility anisotropies from microwave spectroscopy (MS) and molecular-beam (MB) studies*

(From Flygare and Benson[182], reproduced by courtesy of Taylor and Francis Ltd

Molecule	$\chi_{\parallel}-\chi_{\perp}$	*Method*	*Reference*
H_2	0.553 ± 0.02	MB	25, 184
HD	0.535 ± 0.02	MB	185
D_2	0.528 ± 0.04	MB	184
NaF	1.0 ± 0.8	MB	186
KF	-1.8 ± 0.6	MB	187
RbF	-7.2 ± 3.6	MB	187
CsF	-8.9 ± 4.0	MB	188
TlF	-4.5 ± 0.3	MB	189
N_2	-8.56 ± 0.3	other	190, 191
CO	-8.2 ± 0.9	MS	192
CS	-24.0 ± 3.0	MS	192
BrF	-20.0 ± 1.0	MS	182
OCS	-8.0 ± 1.0	MS	193
OCO	-6.32 ± 0.18	other	194
SCS	-14	other	176
OCSe	-10.06 ± 0.18	MS	195
NNO	-10.15 ± 0.15	MS	190
FCCH	-5.19 ± 0.15	MS	196
CH_3CCH	-7.70 ± 0.14	MS	197
CH_3CCCCH	-13.08 ± 0.16	MS	182
CH_3I	-10.98 ± 0.45	MS	198
CH_3Br	-8.50 ± 0.40	MS	198
CH_3Cl	-7.95 ± 0.50	MS	198
CH_3F	-8.2 ± 0.8	MS	182
CH_3NC	-13.5 ± 1.7	MS	199
CH_3CN	-10.20 ± 1.0	MS	199
$CH_3C^{15}N$	-10.5 ± 0.5	MS	199
CD_3CN	-10.2 ± 1.0	MS	199
HCF_3	-1.2 ± 0.6	MS	182
NH_3	1.3 ± 1	MB	108
PH_3	-2.7 ± 0.8	MS	200
NF_3	3.0 ± 1.5	MS	201
PF_3	-1.3 ± 0.5	MS	201
OPF_3	-1.8 ± 0.8	MS	201

*Units are 10^{-6} erg G^{-2} mol^{-1}.

reported for fluoroacetylene, both values being smaller than the INDO values of Table 4.9. The MS value of $\Delta\chi$ for methyl cyanide is considerably larger than the acetylene results and, although this increase is qualitatively reproduced by our INDO method, absolute calculated values are again too large.

Table 4.23 Independent magnetic susceptibility anisotropies in asymmetric top molecules*†‡

(From Flygare and Benson[182], reproduced by courtesy of Taylor and Francis Ltd.)

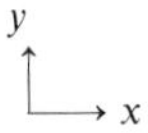 y ↑ → x	$2\chi_{xx}-\chi_{yy}-\chi_{zz}$	$2\chi_{yy}-\chi_{xx}-\chi_{zz}$	*Reference*
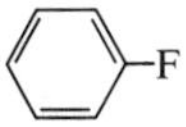 –F	52.9 ± 0.8	63.6 ± 1.5	202
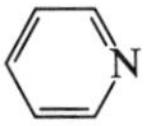 N	54.3 ± 0.6	60.5 ± 0.8	203
	9.1 ± 2.2	5.7 ± 1.6	204
	37.8 ± 0.3	30.7 ± 0.3	205
 N	50.2 ± 1.0	34.6 ± 1.8	206
 O	43.0 ± 0.2	34.4 ± 0.2	207
 S	49.6 ± 1.1	50.6 ± 1.3	207
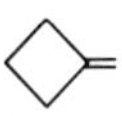	-6.4 ± 0.5	$+4.3 \pm 1.7$	208
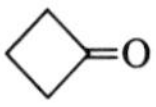 =O	14.8 ± 0.9	-10.6 ± 1.0	209
	-0.9 ± 0.5	$+5.0 \pm 0.7$	208
 O	-20.1 ± 0.5	-13.5 ± 0.8	231
 S	-20.9 ± 1.0	-24.6 ± 1.0	231
	18.3 ± 0.5	14.9 ± 0.6	208
	7.1 ± 0.6	26.8 ± 0.4	228
N	4.6 ± 0.8	16.5 ± 0.7	206
O	0.8 ± 1.0	18.1 ± 0.6	210

Table 4.23 *continued*

y ↑ → x	$2\chi_{xx}-\chi_{yy}-\chi_{zz}$	$2\chi_{yy}-\chi_{xx}-\chi_{zz}$	*Reference*
S (three-membered ring)	12.1 ± 0.9	18.7 ± 0.6	211
O, C=O, O (five-membered ring)	7.2 ± 1.2	21.7 ± 1.4	182
	-5.9 ± 0.3	21.2 ± 0.3	212
	-14.9 ± 1.1	$+7.8 \pm 1.5$	213
C=C(–C)–C=C	16.7 ± 1.2	19.2 ± 1.0	205
C=C–C=O	24.1 ± 0.9	17.1 ± 1.5	182
C=C–C	-0.7 ± 0.3	13.4 ± 0.5	214
C≡C–C–C=O	4.4 ± 0.8	9.0 ± 1.6	215
C–O–C	-10.4 ± 0.7	$+1.2 \pm 0.6$	216
C–S–C	-4.2 ± 0.5	-2.8 ± 0.5	216
C–C=O	8.1 ± 2.5	9.6 ± 1.4	217
H, H >C=O	25.5 ± 0.5	-3.9 ± 0.3	182
H–C(=O)–O–H	9.4 ± 0.3	3.4 ± 0.5	218
H–C(=O)–N(–H)–H	$+8.0 \pm 0.5$	$+2.2 \pm 0.7$	182
H–C(=O)–C(H)(H)–O–H	18.8 ± 2.0	7.1 ± 2.5	219
H–C(=O)–O–CH_3	3.1 ± 0.9	11.0 ± 0.9	219

Table 4.23 *continued*

Molecule (y ↑, $\rightarrow x$)	$2\chi_{xx}-\chi_{yy}-\chi_{zz}$	$2\chi_{yy}-\chi_{xx}-\chi_{zz}$	*Reference*
$H_2C=C=O$	-5.0 ± 0.7	-0.2 ± 0.6	220, 221
H_2CF_2	$+0.8\pm0.4$	-3.9 ± 0.5	222
$O=CF_2$	$+1.6\pm0.9$	$+5.3\pm0.6$	222
$O=C(F)H$	$+6.1\pm0.3$	$+5.9\pm0.3$	223
$H_2C=CHF$	-0.8 ± 0.2	$+9.6\pm0.2$	223
$H_2C=CF_2$	-2.3 ± 0.6	$+7.7\pm0.5$	222
$FHC=CHF$	$+5.7\pm0.4$	-1.6 ± 0.3	222
$FHC=CF_2$	-4.2 ± 0.2	$+7.7\pm0.3$	223
H_3C-CH_2F	-6.6 ± 0.4	-0.7 ± 0.5	223
H_3C-CHF_2	-3.4 ± 0.7	-1.7 ± 0.9	223
O_3 (O–O–O)	$+98.0\pm5.6$	-17.4 ± 4.4	224, 227
SO_2 (O–S–O)	$+6.4\pm0.5$	$+3.1\pm0.3$	224
OF_2 (F–O–F)	-8.8 ± 1.4	-4.4 ± 0.7	224

Table 4.23 *continued*

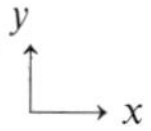 Molecule (y ↑, x →)	$2\chi_{xx}-\chi_{yy}-\chi_{zz}$	$2\chi_{yy}-\chi_{xx}-\chi_{zz}$	*Reference*
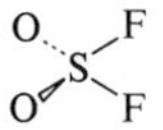 SO_2F_2	-12.6 ± 2.6	$+13.2 \pm 3.5$	201
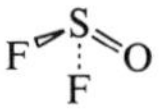 SOF_2	$+0.7 \pm 1.6$	$+0.8 \pm 2.0$	201
 SF_4	-2.9 ± 2.2	-4.9 ± 1.8	225
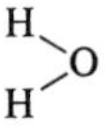 H_2O	-0.199 ± 0.048	$+0.464 \pm 0.024$	226

*Units are 10^{-6} erg G^{-2} mol^{-1}.
†All these results are from Zeeman microwave studies.
‡The molecules are arranged so that the z-axis is perpendicular to the heavy atom plane and the orientation of the heavy atom plane is approximately as shown in the figures.

For the ring compounds in Table 4.23 it is interesting to compare values of $\Delta\chi = \chi_{zz} - \frac{1}{2}(\chi_{xx} + \chi_{yy})$. This anisotropy is equal to one half the negative sum of the two susceptibility anisotropies given in Table 4.23. Many attempts have been made to relate such $\Delta\chi$ values to aromaticity[230]. The striking difference between the values of $\Delta\chi$ for benzene (-59.7) and for cyclohexadiene (-7.4) can be regarded as an indication of large interatomic ring contributions in benzene. Pyridine and fluorobenzene show similar values of $\Delta\chi$ to that of benzene. On going to the five-membered rings there is a reduction in this anisotropy and $\Delta\chi$ is further reduced in the three-membered rings. The positive anisotropies for the four-membered rings are quite striking since the values of $\Delta\chi$ are negative for all other ring molecules that have been studied. Benson *et al.*[231] attribute these positive values to large local contributions to $\Delta\chi$.

Flygare and his collaborators have also attempted to estimate non-local contributions to anisotropies by comparing $\Delta\chi$ for ring compounds with the values for their open-ring analogues. For example, the open-ring compounds propene, dimethyl sulphide, dimethyl ether and isoprene have significantly lower (in magnitude) anisotropies than the respective ring analogues cyclopropene, ethylene sulphide, ethylene oxide and cyclopentadiene. If there is no change in carbon hybridisation as the ring is closed, then the change in $\Delta\chi$ may be assigned to non-local effects. Although the assumption of constant hybridisation on ring closure is rather poor for three- and four-membered rings, it may be more valid for the five-membered rings where there is less strain.

The $\Delta\chi$ values for the fluorocarbons also show some interesting trends. On going from formaldehyde to carbonyl fluoride, the value of $\Delta\chi$ becomes less negative, a similar trend to that found in the corresponding ethylene studies. Addition of a first fluorine gives a change of +4.8. A second fluorine produces on average a change of +2.4 and the change on addition of a third fluorine is +0.6.

The large magnetic susceptibility anisotropy in ozone has been attributed to the existence of low-lying electronic states[224]. Table 4.23 indicates that a rapidly growing volume of experimental magnetic susceptibility data is being amassed from Zeeman microwave studies. Despite the many theories proposed there is still a need for a theory capable of interpreting the many interesting experimental trends. Certainly theoreticians can no longer claim to be hindered by a lack of accurate experimental data.

4.4 MAGNETIC SUSCEPTIBILITIES AND N.M.R. MAGNETIC SHIELDING

Another method which has been used extensively to estimate magnetic anisotropies is based on the close connection between the magnetic susceptibility tensor $\boldsymbol{\chi}$ and the nuclear shielding tensors, $\boldsymbol{\sigma}$, in a molecule. The majority of such studies have been based on equation (4.133) which we reproduce here for convenience

$$\boldsymbol{\sigma}_{\mathrm{A}}^{\mathrm{AB}} = (\tfrac{1}{3})\Delta\chi_{\mathrm{B}}(1 - 3\cos^2\theta_{\mathrm{B}})R_{\mathrm{AB}}^{-3} \qquad (4.133)$$

It should be emphasised again that equation (4.133) is based on the dipole approximation. Many attempts have been made to obtain information on $\Delta\chi$ for various electron groupings using this equation. Conversely, a number of workers have used a knowledge of $\Delta\chi$ from other sources to interpret n.m.r. chemical shifts, particularly for protons. Reviews of such research have been given by Bothner-By and Pople[168], by Jackman and Sternhell[232] and by Bovey[233]. Here we only mention the main points for some important electron groupings and suggest that the books by Jackman and Sternhell[232] and by Bovey[233] be consulted for the extensive literature which now exists.

The usual procedure is to compare two compounds, which ideally are of known geometry and identical except for the presence of the group in one and its absence or different orientation in the other. By measuring the chemical shift and assuming that changes in the shielding arising from other sources are negligible, or can be allowed for, application of equation (4.133) yields information about $\Delta\chi$. This approach suffers from two main difficulties: (a) these neighbour-anisotropy contributions to shielding are small and only really significant for protons and (b) the dipolar approximation is only valid if the nucleus studied is a distance of about 3–4 Å from the electron group[106]. Under such conditions, however, the contributions to the shieldings are small, and other effects such as electric-field effects[234] give rise to shielding contributions of comparable, but, uncertain size. For example, Slomp and McGarvey[235] examined the α and β isomers of a series of 6-methyl steroids and found that the protons of the β-methyl groups were less shielded than those of the α-methyls. Magnetic anisotropy effects cannot explain this shielding trend, and it was suggested that the steric compression to which the β-methyls are subject results in a second-order paramagnetism, which gives a downfield shift. Other cases in which the measured shifts are in a direction opposite to those expected from anisotropy considerations have been discussed by Wellman and Bordwell[236], by Nagata *et al.*[237] and by Cook and Danyluk[238].

The limitations of equation (4.133) are often ignored and this has led to widely differing values of $\Delta\chi$ for the same electron group in various types of molecule. In many cases the values of $\Delta\chi$ obtained are in complete disagreement with those obtained from other, more reliable experiments.

In the following, we will discuss the results for some important electron groups.

4.4.1 Shielding by C—C and C—H single bonds

It is convenient and usual to discuss the anisotropies associated with these two bonds together, as their $\Delta\chi$ values are usually determined from the same systems. For several years it was customary to assume that C—H bonds were magnetically isotropic. Using this assumption, Bothner–By and Naar–Colin[107] found that a $\Delta\chi$(C—C) value of 3.3 would explain the chemical-shift differences between cyclopentane, cyclohexane and open-chain methylene compounds. Jackman[239] noted that the shift between equatorial and axial protons in cyclohexane derivatives could also be accounted for on this basis. Other applications were those of Musher[240], Moritz and Sheppard[241], and Hall[242]. Musher[243] has also pointed out that the unrealistic value for $\Delta\chi$(C—C) of 15 is required to rationalise the chemical shifts in some bicycloheptanols. Bothner–By and Pople[168], however, have suggested that any value for $\Delta\chi$(C—C) larger than 4.5 seems unlikely since the mean susceptibility is -3.0 and larger values of $\Delta\chi$ would require χ along the bond direction to be paramagnetic.

Some of these difficulties may be avoided by ascribing part of the chemical shift to electronegativity differences and allowing a non-zero value for $\Delta\chi$(C—H) as first suggested by Narasimhan and Rogers[244a]. For studies of propane they suggest values for $\Delta\chi$(C—C) of 1.5–3.0 and $\Delta\chi$(C—H) of 0.24–1.5. Zurcher[245] has suggested the alternate values of $\Delta\chi$(C—C) from 1.95 to 3.75 and $\Delta\chi$(C—H) from 0.0 to 1.5. Davies[180] proposed values of $\Delta\chi$(C—C) *c.* 4.0 and $\Delta\chi$(C—H) *c.* 3.0 from a survey of experimental and theoretical data.

In an effort to remove the errors associated with the dipole approximation, Apsimon *et al.*[246] have recently modified equation (4.133) to correct for the deficiencies at small distances. Their modified equation for $\boldsymbol{\sigma}_{\mathrm{A}}^{\mathrm{AB}}$ is

$$\begin{aligned}\boldsymbol{\sigma}_{\mathrm{A}}^{\mathrm{AB}} &= \tfrac{1}{3}\Delta\chi_{\mathrm{B}}(1-3\cos^2\theta_{\mathrm{B}})\,R_{\mathrm{AB}}^{-3}\\ &\quad + S^2\{-\tfrac{1}{2}(\chi_{\mathrm{L}}+2\chi_{\mathrm{T}})+5(\chi_{\mathrm{L}}\cos^2\theta_{\mathrm{B}}+\chi_{\mathrm{T}}\sin^2\theta_{\mathrm{B}})\\ &\quad -35(\chi_{\mathrm{L}}\cos^4\theta_{\mathrm{B}}+\chi_{\mathrm{T}}\sin^4\theta_{\mathrm{B}})/6\}\,R_{\mathrm{AB}}^{-5} \qquad (4.192)\end{aligned}$$

where S is half the length of the induced dipole and χ_{L} and χ_{T} are the longitudinal and transverse components of the susceptibility tensor of the electron group at B.

Apsimon *et al.* obtained one equation for $\Delta\chi$(C—C) and $\Delta\chi$(C—H) by applying equation (4.192) to cyclohexane where the chemical shift difference has been accurately measured. Equation (4.180) was used as a second piece of data with the value of $\Delta\chi(C_2H_6)$ derived from Cotton–Mouton studies. These two relationships allow $\Delta\chi$(C—C) = 8.43 and $\Delta\chi$(C—H) = 6.63 to be deduced.

The correction employed in equation (4.192) assumes that at short distances the field of a magnetically anisotropic bond is that of a magnetic dipole of finite length. No justification is given for the validity of this model. In fact, there is rather poor agreement between the calculated and observed chemical shifts of the methine resonances upon methyl substitution at distant carbon atoms in a series of cycloalkanols[246].

A general expression for calculating the contributions of the magnetically anisotropic X—Y bonds of freely rotating XY_3 groups to the shielding of protons of a freely rotating methyl group elsewhere in the same molecule has recently been given by Raynes[247].

In view of the rather small observed shifts, the contributions of unknown magnitude from other sources and the approximate nature of equations (4.133) and (4.192), it would seem wise to treat the values for $\Delta\chi$(C—C) and $\Delta\chi$(C—H) obtained by the n.m.r. method with some reservation.

4.4.2 Shielding by C=C double bonds

Several models for the susceptibility tensor for a C=C double bond have been suggested[89, 239, 248]. Jackman has suggested a large diamagnetism in the y direction. Pople's theory of susceptibility[89], on the other hand, predicts a paramagnetism in the x direction centred on the carbon atoms. As Bothner–By and Pople[168] have pointed out, the available evidence seems to support

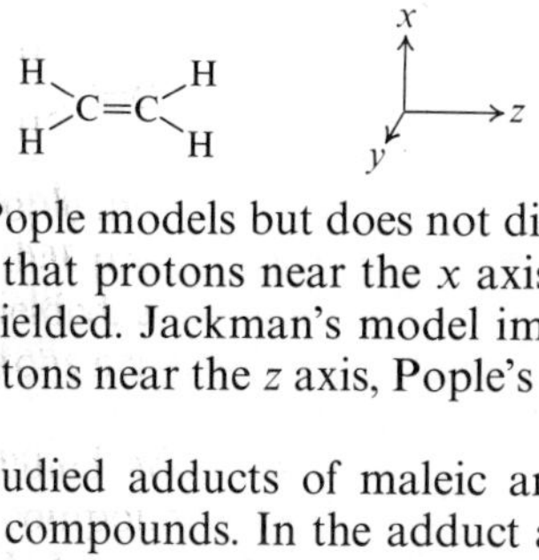

both the Jackman and Pople models but does not distinguish between them. Both suggestions imply that protons near the x axis will be deshielded and those near the y axis shielded. Jackman's model implies a deshielding contribution for remote protons near the z axis, Pople's a shielding contribution for such protons.

Ayer *et al.*[249] have studied adducts of maleic anhydride, maleopimaric acid and several related compounds. In the adduct a double bond is rigidly held in such a position that an angular methyl group is close to the y axis; it is shielded in accord with the Pople and Jackman models. For norbornadienes and norbornenes the bridge hydrogens located *syn* to a double bond are more shielded than those located *anti*[250], again in accord with both Jackman and Pople. Many other examples of shielding by C—C double bonds are given by Jackman and Sternhell[232] and by Bovey[233].

4.4.3 Shielding by C=O double bonds

Pople[89] has calculated the magnetic anisotropy of the carbonyl group and finds a large paramagnetism in the x direction (see ethylene figure) centred on the carbon atom and the largest diamagnetism on the oxygen atom in the *y* direction. Jackman[239] has suggested a large diamagnetism in the direction normal to the nodal plane of the π orbitals, as with ethylene. Again both models agree in predicting deshielding for remote protons close to the *x*-axis and shielding for remote protons close to the *y* axis; they differ in predictions about the contribution to protons close to the *z* axis in the region closer to the carbon.

Jackman[232, 239] has given several examples of substituted acrylic esters which are in accord with both models. Narasimhan and Rogers[244b] suggest the values $\chi_{zz} \approx 0$, $\chi_{yy} \approx 2.5$, $\chi_{xx} = -6.6$ from a study of the different chemical shifts in substituted amides. These models are qualitatively supported by the work of Williams, Bhacca and Djerassi[251] on some ketosteroids.

Using the Jackman model, which predicts deshielding for all protons in the *xz* plane, one would expect that the *N*-substituent *cis* to the carbonyl group in amides would be more deshielded than the *trans* substituent. However, as Bovey[233] points out, the proton *cis* to the carbonyl group in formamide is thought to be more shielded than the *trans* proton. It thus appears that amido and keto carbonyl groups may have somewhat different shielding properties.

Jackman's suggestions have recently been questioned by Apsimon *et al.*[252, 253]. These workers propose a different model in which protons in the *xz* plane and close to the *x* axis are deshielded, but in which protons in the *xz* plane and distant from the *x* axis are shielded. Support for the Apsimon model has been reported by Karabatsos *et al.*[254] who find that their n.m.r. studies of some aldehydes and ketones can be interpreted if H_A resonates at higher magnetic fields than H_B.

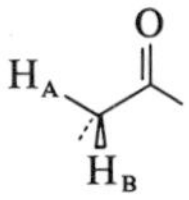

4.4.4 Shielding by C≡C and C≡N triple bonds

The acetylene protons are found to be less shielded than ethane protons but more shielded than the protons in ethylene. Charge-density considerations, however, would predict the acetylene protons to be the least shielded. Pople explained this unusual shielding of the acetylene protons in terms of a contribution from the anisotropic susceptibility of the acetylenic group. Theoretical calculations have given values for $\Delta\chi(C{\equiv}C)$ between -19.4[89] and -1.61[32].

Zeil and co-workers[255, 256] studied the proton shifts in a variety of ethyl- and isopropylacetylenes, haloacetylenes and nitriles. They took into account both the electronegativity of the substituents and the magnetic anisotropy of the triple bonds and found the following anisotropies: —C≡C—H, —34, C≡C—Cl, —37, —C≡N, —34.

Reddy and Goldstein[257] used the observed ^{13}C—H coupling constants to provide an indication of the local shielding in a C—H bond. If all contributions to shielding are local in character, plots of observed chemical shift versus J_{C-H} will be linear for families of compounds with similar structure. They used the vertical displacement of these lines to estimate values of $\Delta\chi$ for the acetylenic and nitrile groups suggesting the value of -16.5 for both groups.

The lack of reliability of this approach is clearly demonstrated by comparing these values with the much smaller results found by Flygare and his group.

4.4.5 Shielding by rings

The shielding of protons by aromatic rings has been interpreted using the classical theories of Pople[139], Waugh and Fessenden[142], and Johnson and Bovey[143] and the quantum mechanical theories of Pople[88], and McWeeny[94]. The latter are usually more successful. Jonathon *et al.*[145] have reproduced the proton chemical shift trends in polycyclic hydrocarbons using Pople's method of estimating ring currents. Differential shielding of non-equivalent methylene protons in a variety of systems containing a benzene ring has been studied by Whitesides *et al.*[258]. They conclude that the shifts can all be explained with reasonable molecular geometries using the Johnson–Bovey calculations for the remote shielding contributions of the benzene ring. Other studies of this type are those of House *et al.*[259], Longone and Chow[260], and Kurland and Wise[261].

In [18]-annulene, bisdehydro-[14]-annulene and related systems[144, 262–264] the inner protons are at extraordinarily high fields in agreement with theoretical predictions for $(4n+2)$ π-electron systems. It should also be noted that in these molecules the outer protons are more deshielded than those of benzene itself. This follows from the fact that the shielding effect of the ring current depends on the area of the ring. A similar pattern of chemical shifts has been observed for porphin and other porphyrins[129, 265–267]. It has been predicted[268] that in very large annulenes ($>$[24]), the ring current will no longer exist and their spectra will resemble linear polyenes. McWeeny's theory indicates that the proton spectrum of cyclic hexaphenyl should resemble biphenyl rather than an annulene and this appears to be the case[269].

The methylene-bridged [10]-annulene[270] illustrates the strongly shielded region above and below the plane of an aromatic ring system. Many other examples of the long-range shielding effects of aromatic rings have been given by Jackman and Sternhell[232]. It should be noted that in many instances local π-electron charge densities play an important role in determining chemical shifts.

It has been suggested by Elvidge and Jackman[271] that if the contribution of the ring current to the shielding of ring protons and ring methyl substituents can be determined it can be used as a measure of the aromaticity of the system. This suggestion has been supported[272–274] and criticised[162, 275]. Dauben *et al.*[276] have attempted to relate aromaticity to a quantity χ_{exalt} defined as $\chi_{meas.} - \chi_{est.}$ where $\chi_{est.}$ is the susceptibility estimated neglecting ring currents. Jones[230] has recently reviewed the various criteria that have been used to measure aromaticity.

The unusually high diamagnetic susceptibilities of cyclopropanes has led a number of workers to suggest the presence of a ring current in cyclopropane. Tadanier and Cole[277] have attributed the lesser shieldings of 6-β(axial) protons as compared with 6-α protons in 3α,5α-cyclosteroids to differential shielding by the cyclopropane ring. Sauers and Sonnet[278] have examined shifts in the isomeric adducts of carbethoxycarbene and bicycloheptadiene, in norbornyl and nortricyclyl derivatives and come to a similar conclusion. Forsen and Norin[279] have noted upfield shifts of protons located near the threefold axis of cyclopropane rings in nortricyclyl derivatives. The magnitude of the ring current in cyclopropane has been discussed by Lacher *et al.*[280], Patel *et al.*[122] and Burke and Lauterbur[281]. It should be noted that the non-local contribution to the susceptibility calculated by Bley[98] is consistent with all these observations.

The shieldings associated with other types of bond and with atoms have also received some attention. The shielding associated with nitrogen atoms has been discussed by Francois[282], Yamaguchi *et al.*[283] and Gil and Murrell[284]. Zurcher[285] has discussed the shifts produced by chlorine substituents and by hydroxyl groups in steroids. The magnetic anisotropy of the nitro group has been studied by Yamaguchi[286] and a similar study for the nitroso group has been reported by Harris and Spragg[287]. Homer and Callaghan[288] have used values of $\Delta\chi$(C—F) and $\Delta\chi$(C—Cl) to determine proton resonance shifts. The ideas on the shielding effects of C—O double bonds have been extended to C—S double bonds by Long and Townsend[289].

It should be noted that many of these studies assume that the susceptibility tensors associated with electron groupings are 'constants' which can be transferred from molecule to molecule. It is likely, however, that such anisotropies vary from molecule to molecule. Theoretical studies[290–293] of the electronic structure of molecular ground states indicate that substituents can have considerable effects on adjacent bonds. This is particularly true for adjacent multiple bonds where the attached group may produce a relatively large polarisation of the charge distribution in such a bond, in addition to donating charge to, or withdrawing charge from it. The susceptibility associated with a multiple bond is likely to be sensitive to such charge effects. Similarly, substituents can have large effects on the excited states associated with electron groups. For example, in substituted carbonyl compounds, XCHO, the position of the low-lying $n \rightarrow \pi^*$ state relative to the ground state is quite sensitive to the nature of X[294]. Since the contribution of this excited state to $\boldsymbol{\chi}^p$ is largest along the C—O bond direction, this suggests that substituent effects on excited states also can lead to changes in bond susceptibility anisotropies.

4.5 CONCLUSIONS

It is over 40 years since Van Vleck derived the equation which provided the basic understanding of the diamagnetism of diamagnetic molecules. Applications of molecular orbital theory have furthered this understanding by suggesting various relationships between magnetic susceptibility and electronic structure. However, these relationships, which are based on very approximate theories, are still very qualitative and are often regarded with reservations. Despite the many theoretical studies of the last 10 years little additional progress has been made towards a more quantitative theory of magnetic susceptibility which would not only allow a more critical appraisal of the existing qualitative ideas but may also suggest additional correlations with electronic structure. Perhaps both the *ab initio* and semi-empirical methods which have been developed in the last few years will prove to be as useful for studying magnetic susceptibility as they have been for properties which depend solely on the electronic structure of the ground state.

Accurate experimental magnetic susceptibility anisotropy data is now fairly commonplace. If the problems associated with the temperature-independent contributions to the Cotton–Mouton constant can be solved, thereby increasing the precision of such anisotropy results, the magnetic birefringence method promises to be an important complementary technique to the Zeeman microwave method, since it is applicable to non-polar molecules.

Interpretation of chemical shifts in terms of changes in magnetic anisotropy requires both more accurate values of group susceptibility anisotropies and improvement of the models that are used to relate $\boldsymbol{\chi}$ and $\boldsymbol{\sigma}$. There are, however, some indications that the idea of transferable bond susceptibilities needs to be re-examined.

Acknowledgements

It is a pleasure to thank J. A. Pople, J. W. McIver, Jr. and D. P. Miller for stimulating discussions of diamagnetism whilst I was a member of Dr. Pople's group (1968–1970). I would also like to thank P. Ellis, G. Maciel and J. McIver for communicating their results prior to publication. Thanks are also due to C. M. Ditchfield for her help in preparing this manuscript.

References

1. Selwood, P. W. (1956). *Magnetochemistry*. (New York: Interscience)
2. Pascal, P. (1910). *Ann. Chim. et. Phys.*, **19**, 5; (1912) **25**, 289; (1913) **29**, 218
3. Pacault, A. (1948). *Rev. Sci.*, **86**, 38
4. Lonsdale, K. and Krishnan, K. S. (1963). *Proc. Roy. Soc. (London)*, **A156**, 597
5. Beams, J. W. (1932). *Rev. Mod. Phys.*, **4**, 131
6. Chinchalkar, S. (1931). *Ind. J. Phys.*, **6**, 1
7. Buckingham, A. D., Prichard, W. H. and Whiffen, D. H. (1965). *Chem. Commun.*, 51; (1967). *Trans. Faraday. Soc.*, **63**, 1057
8. Huttner, W. and Flygare, W. H. (1967). *J. Chem. Phys.*, **47**, 4137
9. Huttner, W., Lo, M-K. and Flygare, W. H. (1968). *J. Chem. Phys.*, **48**, 1206
10. McConnell, H. (1957). *J. Chem. Phys.*, **27**, 226
11. Pople, J. A. (1962). *J. Chem. Phys.*, **37**, 53
12. Van Vleck, J. H. (1932). *Electric and Magnetic Susceptibilities*. (New York: Oxford University Press)
13. For a recent review see Lipscomb, W. N. (1966). *Advan. Mag. Res.*, **2**, 137
14. Hameka, H. F. (1965). *Advanced Quantum Chemistry*. (Massachusetts: Addison-Wesley Publishing Co.)
15. Hirschfelder, J. O., Byers-Brown, W. and Epstein, S. T. (1964). *Advan. Quant. Chem.*, **1**, 255
16. Langevin, P. (1905). *J. Phys.* **4**, 678; (1905). *Ann. Chim. Phys.*, **5**, 70

17. Ramsey, N. F. (1950). *Phys. Rev.,* **77,** 567; (1950) **78,** 699
18. Lamb, W. E. (1941). *Phys. Rev.,* **60,** 817
19. Snyder, L. C. and Parr, R. G. (1961). *J. Chem. Phys.,* **34,** 827
20. Hameka, H. F. (1964). *J. Chem. Phys.,* **40,** 3127
21. See for example, Pauling, L. and Wilson, E. B. (1965). *Introduction to Quantum Mechanics* 204 (New York: McGraw-Hill)
22. Van Vleck, J. H. and Frank, A. (1929). *Proc. Nat. Acad. Sci. U.S.,* **15,** 539
23. Witmer, E. (1935). *Phys. Rev.,* **48,** 380; (1937) **51,** 383
24. James, H. M. and Coolidge, A. S. (1933). *J. Chem. Phys.,* **1,** 825
25. Barnes, R. G., Bray, P. J. and Ramsey, N. F. (1954). *Phys. Rev.,* **94,** 893
26. Bonet, G. V. and Bushkovitch, A. V. (1953). *J. Chem. Phys.,* **21,** 1299
27. Buckingham, R. A., Massey, H. S. W. and Tibbs, S. R. (1941). *Proc. Roy. Soc. (London),* **A178,** 119
28. Carter, C. (1956). *Proc. Roy. Soc. (London),* **A235,** 321
29. Pacault, A., Hoarau, J. and Marchand, A. (1961). *Advan. Chem. Phys.,* **3,** 184
30. Weltner, W. (1958). *J. Chem. Phys.,* **28,** 477
31. Ishiguro, E. and Koide, S. (1954). *Phys. Rev.,* **94,** 350
32. Tillieu, J. and Guy, J. (1954). *Compt. Rend.,* **239,** 1203
33. Gans, R. and Mrowka, B. (1935). *Konigsberger Gelehrte Gesellschaft.,* **12,** 1, 30
34. McLachlan, A. D. and Baker, M. R. (1961). *Molec. Phys.,* **4,** 255
35. Das, T. P. and Bersohn, R. (1956). *Phys. Rev.,* **104,** 849; (1956). **115,** 897
36. Stephen, M. J. (1957). *Proc. Roy. Soc. (London),* **A243,** 264
37. Guy, J., Tillieu, J. and Baudet, J. (1958). *Compt. Rend.,* **246,** 574
38. See for example, Slater, J. C. (1960). *Quantum Theory of Atomic Structure,* Vol. 1. (New York: McGraw-Hill), 291
39. For a discussion of the role of the gauge see Reference 12, p. 162
40. Epstein, S. T. (1965). *J. Chem. Phys.,* **42,** 2897
41. Karplus, M. and Kolker, H. J. (1961). *J. Chem. Phys.,* **35,** 2235; (1963) **38,** 1263
42. Das, T. P. and Karplus, M. (1962). *J. Chem. Phys.,* **36,** 2275; Hurst, R. P., Karplus, M. and Das, T. P. (1962). *J. Chem. Phys.,* **36,** 2786
43. See Reference 12, p. 172
44. Sadlej, A. J. (1971). *Molec. Phys.,* **20,** 593
45. Rebane, T. K. (1960). *J. Exptl. Theoret. Phys. (U.S.S.R),* **38,** 963
46. Hoyland, J. R. and Parr, R. G. (1963). *J. Chem. Phys.,* **38,** 2991
47. Hoyland, J. R. (1964). *J. Chem. Phys.,* **41,** 3135
48. Brillouin, L. (1933–1934). *Actualities Sci. et. Ind.* Nos. 71, 159
49. Pople, J. A., McIver, J. W., Jr. and Ostlund, N. S. (1968). *J. Chem. Phys.* **49,** 2960
50. Ditchfield, R., Miller, D. P. and Pople, J. A. (1970). *J. Chem. Phys.,* **53,** 613
51. Stevens, R. M., Pitzer, R. M. and Lipscomb, W. N. (1963). *J. Chem. Phys.,* **38,** 550
52. Blizzard, A. C. and Santry, D. P. (1970). *Chem. Commun.,* 87
53. Ditchfield, R., Miller, D. P. and Pople, J. A. (1971). *J. Chem. Phys.,* **54,** 4861
54. Stevens, R. M. and Lipscomb, W. N. (1964). *J. Chem. Phys.,* **40,** 2238
55. Stevens, R. M. and Lipscomb, W. N. (1964). *J. Chem. Phys.,* **41,** 184
56. Stevens, R. M. and Lipscomb, W. N. (1964). *J. Chem. Phys.,* **41,** 3710
57. Stevens, R. M. and Lipscomb, W. N. (1965). *J. Chem. Phys.,* **42,** 3666
58. Stevens, R. M. and Lipscomb, W. N. (1965). *J. Chem. Phys.,* **42,** 4302
59. Hegstrom, R. A. and Lipscomb, W. N. (1968). *J. Chem. Phys.,* **48,** 809
60. Stevens, R. M. and Karplus, M. (1968). *J. Chem. Phys.,* **49,** 1094
61. Laws, E. A., Stevens, R. M. and Lipscomb, W. N. (1969). *Chem. Phys. Lett.,* **4,** 159
62. Arrighini, G. P., Maestro, M. and Moccia, R. (1968). *J. Chem. Phys.,* **49,** 882
63. Arrighini, G. P., Guidotti, C., Maestro, M., Moccia, R. and Salvetti, O. (1968). *J. Chem. Phys.,* **49,** 2224
64. Arrighini, G. P., Guidotti, C., Maestro, M., Moccia, R. and Salvetti, O. (1969). *J. Chem. Phys.,* **51,** 480
65. Arrighini, G. P., Giovanni, P., Guidotti, C. and Salvetti, O. (1970). *J. Chem. Phys.,* **52,** 1037
66. Arrighini, G. P., Maestro, M. and Moccia, R. (1970). *J. Chem. Phys.,* **52,** 6411
67. Arrighini, G. P., Maestro, M. and Moccia, R. (1970). *Chem. Phys. Letts.,* **7,** 351
68. Arrighini, G. P., Tomasi, J. and Petrongolo, C. (1970). *Theoret. Chim. Acta,* **18,** 341
69. Ditchfield, R., Miller, D. P. and Pople, J. A. (1970). *Chem. Phys. Lett.,* **6,** 573

70. Hehre, W. J., Stewart, R. F. and Pople, J. A. (1969). *J. Chem. Phys.*, **51,** 2657
71. Ditchfield, R., Hehre, W. J. and Pople, J. A. (1970). *J. Chem. Phys.*, **52,** 5001
72. Ditchfield, R., Hehre, W. J. and Pople, J. A. (1971). *J. Chem. Phys.*, **54,** 724
73. Not allocated
74. Amos, A. T. and Roberts, H. G. (1969). *J. Chem. Phys.*, **50,** 2375
75. London, F. (1937). *J. Phys. Radium,* **8,** 397
76. Hameka, H. F. (1959). *Z. Naturforsch,* **14a,** 599
77. Hameka, H. F. (1959). *Physica,* **25,** 626
78. Hameka, H. F. (1961). *J. Chem. Phys.*, **34,** 366, 1996
79. Hameka, H. F. (1962). *Physica,* **28,** 908
80. Hameka, H. F. (1962). *Rev. Mod. Phys.*, **34,** 87
81. Hamano, H., Kim, H. and Hameka, H. F. (1963). *Physica,* **29,** 111; (1963). ibid, **29,** 117
82. de la Vega, J. R. and Hameka, H. F. (1964). *J. Chem. Phys.*, **40,** 1929
83. de la Vega, J. R. and Hameka, H. F. (1965). *Physica,* **35,** 313
84. de la Vega, J. R., Fang, Y. and Hameka, H. F. (1967). *Physica,* **36,** 577
85. de la Vega, J. R., Ziobro, D. and Hameka, H. F. (1967). *Physica,* **37,** 265
86. Pan, Y-C. and Hameka, H. F. (1970). *J. Chem. Phys.*, **53,** 1265
87. Pople, J. A. (1957). *Proc. Roy. Soc. (London),* **A239,** 541, 550
88. Pople, J. A. (1958). *Molec. Phys.*, **1,** 175
89. Pople, J. A. (1962). *J. Chem. Phys.*, **37,** 60
90. Pople, J. A. (1962). *Discuss. Faraday Soc.*, **34,** 7
91. Pople, J. A. (1963). *J. Chem. Phys.*, **38,** 1276
92. Pople, J. A. (1964). *J. Chem. Phys.*, **41,** 2559
93. Ferguson, A. F. and Pople, J. A. (1965). *J. Chem. Phys.*, **42,** 1560
94. McWeeny, R. (1958). *Molec. Phys.*, **1,** 311
95. Hall, G. G. and Hardisson, A. (1962). *Proc. Roy. Soc. (London),* **A268,** 328
96. Davis, D. W. (1961). *Trans. Faraday Soc.*, **57,** 2081
97. Bley, W. R. (1969). *Molec. Phys.*, **16,** 303
98. Bley, W. R. (1971). *Molec. Phys.*, **20,** 491
99. Ellis, P. D., Maciel, G. E. and McIver, J. W. Jr. *J. Amer. Chem. Soc.* (to be published)
100. Ditchfield, R. *J. Chem. Phys.*, (to be published)
101. Pariser, R. and Parr, R. G. (1953). *J. Chem. Phys.*, **21,** 466
102. Pople, J. A. (1953). *Trans. Faraday Soc.*, **49,** 1375
103. Moccia, R. (1970). *Chem. Phys. Lett.*, **5,** 260
104. Ditchfield, R. (unpublished work)
105. Saika, A. and Slichter, C. P. (1954). *J. Chem. Phys.*, **22,** 26
106. Didry, J. R. and Guy, J. (1961). *Compt. Rend.*, **253,** 422
107. Bothner-By, A. A. and Naar-Colin, C. (1958). *Ann. N.Y. Acad. Sci.*, **70,** 833
108. Kukolich, S. G. (1970). *Chem. Phys. Lett.*, **5,** 401; erratum (1971). *Chem. Phys. Lett.*, **12,** 216
109. Kukolich, S. G. (1969). *J. Chem. Phys.*, **50,** 3751
110. Taft, H. and Dailey, B. P. (1969). *J. Chem. Phys.*, **51,** 1002
111. Verhoeven, J. and Dymanus, A. (1970). *J. Chem. Phys.*, **52,** 3222
112. Wiberg, K. B. and Nist, B. J. (1961). *J. Amer. Chem. Soc.*, **83,** 1226
113. Hameka, H. F. (1962). *J. Chem. Phys.*, **37,** 3008
114. For a reply to the criticisms of Reference 113 see (1962). Pople, J. A. *J. Chem. Phys.*, **37,** 3009
115. O'Sullivan, P. S. and Hameka, H. F. (1970). *J. Amer. Chem. Soc.*, **92,** 25, 1281
116. Ditchfield, R., McIver, Jr., J. W., Miller, D. P. and Pople, J. A. (unpublished results)
117. Pople, J. A., Santry, D. P. and Segal, G. A. (1965). *J. Chem. Phys.*, **43,** S129
118. Pople, J. A. and Segal, G. A. (1965). *J. Chem. Phys.*, **43,** S136; (1966). **44,** 3289
119. Pople, J. A., Beveridge, D. L. and Dobosh, P. A. (1967). *J. Chem. Phys.*, **47,** 2026
120. Pople, J. A. and Beveridge, D. L. (1971). *Approximate Molecular Orbital Theory.* (New York: McGraw-Hill)
121. Similar general statements have been made by Amos, A. T. and Musher, J. (1968). *J. Chem. Phys.*, **49,** 2158
122. Patel, D. J., Howden, M. E. H. and Roberts, J. D. (1963). *J. Amer. Chem. Soc.*, **85,** 3218
123. Pauling, L. (1936). *J. Chem. Phys.*, **4,** 673
124. Hoarau, J. Lumbroso, N. and Pacault, A. (1956). *Compt. Rend.*, **242,** 1702
125. London, F. (1937). *J. Chem. Phys.*, **5,** 837
126. Hückel (1931). *Z. Physik,* **70,** 204

127. McWeeny, R. (1951). *Proc. Phys. Soc. (London)*, **64A,** 261 921; (1952). **65A,** 839; (1953). **66A,** 714
128. Berthier, G., Mayot, M., Pullman, B. and Pullman, A. (1952). *J. Phys. Rad.*, **13,** 15
129. Abraham, R. J. (1961). *Molec. Phys.*, **4,** 145
130. Craig, D. P., Heffernan, M. L., Mason, R. and Paddock, N. L. (1961). *J. Chem. Soc.*, 1376
131. Itoh, T., Ohno, K. and Yoshizumi, H. (1955). *J. Phys. Soc. (Japan)*, **10,** 103
132. Caralp, L. and Hoarau, J. (1963). *J. Chim. Phys.*, **60,** 889
133. Dailey, B. P. (1964). *J. Chem. Phys.*, **41,** 2304
134. Davies, D. W. (1961). *Nature (London)*, **190,** 1102
135. Davies, D. W. (1967). *The Theory of the Electric and Magnetic Properties of Molecules.* (New York: Wiley)
136. Hoarau, J. (1956). *Ann. Chim.*, **1,** 544
137. Pople, J. A. and Untch, K. G. (1966). *J. Amer. Chem. Soc.*, **88,** 4811
138. Berthier, G., Mayot, M. and Pullman, B. (1951). *J. Phys. Rad.*, **12,** 717; see also Salem, L., *The Molecular Orbital Theory of Conjugated Systems*, 194–210. (New York: W. A. Benjamin, Inc.)
139. Pople, J. A. (1956). *J. Chem. Phys.*, **24,** 1111
140. See Reference 12, p. 22
141. Bernstein, H. J., Scheider, W. G. and Pople, J. A. (1956). *Proc. Roy. Soc. (London)*, **A236,** 515
142. Waugh, J. S. and Fessenden, R. W. (1957). *J. Amer. Chem. Soc.*, **79,** 846
143. Johnson, C. E. and Bovey, F. A. (1958). *J. Chem. Phys.*, **29,** 1012
144. Sondheimer, F. (1963). *Pure appl. Chem.*, **7,** 363
145. Jonathon, H. Gordon, S. and Dailey, B. P. (1962). *J. Chem. Phys.*, **36,** 2443
146. Maddox, I. J. and McWeeny, R. (1962). *J. Chem. Phys.*, **36,** 2353
147. Memory, J. D. (1963). *J. Chem. Phys.*, **38,** 1341
148. Figeys, H. P. (1966). *Tetrahedron Lett.*, **38,** 4625
149. Jung, D. E. (1969). *Tetrahedron*, **25,** 129
150. Memory, J. D., Parker, G. W. and Halsey, J. C. (1966). *J. Chem. Phys.*, **45,** 3567
151. Memory, J. D. and Cobb, T. B. (1967). *J. Chem. Phys.*, **47,** 2020
152. Haigh, C. W., Mallion, R. B. and Armour, E. A. G. (1970). *Molec. Phys.*, **18,** 751
153. Black, P. J., Brown, R. D. and Heffernan, H. L. (1967). *Aust. J. Chem.*, **20,** 1305, 1325
154. Hall, G. G., Hardisson, A. and Jackman, L. M. (1963). *Tetrahedron Suppl.*, 2, **19,** 101
155. Roberts, H. G. Ff. (1971). *Theoret. Chim. Acta*, **22,** 105
156. For a discussion of the various types of coupled perturbation methods see Langhoff, P. W., Karplus, M. and Hurst, R. P. (1966). *J. Chem. Phys.*, **44,** 505 and Dalgarno, A. (1959). *Proc. Roy. Soc. (London)*, **A251,** 282
157. McWeeny, R. (1971). *Chem. Phys. Lett.*, **9,** 341
158. Parker, G. W. and Memory, J. D. (1965). *J. Chem. Phys.*, **43,** 1388 see also Edwards, T. G. and McWeeny, R. (1971). *Chem. Phys. Lett.*, **10,** 283
159. Amos, A. T. and Roberts, H. G. Ff. (1971). *Molec. Phys.*, **20,** 1073
160. Roberts, H. G. Ff. and Amos, A. T. (1971). *Molec. Phys.*, **20,** 1081, 1089
161. Roberts, H. G. Ff. (1971). *Chem. Phys. Lett.*, **11,** 259
162. Musher, J. I. (1965). *J. Chem. Phys.*, **43,** 4081
163. Musher, J. I. (1966). *Advan. Mag. Res.*, **2,** 177
164. Musher, J. I. (1967). *J. Chem. Phys.*, **46,** 1219 see however Gaidis, J. M. and West, R. (1967). *J. Chem. Phys.*, **46,** 1218
165. Bitter, F. (1929). *Phys. Rev.*, **33,** 389
166. Barter, C., Meisenheimer, R. G. and Stevenson, D. P. (1960). *J. Phys. Chem.*, **64,** 1312
167. Mulay, L. N. and Mulay, I. L. (1970). *Analyt. Chem.*, **42,** 325R
168. Bothner-By, A. A. and Pople, J. A. (1965). *Ann. Rev. Phys. Chem.*, **16,** 43
169. Lasheen, M. A. (1964). *Phil. Trans. Roy. Soc. (London)*, **A256,** 357
170. Krishnan, K. S. and Banerjee, S. (1935). *Phil. Trans. Roy. Soc. (London)*, **A234,** 265
171. Lasheen, M. A. (1969). *Acta. Crystallogr.*, **A25,** 581
172. Lasheen, M. A. (1970). *Acta Crystallogr.*, **A26,** 681
173. Le Fevre, R. J. W. and Murthy, D. S. N. (1966). *Aust. J. Chem.*, **19,** 179
174. Le Fevre, R. J. W. and Murthy, D. S. N. (1969). *Aust. J. Chem.*, **22,** 1415
175. Le Fevre, R. J. W. and Murthy, D. S. N. (1970). *Aust. J. Chem.*, **23,** 193
176. Le Fevre, R. J. W., Murthy, D. S. N. and Stiles, P. J. (1968). *Aust. J. Chem.*, **21,** 3059
177. Buckingham, A. D. and Pople, J. A. (1956). *Proc. Phys. Soc. (London) B*, **69,** 1133

178. Pople, J. A. (1962). *Discuss. Faraday Soc.,* **34,** 68
179. Lonsdale, K. (1939). *Proc. Roy. Soc. (London),* **A171,** 541
180. Davies, D. W. (1963). *Molec. Phys.,* **6,** 489
181. Flygare, W. H. (1967). *Rec. Chem. Prog.,* **28,** 63
182. Flygare, W. H. and Benson, R. C. (1971). *Molec. Phys.,* **20,** 225 and references therein
183. Huttner, W. and Flygare, W. H. (1967). *J. Chem. Phys.,* **47,** 4137
184. Harrick, N. J. and Ramsey, N. F. (1952). *Phys. Rev.,* **88,** 228
185. Ramsey, N. F. and Lewis, H. R. (1957). *Phys. Rev.,* **108,** 1246; Quinn, W. E., Baker, M. J., La Tourrette, J. T. and Ramsey, N. F. (1958). *Phys. Rev.,* **112,** 1929
186. Graff, G. and Werth, G. (1965). *Z. Phys.,* **183,** 223
187. Graff, G., Schonwasser, R. and Tonutti, M. (1967). *Z. Phys.,* **199,** 157
188. Graff, G. and Runolfsson, O. (1965). *Z. Phys.,* **187,** 140
189. Drechsher, W. and Graff, G. (1961). *Z. Phys.,* **163,** 165; Van Boeckh, R., Graff, G. and Ley, R. (1964). *Z. Phys.,* **179,** 285
190. Flygare, W. H., Shoemaker, R. L. and Huttner, W. (1969). *J. Chem. Phys.,* **50,** 2414
191. Buckingham, A. D., Disch, R. L. and Dunmur, D. A. (1968). *J. Amer. Chem. Soc.,* **90,** 3104
192. Gustafson, S. and Gordy, W. (1970). *J. Chem. Phys.,* **52,** 579
193. Huttner, W. and Morgenstern, K. (1970). *Naturforsch,* **25a,** 547
194. Flygare, W. H., Huttner, W., Shoemaker, R. L. and Foster, P. D. (1969). *J. Chem. Phys.,* **50,** 1714
195. Shoemaker, R. L. and Flygare, W. H. (1970). *Chem. Phys. Lett.,* **6,** 576
196. Shoemaker, R. L. and Flygare, W. H. (1968). *Chem. Phys. Lett.,* **2,** 610
197. Shoemaker, R. L. and Flygare, W. H. (1969). *J. Amer. Chem. Soc.,* **91,** 5417
198. Vanderhart, D. and Flygare, W. H. (1969). *Molec. Phys.,* **18,** 77
199. Pochan, J. M., Shoemaker, R. L., Stone, R. G. and Flygare, W. H. (1970). *J. Chem. Phys.,* **52,** 2478
200. Kukolich, S. G. and Flygare, W. H. (1970). *Chem. Phys. Lett.,* **6,** 45
201. Stone, R. G., Pochan, J. M. and Flygare, W. H. (1969). *Inorg. Chem.* **8,** 2647
202. Huttner, W. and Flygare, W. H. (1969). *J. Chem. Phys.,* **50,** 2867
203. Wang, J. H. S. and Flygare, W. H. (1970). *J. Chem. Phys.,* **52,** 5636
204. Pochan, J. M. and Flygare, W. H. (1969). *J. Amer. Chem. Soc.,* **91,** 5928
205. Benson, R. C. and Flygare, W. H. (1970). *J. Amer. Chem. Soc.,* **92,** 7523
206. Sutter, D. H. and Flygare, W. H. (1969). *J. Amer. Chem. Soc.,* **91,** 6895
207. Sutter, D. H. and Flygare, W. H. (1969). *J. Amer. Chem. Soc.,* **91,** 4063
208. Benson, R. C. and Flygare, W. H. (1970). *J. Chem. Phys.,* **53,** 4470
209. Tigelaar, H. L. and Flygare, W. H. (1970). *J. Chem. Phys.,* **53,** 3943
210. Sutter, D. H., Huttner, W. and Flygare, W. H. (1969). *J. Chem. Phys.,* **50,** 2869
211. Sutter, D. H. and Flygare, W. H. (1969). *Molec. Phys.,* **16,** 153
212. Gierke, T. D., Benson, R. C. and Flygare, W. H. (19). *J. Chem. Phys.,*
213. Hsu, S. L., Andrist, A. H., Gierke, T. D., Benson, R. C., Flygare, W. H. and Baldwin, J. E. (1970). *J. Amer. Chem. Soc.,* **92,** 5250
214. Benson, R. C. and Flygare, W. H. (1969). *Chem. Phys. Lett.,* **4,** 141
215. Benson, R. C., Scott, R. S. and Flygare, W. H. (1969). *J. Phys. Chem.,* **73,** 4359
216. Benson, R. C. and Flygare, W. H. (1970). *J. Chem. Phys.,* **52,** 5291
217. Huttner, W. and Flygare, W. H. (1969). *Trans. Faraday Soc.,* **65,** 1953
218. Kukolich, S. G. and Flygare, W. H. (1969). *J. Amer. Chem. Soc.,* **91,** 2433
219. Wang, J. H. S. and Flygare, W. H. (1970). *J. Chem. Phys.,* **53,** 4479; see erratum in *J. Chem. Phys.,* (1971). **55,** 3616
220. Huttner, W., Foster, P. D. and Flygare, W. H. (1969). *J. Chem. Phys.,* **50,** 1710
221. Lo, M-K., Foster, P. D. and Flygare, W. H. (1968). *J. Chem. Phys.,* **48,** 948
222. Blickensderfer, R., Wang, J. H. S. and Flygare, W. H. (1969). *J. Chem. Phys.,* **51,** 3196
223. Hancock, J. K., Rock, S. L. and Flygare, W. H. (1971). *J. Chem. Phys.,* **54,** 3450
224. Pochan, J. M., Stone, R. G. and Flygare, W. H. (1969). *J. Chem. Phys.,* **51,** 4278
225. Stone, R. G., Tigelaar, H. L. and Flygare, W. H. (1970). *J. Chem. Phys.,* **53,** 3947
226. Kukolich, S. G. (1969). *J. Chem. Phys.,* **50,** 3751
227. Burrus, C. A. (1959). *J. Chem. Phys.,* **30,** 980
228. Benson, R. C. and Flygare, W. H. (1969). *J. Chem. Phys.,* **51,** 3087
229. Bridge, N. J. and Buckingham, A. D. (1966). *Proc. Roy. Soc. (London),* **A295,** 334
230. For a recent critical review of the criteria used for aromaticity see Jones, A. J. (1968). *Rev. Pure Appl. Chem.,* **18,** 253

231. Benson, R. C., Tigelaar, H. L., Rock, S. L. and Flygare, W. H. (1970). *J. Chem. Phys.*, **52,** 5628
232. Jackman, L. H. and Sternhell, S. (1969). *Applications of Nuclear Magnetic Resonance Spectroscopy in Organic Chemistry* (London: Pergamon)
233. Bovey, F. A. (1969). *Nuclear Magnetic Resonance Spectroscopy* (New York: Academic Press)
234. Buckingham, A. D. (1960). *Can. J. Chem.*, **38,** 300
235. Slomp, G. and McGarvey, B. R. (1959). *J. Amer. Chem. Soc.*, **81,** 2200
236. Wellman, J. M. and Bordwell, F. G. (1963). *Tetrahedron Lett.*, 1703
237. Nagata, W., Terasawa, T. and Tori, K. (1964). *J. Amer. Chem. Soc.*, **86,** 3746
238. Cook, C. D. and Danyluk, S. S. (1963). *Tetrahedron*, **19,** 177
239. Jackman, L. M. (1959). *Nuclear Magnetic Resonance Spectroscopy* (London: Pergamon)
240. Musher, J. I. (1961). *J. Chem. Phys.*, **35,** 1159
241. Moritz, A. G. and Sheppard, N. (1962). *Molec. Phys.*, **5,** 361
242. Hall, L. D. (1964). *Tetrahedron Lett.*, 1457
243. Musher, J. I. (1963). *Molec. Phys.*, **6,** 93
244. Narasimhan, P. T. and Rogers, M. T. (a) (1959). *J. Chem. Phys.*, **31,** 1302; (b) *J. Phys. Chem.*, (1959) **63,** 1388
245. Zurcher, R. F. (1961). *Helv. Chim. Acta*, **44,** 1755
246. Apsimon, J. W., Craig, W. G., Demarco, P. V., Mathieson, D. W., Saunders, L. and Whalley, W. B. (1967). *Chem. Commun.*, 359; *Tetrahedron*, (1967) **23,** 2339, 2357, 2375
247. Raynes, W. T. (1971). *Molec. Phys.*, **20,** 321
248. Conroy, H. (1960). *Advances in Organic Chemistry: Methods and Results* (New York: Interscience) 265–328
249. Ayer, W. A. McDonald, C. E. and Stothers, J. B. (1963). *Can. J. Chem.*, **41,** 1113
250. Snyder, E. I. and Franzus, B. (1964). *J. Amer. Chem. Soc.*, **86,** 1166
251. Williams, D. H., Bhacca, N. S. and Djerassi, C. (1963). *J. Amer. Chem. Soc.*, **85,** 2811
252. Apsimon, J. W., Craig, W. G., Demarco, P. V., Mathieson, D. W., Nasser, A. K. G., Saunders, L. and Whalley, W. B. (1967). *Chem. Commun.*, 754
253. Apsimon, J. W., Demarco, P. V., Mathieson, D. W., Craig, W. G., Harim, A., Saunders, L. and Whalley, W. B. (1970). *Tetrahedron*, **26,** 119
254. Karabatsos, G. J., Sonnichsen, G. C., Hsi, N. and Fenoglio, D. J. (1967). *J. Amer. Chem. Soc.*, **89,** 5067
255. Heel, H. and Zeil, W. (1960). *Z. Elektrochem*, **64,** 962
256. Zeil, W. and Buchert, H. (1963). *Z. Physik. Chem. (Frankfurt)*, **38,** 47
257. Reddy, G. S. and Goldstein, J. H. (1962). *J. Chem. Phys.*, **36,** 2644; (1963). **39,** 3509
258. Whitesides, G. M., Holtz, D. and Roberts, J. D. (1964). *J. Amer. Chem. Soc.*, **86,** 2628
259. House, H. O., Magin, R. W. and Thompson, H. W. (1963). *J. Org. Chem.*, **28,** 2403
260. Longone, D. T. and Chow, H. S. (1964). *J. Amer. Chem. Soc.*, **86,** 3898
261. Kurland, R. J. and Wise, W. B. (1964). *J. Amer. Chem. Soc.*, **86,** 1877
262. Jackman, L. M., Sondheimer, F., Amiel, Y., Ben Efraim, D. A., Gaoni, Y., Wolovsky, R. and Bothner-By, A. A. (1962). *J. Amer. Chem. Soc.*, **84,** 4307
263. Sondheimer, F., Gaoni, Y., Jackman, L. M., Bailey, N. A. and Mason, R. (1962). *J. Amer. Chem. Soc.*, **84,** 4595
264. Gaoni, Y., Melera, A., Sondheimer, F. and Wolovsky, R. (1964). *Proc. Chem. Soc.*, 397
265. Becker, E. D., Bradley, R. B. and Watson, C. J. (1961). *J. Amer. Chem. Soc.*, **83,** 3743
266. Abraham, R. J., Jackson, A. H., Kenner, G. W. and Warburton, D. (1963). *J. Chem. Soc.*, 853
267. Caughey, W. S. and Iber, P. K. (1963). *J. Org. Chem.*, **28,** 269
268. Longuet-Higgins, H. G. and Salem, L. (1960). *Proc. Roy. Soc. (London)*, **A257,** 445
269. Staab, H. A. and Binnig, F. (1964). *Tetrahedron Lett.*, 319
270. Vogel, E., Pretzer, W. and Boll. W. A. (1965). *Tetrahedron Lett.*, 3613
271. Elvidge, J. A. and Jackman, L. M. (1961). *J. Chem. Soc.*, 859
272. De. Jongh, H. A. P. and Wynberg, H. W. (1965). *Tetrahedron*, **21,** 515
273. Davies, D. W. (1965). *Chem. Commun.*, 258
274. Elvidge, J. A. (1965). *Chem. Commun.*, 160
275. Abraham, R. J., Sheppard, R. C., Thomas, W. A. and Turner, S. (1965). *Chem. Commun.*, 43
276. Dauben, H. J., Wilson, J. D. and Laity, J. L. (1968). *J. Amer. Chem. Soc.*, **90,** 811; (1969). **91,** 199

277. Tadanier, J. and Cole, W. (1962). *J. Org. Chem.,* **27,** 4610
278. Sauers, R. R. and Sonnet, P. E. (1963). *Chem. Ind. (London),* 786
279. Forsen, S. and Norin, T. (1964). *Tetrahedron Lett.,* 2845
280. Lacher, J. R., Pollock, J. W. and Park, J. D. (1952). *J. Chem. Phys.,* **20,** 1047
281. Burke, J. J. and Lauterbur, P. C. (1964). *J. Amer. Chem. Soc.,* **86,** 1870
282. Francois, H. (1962). *Bull. Soc. Chim. Fr.,* 506
283. Yamaguchi, S., Kuda, S. O. and Nakagawa, N. (1963). *Chem. Pharm. Bull.,* **11,** 1465
284. Gil, V. M. S. and Murrell, J. N. (1964). *Trans. Faraday Soc.,* **69,** 248
285. Zurcher, R. F. (1965). *Nuclear Magnetic Resonance in Chemistry,* (Biagio Pesce, editor) (New York: Academic Press)
286. Yamaguchi, I. (1963). *Molec. Phys.,* **6,** 105
287. Harris, R. K. and Spragg, R. A. (1967). *J. Molec. Spectrov.,* **23,** 158
288. Homer, J. and Callaghan, D. (1968). *J. Chem. Soc. A,* 439, 518
289. Long, R. A. and Townsend, L. B. (1970). *J. Chem. Commun.,* 1087
290. Pople, J. A. and Gordon, M. S. (1967). *J. Amer. Chem. Soc.,* **89,** 4253
291. Newton, M. D. and Lipscomb, W. N. (1967). *J. Amer. Chem. Soc.,* **89,** 4261
292. Hehre, W. J. and Pople, J. A. (1970). *J. Amer. Chem. Soc.,* **92,** 2191
293. Radom, L., Hehre, W. J. and Pople, J. A. (1971). *J. Amer. Chem. Soc.,* **93,** 289
294. For a theoretical discussion of such excited states see Ditchfield, R., Del Bene, J. and Pople, J. A. (1972). *J. Amer. Chem. Soc.,* **94,** 703

5
Acoustic Studies of Molecular Conformational Changes

A. M. NORTH
and
R. A. PETHRICK
University of Strathclyde

5.1 INTRODUCTION AND THEORY

In the last two decades considerable advances have been made towards our understanding of the detailed stereochemistry of organic molecules[1]. Such studies fall broadly into two classes. The first comprises those investigations in which detailed examination of various spectroscopic and diffraction experiments leads to the exact determination of the spatial geometry within a molecule. The second class consists of dynamic measurements, which involve determination of the energy associated with each of the possible stable conformations and of their rates of interconversion. The data obtained from the former study provide either a time-averaged structure or else a superposition of the structures of several conformations. In this case the observed behaviour of a particular molecular system will depend upon the relative magnitudes of the characteristic interaction time for the experiment and that required for conformational change[2].

Molecular acoustics studies fall in the second class of experiments and are concerned with the way in which an oscillating pressure wave interacts with an equilibrium ensemble of conformational states.

It is of historical interest that the first reported conformational dynamic studies using acoustics were contemporary with those using nuclear magnetic resonance. However, it is only recently that the scope of this technique for conformational studies has come to be generally appreciated.

In this review is presented a survey of the data on the dynamics of conformational equilibria as available from acoustic studies and the scope and limitations of such experiments are delineated.

Generally, acoustic measurements of conformational change are, for practical reasons, restricted to studies of liquids or to solutions in an acoustically inactive solvent. Strictly, of course, the energetics associated with purely intramolecular conformational change should be determined in the absence of intermolecular forces. This would necessitate the molecule being studied in the gas phase and so would restrict observations to that relatively small number of molecules possessing adequate vapour pressures. It can be appreciated that the presence of intermolecular forces in liquids may modify significantly the energetics of conformational change, as has been shown for the enthalpy difference between stable conformations[3]. On the other hand, the activation energy opposing conformer interconversion is less susceptible to intermolecular forces and useful data may be obtained from liquid-phase studies. Correlation of such measurements with the predictions made by theoretical calculations for isolated molecules is obviously not easy, but it should be remembered that data on the energetics as measured in the liquid phase are of wider practical application.

5.1.1 The concept of sound propagation in a liquid

A small-amplitude sound wave propagates as an elastic sinusoidal displacement and observation of its amplitude and phase as a function of distance may be related to the energy-transfer processes occurring in the propagation medium. Dispersion phenomena associated with failure of the system to adjust successfully to an applied perturbation can be described in terms of either a fluctuating density or the response of the system to a pressure-wave perturbation. It is usual to use the former description for photon scattering by the equilibrium phonon distribution in a liquid[4]. Sound propagation, however, is normally formulated on the basis of a pressure perturbation and it is this approach which will be followed in this article.

We associate the passage of a pressure wave with an alternating unidirectional compression and rarefaction of a volume element within a medium. Since the period of pressure fluctuation is short compared with the time required for thermal equilibration with the surroundings, the process is essentially adiabatic and can therefore be visualised correctly as a sinusoidal temperature wave (excluding substances such as water at 4 °C). Consequently, a system may react either to the pressure fluctuation, which corresponds to a volume response, or to the temperature alternation, which may be described as a specific-heat response. This variation of the translational temperature about its ambient value leads to the possibility of energy exchange in the system and this in turn leads to reduction of amplitude of the perturbation with distance propagated into the medium.

Although our interest in energy-transfer processes leads naturally to studies of energy absorption, it is important to realise that absorption can arise from other 'classical' examples of molecular behaviour. Thus thermal conduction from the high-temperature 'crests' to the low-temperature 'troughs' of the wave leads to a diminution in wave amplitude and so to sound absorption. Furthermore a sound wave, which is unidirectional compression, can be resolved into a three-dimensional compression plus a shear deformation of the propagation medium. Since shear deformation leads to energy dissipation as viscosity, this aspect of elastic deformation also leads to sound absorption. These two processes are referred to as the 'classical' or viscothermal-loss processes, and of course must be accounted for before absorption due to energy-transfer processes can be measured. In addition to these 'classical' losses, there are a number of other processes which may lead to attenuation of the propagating pressure wave. The molecular origins of these have been described by a number of workers[5,6] and the important processes may be described in the following sub-sections.

5.1.1.1 Translational to rotational–vibrational energy relaxation

If a molecule undergoes a collision with another, it is possible that during the instant of impact, part of the kinetic energy of the pair may be converted from translational energy into internal energy of the molecules. This excess internal energy may then be returned to the system as translational energy after additional collisions between the vibrationally–rotationally excited species and other suitably orientated molecular species.

It is possible to formulate a kinetic scheme describing the rate of conversion of internal energy between various modes and also its exchange with the total translational motion of the system[7]. When the interchange frequency between the normal and the excited rotational and vibrational states becomes of the order of the sound perturbation frequency, a phase shift and attenuation of the sound wave will be observed. In liquids this gives rise to dispersion phenomena in the high gigahertz frequency range.

Relaxation studies of vibrational and rotational excited molecules have been the subject of a number of reviews[8,9] and will not be considered further in this article.

5.1.1.2 Relaxation of liquid structure

The simple classical theory of sound propagation does not allow for the perturbation produced by a sound wave becoming out-of-phase with the spatial rearrangement of the structure of a liquid. The pressure wave produces a successive compression and rarefaction which causes the volume of a given element to alternate periodically about its mean equilibrium or static value. Associated with this fluctuation of the volume will be the flow of molecules between a more and a less compact structure. When this movement becomes out-of-phase with the fluctuations, a so-called 'structural' relaxation will be observed. In most liquids of low viscosity these relaxation processes occur at high megahertz and low gigahertz frequencies. In liquids with viscosities greater than 1 poise it is possible that the absorption at lower frequencies may be modified by the effect of this structural relaxation.

5.1.1.3 Internal rotation and molecular conformational change

A conformation change within a molecule may correspond to a gain or loss of energy, although an energy barrier almost always exists between two discernible conformers. The process of rotation of one end of a molecule relative to another, as in the *gauche–trans* isomerisation in 1,2-disubstituted ethanes, may be associated with excitation of the 'torsional' modes in the molecule. This excitation occurs during molecular collisions, and if not effectively deactivated may lead to conversion of one molecular conformation into another. When the frequency of the energy perturbation matches that of the characteristic time for the interconversion of states, an acoustic loss and dispersion associated with this process will be observed. The observed phenomenon may then be related to the thermodynamic and kinetic parameters describing the interconversion process.

5.1.1.4 Chemical equilibria and fast reactions

The criteria for the observation of an acoustic relaxation are that there should exist two or more thermodynamically stable states of unequal energy with finite populations at ambient temperature. These conditions are fulfilled by many reacting systems and dispersion phenomena have been identified with solvation of charged and uncharged species, with proton transfer and with association and dimerisation equilibria[7].

In the context of this review the third and fourth processes above are of major importance.

5.1.2 The interaction of a periodic pressure perturbation with a system capable of relaxation

The formulation of the perturbation of a chemical system by a sound wave has been described by a number of authors[5–7] and only an abridged treatment will be presented here.

The pressure amplitude p, of a one-dimensional sound wave travelling in the positive x-direction is given by:

$$p = p_0 \exp(-\alpha x) \exp i\omega(t - x/c) \tag{5.1}$$

where p_0 is the pressure amplitude at $x = 0$ and time $t = 0$, α is the absorption coefficient related to the attenuation of the sound wave, c is the phase velocity, and $\omega = 2\pi f$, where f is the frequency of the sound wave. This equation is analogous to the Beer–Lambert law governing light absorption where x is the distance between the point of observation and sound source. The above equation may be rewritten in the form:

$$p = A'e^{i\omega t} \tag{5.2}$$

where A' is an amplitude factor ($= p_0 \exp(-\alpha x) \exp i\omega(-x/c)$). It is assumed that the pressure and temperature variation are synchronous, when the perturbation driving equilibrium away from its static value may be given by equation (5.2).

For a two-state equilibrium of the form:

$$\mathrm{A} \underset{k_{21}}{\overset{k_{12}}{\rightleftharpoons}} \mathrm{B} \tag{5.3}$$

it is possible to define a relaxation time (τ) as the time required for the extent by which the concentration of one reagent differs from its equilibrium value concentration to decrease to 1/e of its original value. From this, the relationship between the rate constants and the relaxation time may be obtained

$$\tau = 1/(k_{12} + k_{21}) \tag{5.4}$$

If we let y be the difference in actual concentrations and the initial values ($y = a - a_0$, $y = b_0 - b$) and $\bar{y}$ be the difference in equilibrium concentrations and the initial values ($\bar{y} = \bar{a} - a_0$, $\bar{y} = b_0 - \bar{b}$), then

$$-dy/dt = (y - \bar{y})(k_{12} + k_{21}) \tag{5.5}$$

thus

$$\tau(dy/dt) + y = \bar{y} \tag{5.6}$$

The above equations apply to a system receiving a sudden impulse and then undergoing a relaxation process. In acoustics, one is usually concerned with a system being subjected to a periodic pressure change, as given by equation (5.2).

If the pressure variation described by equation (5.2) is correlated with y, then

$$\bar{y} = Ae^{i\omega t} \tag{5.7}$$

and, due to the finite time of exchange,

$$y = Be^{i\omega t} \tag{5.8}$$

Since A, as defined by equation (5.7), is complex, then B is also complex and

$$\begin{aligned} dy/dt &= Be^{i\omega t} i\omega \\ &= i\omega y \end{aligned} \tag{5.9}$$

Using equation (5.6), one obtains

$$i\omega y\tau + y = \bar{y} \tag{5.10}$$

and so

$$\begin{aligned} y &= \frac{\bar{y}}{1 + i\omega\tau} \\ &= \bar{y}\,\frac{1}{1+\omega^2\tau^2} - \bar{y}\,\frac{i\omega\tau}{1+\omega^2\tau^2} \\ &= G\bar{y} \end{aligned} \tag{5.11}$$

where G is a complex function given by

$$G = G_r + iG_i$$

with

$$G_r = \frac{1}{1+\omega^2\tau^2} \text{ and } G_i = \frac{-\omega\tau}{1+\omega^2\tau^2} \tag{5.12}$$

Alternatively, equation (5.10) can be expressed in the form:

$$d\bar{y} = dy(1 + i\omega\tau) \tag{5.13}$$

The function G is known as the 'transfer function'. The real (G_r) and imaginary (G_i) parts can be related to the dispersion and absorption of the sound wave respectively. It is normal to use the well-known trigonometric relationships to express G in terms of the phase angle ϕ.

When $\tan\phi = -G_i/G_r = \omega\tau$

the driving force is given by equation (5.7), and using equation (5.11) one obtains,

$$y = \frac{e^{-i\phi}}{(1+\omega^2\tau^2)^{\frac{1}{2}}} A e^{i\omega t} = \frac{A e^{i(\omega t-\phi)}}{(1+\omega^2\tau^2)^{\frac{1}{2}}} \tag{5.14}$$

The physical meaning of this equation is that y lags behind $\bar{y}$ with a phase angle ϕ. Initially, at low frequency, the two are in phase, but as the periodicity approaches that of the relaxation time τ, then a phase lag is introduced which reaches a maximum at $\omega\tau = 1$ or $\phi = 45$ degrees.

The two general methods of measurement follow from consideration of the stationary state solutions of equation (5.14)[10]:

(a) One method is concerned with the real part of the transfer function $G_r = 1/(1+\omega^2\tau^2)$ or its square root, ρ. G_r and ρ can be related to an observed dispersion in absorption coefficient (α/f^2), and in velocity ($(c_0^2-c_\infty^2)/c_0^2$), which

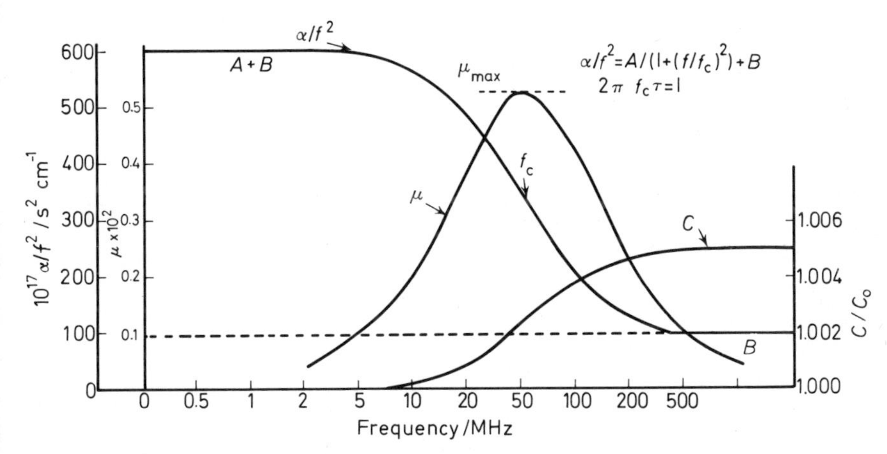

Figure 5.1 Variation of typical ultrasonic parameters with frequency for a single-relaxation process

have the form shown in Figure 5.1. All the methods utilising the frequency or time-dependence of the real part of the transfer function G_r are termed 'dispersion' methods.

(b) The other methods are related to the imaginary part of the transfer function,

$$G_i = (-\omega\tau/(1+\omega^2\tau^2)).$$

Any phase shift between two conjugate variables, such as y and $\bar{y}$, will result in a loss of energy. The loss of energy per wavelength ($\alpha\lambda$) will show a maximum, as indicated previously, at $\omega = 1/\tau$. Methods are available which can measure this loss directly and the relaxation time may be evaluated using standard equations. Since the basis for all methods observing G_i is the measurement of energy absorption, this group can be termed 'absorption' methods.

In order to proceed further it is necessary to relate the various thermodynamic functions of the system to the appropriate parameters of the frequency-dependent perturbation. For example, the acoustic loss, $\mu = \alpha\lambda$, can be related[5,6] to the adiabatic compressibility, β, by (for small μ)

$$\frac{\mu}{\pi} = \frac{\omega\tau(\beta_0 - \beta_\infty)}{\beta_0 + \omega^2\tau^2\beta_\infty} \tag{5.15}$$

where β_0 and β_∞ represent the adiabatic compressibility measured at frequencies respectively far below and far above those within which the dispersion is occurring. When $\omega\tau = 1$, μ reaches its maximum value, μ_{max}, given by (for $\beta_0 - \beta \ll \beta_0$) and $\omega = 2\pi f_c$

$$\mu_{max} = \frac{\pi(\beta_0 - \beta_\infty)}{2\beta_0} \tag{5.16}$$

The sound velocity, too, is related to the adiabatic compressibility by $c^2 = 1/\beta\rho$ where ρ is the density of the medium.

Alternatively, the adiabatic compressibility can be related[5,6] to the specific heat at constant pressure, C_p, so that

$$\frac{2\mu_{max}}{\pi} = \frac{(\gamma - 1)\Delta C_p}{(C_p - \Delta C_p)}\left\{1 - \frac{\Delta V^0}{V} \cdot \frac{C_p}{\Delta H^0 \theta}\right\} \tag{5.17}$$

where ΔC_p is the total increment in the specific heat due to the process of interest, γ is the ratio of the specific heat at constant pressure to that at constant volume, ΔV^0 and ΔH^0 are respectively the volume and enthalpy changes associated with the process and θ is the coefficient of thermal expansion.

Since ΔC_p can be expressed in terms of the thermodynamic equilibrium parameters for the process of interest, ΔG^0, ΔH^0, ΔS^0, ΔV^0

$$\Delta C_p = R\left(\frac{\Delta H^0}{RT}\right)^2 \frac{\exp(-\Delta G^0/RT)}{[1 + \exp(-\Delta G^0/RT)]^2} \tag{5.18}$$

we are able to relate the acoustic observations to the appropriate thermodynamic quantities.

The value of α/f^2 for the relaxation process is

$$\frac{\alpha}{f^2} = \frac{A}{1 + (f/f_c)^2} \tag{5.19}$$

where $A = (\beta_0 - \beta)/2\beta_0 c f_c$; c is the sound velocity, or $\mu_{max} = \frac{1}{2}Acf_c$. In practice, the 'classical' contributions to α/f^2 from shear and volume viscosities (and also from other relaxation processes occurring at frequencies much higher than f_c) are denoted by a constant, B. Thus

$$(\alpha/f^2) = A/(1 + (f/f_c)^2) + B \tag{5.20}$$

Similarly, the dispersion in the velocity can be shown to be

$$c^2 - c_0^2 = (2\mu_{max}/\pi)c_0 c_\infty(\omega\tau)^2/(1 + (\omega\tau)^2) \tag{5.21}$$

c_0 and c_∞ refer to the sound velocity at frequencies well below and well above the relaxation frequency.

A number of reviews have appeared outlining the methods which may be used for the investigation of the absorption and dispersion properties of a system subjected to a periodic sound wave[5, 7, 11].

Absorption and velocity can usually be measured to an accuracy of 2% over a frequency range from 100 KHz to 1000 MHz. Methods do exist which enable either the velocity or the absorption to be determined to a higher degree of accuracy, but there is usually a limitation as to the available frequency range for such measurements. In conformational studies, the most popular range of frequency has been that between 5 and 150 MHz. In this range the rate constant for the isomer interchange is of the order of 10^{+6} to 10^{+9} s^{-1}, and the corresponding activation energies are 10–50 kJ mol^{-1}.

5.1.3 Experimental determination of energy parameters

It is possible, using various computational techniques, to fit equations (5.20) and (5.21) to experimental data. If the data do not reduce to a single relaxation process, more complex procedures may be invoked, although the significance of the fit decreases approximately as the square of the number of parameters to be fitted[10, 17]. The relevant equations for three-state equilibria[5, 6] and multiple relaxations[10] have been outlined elsewhere and will not be considered here. In many systems, especially those exhibiting simple rotational isomeric equilibria, analysis of the dispersion as a single relaxation process is quite justified.

5.1.3.1 Thermodynamic parameters

The amplitude of an observed dispersion is directly related to the thermodynamic parameters defining the differences between the two stable conformers. In order to simplify the analysis of the observed data it is usual to invoke a relation between the velocity and the specific heat. This is

$$c^2 = (\gamma - 1)\frac{JC_p}{\theta^2 V \rho T} \tag{5.22}$$

where J is the joule. The experimentally observable parameters in the above equations are μ_{max} and $d\mu_{max}/dT$. The unknowns are ΔH^0, ΔS^0 and ΔV^0. In order to solve for ΔH^0 and ΔS^0 the major assumptions are usually made that

$$(\Delta V^0/V)(C_p/\theta) \ll \Delta H^0 \text{ and } \Delta C_p \ll C_p \tag{5.23}$$

These lead to the simplified forms:

$$\frac{\Delta G^0}{RT} = \frac{2 \ln \xi_i + 1 + (F_i^2 + (\xi_i + 1)^2)^{\frac{1}{2}}}{F_i} \tag{5.24}$$

and

$$\frac{\Delta H^0}{RT} = 2(F_i^2 + (\xi_i + 1)^2)^{\frac{1}{2}} \tag{5.25}$$

where

$$F_i = (2\mu_{max} J C_p^2/\pi R\theta^2 \rho c^2 V T)^2$$
$$= \frac{\Delta H^0 \exp(\Delta G^0/2RT)}{RT(1+\exp(\Delta G^0/RT))}$$

and $\xi_i = (T_i/F_i)(dF_i/dT_i)$. The subscript i refers to measurements at a particular temperature.

In many examples, the values of F_i in equation (5.25) cannot be determined experimentally because there are no data available on the 'static' parameters C_p and θ. In these cases a further assumption, that C_p/θ is independent of temperature, is made[6] and the temperature variation of F_i in equation (5.25)

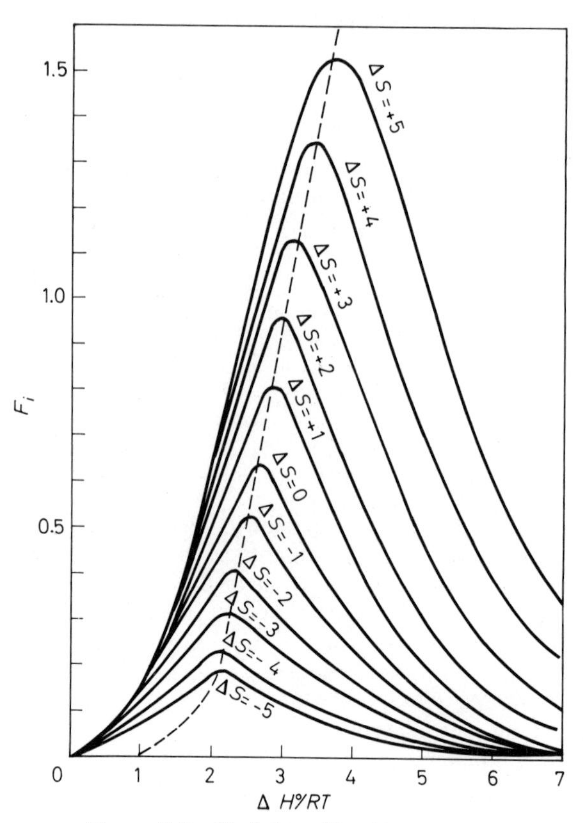

Figure 5.2 Variation of F_i with temperature

is now governed by the experimental quantity $(\mu_{max}/T^2)^{\frac{1}{2}}$. The theoretical variation of F_i with temperature can be evaluated by plotting the right-hand side of equation (5.25) against $\Delta H^0/RT$. A family of curves is obtained depending upon the value used for ΔS^0, the entropy difference. In all curves F_i reaches a maximum at a certain value of ΔH^0/RT. The variation of both F_{max} and the corresponding value of $\Delta H^0/RT$ is shown in Figure 5.2. Information about the equilibrium thermodynamic parameter ΔH^0 can now be obtained by comparing the experimental variation of $(\mu_{max}/T^2)^{\frac{1}{2}}$ with the curves shown in Figure 5.2. Three cases can be distinguished.

(a) If $(\mu_{max}/T^2)^{\frac{1}{2}}$ increases with temperature rise, then it is evident from Figure 5.2 that $\Delta H^0 > (2.3\text{–}3.2)RT$ and thus it is safe to assume that ΔG^0 is so large that the term $\exp(-\Delta G^0/RT)$ is small compared to unity. In this case an approximate value of $(\Delta H^0/RT)$ can be obtained from the slope of the plot $\log(T\mu_{max}/c^2)$ against reciprocal temperature.

(b) If the plot of (μ_{max}/T^2) against temperature reaches a maximum, then it is evident that $\Delta H^0 \approx (2.3\text{–}3.2)RT$ where T is the maximum temperature. The exact value of ΔH^0 can be determined if ΔS^0 is known.

(c) If μ_{max}/T^2 decreases with increasing temperature, then it follows that $\Delta H^0 \leqslant (2.3\text{–}3.5)RT$ where T corresponds to the lowest temperature of experimental observation.

5.1.3.2 *Kinetics of isomerisation*

The rate constants for the forward and reverse processes have the form:

$$k_{12} = 2\pi f_c(1+K^{-1})$$

and (5.26)

$$k_{21} = 2\pi f_c/(1+K)$$

where K (the equilibrium constant) $= k_{12}/k_{21}$. The temperature dependence of the rate constants is governed by the Arrhenius equation:

$$k_{12} = A_{12}\exp(-E_{12}/RT) \tag{5.27}$$

or the Eyring absolute rate equation

$$k_{12} = \mathscr{K}(\boldsymbol{k}T/\boldsymbol{h})\exp(-\Delta H^{\ddagger}_{12}/RT)\exp(\Delta S^{\ddagger}_{12}/R) \tag{5.28}$$

In these equations A is the Arrhenius frequency factor, E the activation energy, $\Delta H^{\ddagger}$ the activation enthalpy or potential barrier, $\Delta S^{\ddagger}$ the entropy of activation, $\mathscr{K}$ the transition probability. $\boldsymbol{k}$ and $\boldsymbol{h}$ have their usual meaning. In equation (5.28), $\Delta H^{\ddagger}_{12}$ can be determined from the slope of the plot of $\log(k_{12}/T)$ against $1/T$. Even when the equilibrium constant K is not known, the slope of the plot of $\log(f_c/T)$ against reciprocal temperature can still yield an accurate estimate of $\Delta H^{\ddagger}_{12}$.

An interesting point regarding the curves drawn in Figure 5.2 is that very small values of F can be measured, and so one-sided equilibria can be observed. This is in contrast with, for example, standard spectroscopic measurements, which ideally should be carried out on systems containing states not too dissimilar in population. A number of systems in which the energy difference is 25 kJ mol^{-1} or more have been studied using acoustics.

If the system studied can be described as a two-state equilibrium, it is possible to gain some insight as to the mechanism of the internal rotational processes from the magnitude of the Arrhenius factor A, although it should be pointed out that the error in this quantity is usually quite large. The rotational isomeric conversion is usually considered to be a thermally activated process. It is then possible to identify two distinct mechanisms.

The first involves one isomer becoming thermally activated to an eclipsed state, which corresponds to a maximum in the potential energy curve, and then immediately being deactivated to the stable isomer in the adjacent potential minimum. The alternative possibility is that the thermally activated molecule reaches the transition state and then undergoes essentially 'free internal rotation' for some time before being deactivated to any of the available potential minima with equal probability. These two distinct mechanisms were originally proposed by Newmark and Sederholm[12, 13] and have been used by Gutowsky and his co-workers[14] in their analysis of the n.m.r. exchange rates in fluorinated ethanes in the liquid phase.

The first mechanism is the simplest possible, and corresponds to the case where the transition probability, $\mathscr{K}$ in the Eyring rate equation (5.28) is unity. From this description of internal rotation, Alger, Gutowsky and Vold[14] have considered the isomeric transition in substituted ethanes and have argued that the major contribution to the entropy of activation $\Delta S^{\ddagger}_{12}$ will come from the difference in vibrational terms between the eclipsed transition state and the lower energy isomer. The only difference in these vibrational terms is, to a first approximation, the absence of torsional modes in the eclipsed state. Obviously this cannot be strictly true, since the achieving of the eclipsed state must lead to some distortion of the molecular framework and so to shifts in the bending, deformation and rocking modes. However, neglecting these secondary effects, it is possible to denote the free energy associated with the torsional oscillations as

$$\Delta G^{\ddagger}_{\mathrm{osc}} = T\Delta S^{\ddagger}_{12}$$

Substitution in the Eyring rate equation leads to:

$$k_{12} = (\boldsymbol{k}T/\boldsymbol{h}Q_{\mathrm{osc}})\exp(-\Delta H^{\ddagger}_{12}/RT) \tag{5.29}$$

where Q_{osc} is the partition function of the torsional oscillations of the lowest energy isomer. If it is assumed that during the torsional oscillations the ends of the molecule twist back and forward about covalent single bonds, and also that the molecules are in torsional energy levels well below the barrier to interconversion, then

$$Q_{\mathrm{osc}} = \mathrm{e}^{-U/2}/(1-\mathrm{e}^{-U}) \tag{5.30}$$

where $U = (\boldsymbol{h}/\boldsymbol{k}T)(3/2\pi)(\Delta H^{\ddagger}_{12}/2I_{\mathrm{r}})^{\frac{1}{2}}$, and I_{r} is the appropriate value of the reduced moment of inertia for the axis of the bond about which internal rotation is assumed to occur. It also follows that U is small enough such that Q_{osc} has essentially the high temperature value of $\boldsymbol{k}T/\boldsymbol{h}\nu_0$ and so

$$k_{12} = \nu_0 \exp(-\Delta H^{\ddagger}_{12}/RT) \tag{5.31}$$

where ν_0 is the torsional frequency of the isomer with the lowest energy. In the second description of internal rotation, the molecule has a probability of $1/n$ of returning at random to one of the n potential energy minima. For rotation about bonds between sp^3 carbon atoms this model leads to a calculated frequency factor of $\nu_0/3$ in equation (5.31). In this model of delayed deactivation the experimental relaxation frequencies must be independent of the magnitude of the highest barrier to rotation because, if the energy of a molecule exceeds the second highest barrier, the molecule can rotate to an angle corresponding to any of the three potential minima without passing over the highest barrier.

In suitable cases, the Arrhenius factor A may be determined from the analysis of the acoustic data, and these values compared with those obtained from theory. The theory requires that the torsional frequency of the isomer with the lowest energy be determined from examination of either the far-infrared or Raman spectrum of the molecule. In Table 5.1 some experimentally determined A factors and measured torsional frequencies are listed for a

Table 5.1 Comparison of spectroscopic torsional frequencies and acoustic Arrhenius factors A

Molecule	10^{-12} × *Torsional frequency*/s^{-1}	10^{-12} *A factors*/s^{-1}
1,2-Dichloroethane	3.8	0.1–1.0
1,2-Dibromoethane	2.7	1–10
1,2-Dibromo-2-methylpropane	4.0	1.5
1,2-Dichloro-2-methylpropane	3.3	0.4
meso-2,3-Dichlorobutane	3.0	4.0
meso-2,3-Dibromoethane	3.3	30
2-Chloro-2-methylbutane	3.4	0.7
2-Bromo-2-methylbutane	3.4	23
2-Iodo-2-methylbutane	3.5	1170
1-Fluoro-1,1,2,2-tetrachloroethane	2.58	2.2
		0.8*
1,2-Difluoro- 1,1,2,2-tetrachloroethane	2.4	0.5*

*Value obtained from an n.m.r. study[14].

number of substituted ethanes. Certain of the A factors and the torsional frequencies are the same, indicating that there is some significance in this formulation of the internal rotational process. Where large discrepancies are observed, this may be due to large distortion effects modifying the description of the partition function for the transition state, or to incorrect identification of the appropriate torsional mode from a vibrational spectrum.

5.2 CRITICAL EVALUATION OF THE DETERMINATION OF CONFORMER ENERGY DIFFERENCES BY ACOUSTIC METHODS

In Table 5.2 some of the enthalpy differences obtained from ultrasonic investigations are compared with values from infrared studies. It will be seen that a number of rather large discrepancies exist. In general, the assumptions involved in the determination of the enthalpy differences from infrared studies are fewer than those involved in the ultrasonic measurements.

The development of equations (5.17) and (5.18) requires several approximations which are based on the fact that μ_{max} is small and the dispersion in velocity is negligible. These equations are based also on the assumption that the observed relaxation process is entirely due to a two-state isomeric equilibrium which obeys first-order kinetics and occurs in ideal solution.

Table 5.2 Enthalpy differences between rotational isomers

Molecule	*Infrared* $kJ\ mol^{-1}$	*Reference*	*Ultrasonic* $kJ\ mol^{-1}$	*Reference*
n-Propyl bromide	1.67, 2.09	3	5.4	15
1,1,2-Trichloroethane	0.92, 1.2	3	8.8	15
1,1,2-Tribromoethane	2.1	3	6.7	15
2-Methylbutane	0.0	3	3.75	16
Isobutyl chloride	1.5	17	5.5	17
Isobutyl bromide	3.18	18	2.90	17
1,2-Dichloro-2-methylpropane	0.00	17	8.4	17
1,2-Dibromo-2-methylpropane	3.1	17	4.05	17
1,1,2,2-Tetrabromoethane	3.14, 3.76	3	3.75	19
2,3-Dimethylbutane	0.42	3	4.2	16
1,2-Dibromo-1,1,2,2-tetrafluoroethane	3.89	3	6.7	20

Table 5.3 Volume change during rotational isomerism[21,22,145]

Molecule	*Solvent*	*Molar volume change* $\Delta V^{\circ}/V$	*Method*
1,1,2-Trichloroethane	Pure liquid	−0.06	
	p-xylene	−0.03	
	mesityl oxide	−0.01	
	n-heptane	−0.01	
	ethyl acetate	−0.01	
	methyl cyanide	−0.03	
	trichloroethylene	−0.01	
	nitrobenzene	−0.02	
1,1,2-Tribromoethane	pentachloroethane	0.004	
	trichloroacetonitrile	0.013	
	acetonitrile	−0.015	Acoustic and spectroscopic measurements
4-Methyl-1,3-dioxane	cyclohexane	0.02	
1,2-Dibromo-1,1,2,2-tetrafluoroethane	hexane	0.02	
meso-2,3-Dibromobutane	hexane	0.03	
	cyclohexane	0.01	
	carbon tetrachloride	0.03	
	trichloroethylene	0.01	
	chloroform	0.01	
	dichloromethane	−0.01	
	ethylene dichloride	−0.01	
	dimethylsulphoxide	−0.02	
1,2-Dimethylcyclohexane		0.026	
1,3-Dimethylcyclohexane		0.022	
1,4-Dimethylcyclohexane		0.033	
4-Methyl-2-methoxy-1,3-dioxane		0.04	
4,6-Dimethyl-2-methoxy-1,3-dioxane		0.02	
1,3-Dioxane		0.02	Thermodynamic measurements
Dichloroethane		0.062	
Dichloroethylene		0.021	
Dibromoethylene		0.008	
Di-iodoethylene		0.073	
n-Butane		0.039	

Provided $\Delta C_p \ll C_p$ (an assumption which is reasonable since μ_{max} is very small)

$$\frac{2\mu_{max}C_p}{(\gamma-1)} = nR\left(\frac{\Delta H^0}{RT}\right)\left\{\frac{\exp(-\Delta G^0/RT)}{(1+\exp(-\Delta G^0/RT)^2}\right\}\times\left\{1-\frac{\Delta V^0}{V}\cdot\frac{C_p}{\theta\Delta H^0}\right\}^2 \tag{5.32}$$

where n is the mole fraction of solute.

If it may be assumed that the equilibrium parameters ΔH^0, ΔS^0 and ΔV^0 are essentially independent of concentration, and ΔH^0, ΔS^0 can be measured by an independent technique, it is possible to obtain information on the magnitude of the volume change accompanying the internal rotational process. A plot of $2\mu_m(\gamma-1)C_p\pi$ against concentration has a slope of

$$R\left(\frac{\Delta H^0}{RT}\right)^2\frac{\exp(-\Delta G^0/RT)}{(1+\exp(-\Delta G^0/RT))^2}\left\{1-\frac{\Delta V^0}{V}\frac{C_p}{\theta\Delta H^0}\right\}^2 \tag{5.33}$$

A recent study of a number of molecules exhibiting rotational isomerism has shown that such linear plots are observed at low concentrations[21, 22].

At concentrations greater than 0.3 mole fractions, deviations from linearity indicate the departures from ideality of the binary mixture. Values of $\Delta V^0/V$ so obtained are listed in Table 5.3. The ΔH^0 values were obtained from n.m.r. studies with $\Delta S^0 = R\ln 2$. Inspection of Table 5.3 indicates that the inequality

$$(\Delta V^0/V)(C_p/\theta)\ll\Delta H^0$$

is not valid for these molecules.

In principle, the application of hydrostatic pressure might be used in the evaluation of ΔV^0, and thus allow acoustic evaluation of ΔH^0 and ΔS^0. Whilst there would appear to be some reason for sceptiscism with regard to the current use of ultrasonic measurements in the determination of energy-difference parameters, the kinetic activation parameters are not liable to such uncertainties, and correlate well with data obtained using other techniques[23].

In the context of volume effects and their determination, it is interesting to consider the effects of internal pressure[24]. In studies of 4,4-dimethyl-2-silapentane and 2,3-dimethyl-2-silabutane it was found that increasing internal pressure of the solvent increases the population of the *gauche* with respect to the *trans* conformation. In these cases the differences in molar volumes for the *trans–gauche* equilibria are respectively -3.7 and $-4.1\ \text{cm}^3\ \text{mol}^{-1}$ for the 4,4-dimethyl-2-silapentane and 2,3-dimethyl-2-silabutane. These studies indicate that solvent internal pressure may play a role in controlling equilibria and isomerisation kinetics in systems where solvent effects normally are considered to be negligible. Similar conclusions have been reached from comparative studies of hydrocarbons in the liquid and gas state[25].

5.3 SIMPLE MOLECULES

For convenience, the molecules will be considered under sections descriptive of the bond about which internal rotation occurs.

5.3.1 Ethane-like molecules (sp^3–sp^3)

For molecules such as 1,2-dibromoethane the two equivalent higher energy minima in the angular dependence of potential energy correspond to the optically active *gauche* isomers. This state is degenerate with an entropy $R \ln 2$ in excess of that of the more stable *trans* isomer. On the other hand, in 1,1,2-trichloroethane, the more stable isomeric state is degenerate.

The potential energy barriers for the less-stable to more-stable isomeric transitions in several ethane derivatives which exist as two-state equilibria are listed in Table 5.4. The second column in this table indicates the sign of the statistical entropy differences, a + sign indicating that the higher energy isomer is the optically active pair.

In a particular series of related molecules the energy barrier opposing interconversion increases as the size of the halogen atom increases.

The importance of electrostatic forces can be seen if one considers 1,1,2-trichloroethane and 2-methylbutane, where the barriers are respectively 24.3 and 19.7 kJ mol^{-1}. If the barrier were entirely determined by steric forces, a higher value would be expected for 2-methylbutane.

A study of 1-fluoro-1,1,2,2-tetrachloroethane[23] yielded activation enthalpies of 35 kJ mol^{-1} in reasonable agreement with the value, 31 kJ mol^{-1}, obtained from n.m.r. Infrared and Raman spectra, too, give 37 kJ mol^{-1} for the activation energy to internal rotation. It should be pointed out that a number of assumptions are involved in the determination of the barrier heights from torsional data; the most significant of these is the assumption that this vibrational mode is pure.

In general, the larger the number of halogen atoms contained by a single molecule, the higher the barrier. In the molecules studied by n.m.r.[28], it was suggested that all barriers influence the spectral line shapes, indicating that each rotational interconversion in these molecules will be followed by an immediate deactivation. A detailed correlation of the rotational barriers with pair-wise interaction constants[12] or with a complete Westheimer type calculation[33] does not appear to be possible.

Propanes and butanes may be considered as methyl substituted ethanes and represented as simple *trans* and *gauche* conformations with potential energy curves similar to those of halogen substituted ethanes[32]. The data obtained from acoustics are in good agreement with those obtained from n.m.r. line-shape analysis.

In n-butane, 1,2-dichloroethane and 1-chloropropane, *gauche* halogen–halogen and *gauche* methyl–methyl interactions are of higher energy than *gauche* halogen–methyl interactions. This has been explained by Szasz[34] in terms of an electrostatic attraction between the alkyl group and the electronegative halogen atoms. In 2,2,3,3-tetrachlorobutane and 2,2,3,3-tetrabromobutane[35] the methyl–halogen interactions lead to a loss in stability of the methyl *trans* state.

In the case of the butanes, substitution of bromine for chlorine results in an increased activation free energy $\Delta G^‡$, Table 5.4. This is in contrast to the observations on the simple halogenated ethanes. The same is true when halogen is substituted for methyl. However, in some cases changes in $\Delta S^‡$ cause opposite changes in E_a and $\Delta G^‡$.

Table 5.4 Ethane derivatives

Molecule	$\Delta G^{\ddagger}$/ kJ mol^{-1}	*Entropy difference*	*Activation barrier* $\Delta H^{\ddagger}_{12}$/ kJ mol^{-1}	*Method*	*State*	*Reference*
1,2-Dichloroethane	—	+	13.32	Acoustics	Soln. in acetone	26
1,2-Dibromoethane	—	+	17.55	Acoustics	Soln. in acetone	26
1,1,2-Trichloroethane	—	–	24.24	Acoustics	Liquid	15
1,1,2-Tribromoethane	—	–	26.75	Acoustics	Liquid	15
1,1,2,2-Tetrabromoethane	—	+	17.97	Acoustics	Liquid	19
n-Propyl bromide	—	–	15.05	Acoustics	Liquid	15
1-Fluoro-1,1,2,2-tetrachloroethane	—	–	34.69	Acoustics	Liquid	23
	—	–	31.0	n.m.r.	Soln. in $CFCl_3$	14
1,2-Difluoro-1,1,2,2-tetrachloroethane	—	–	38.24	n.m.r.	Soln. in $CFCl_3$	13
1,2-Dibromo-1,1-dichloroethane	—	+	32.18	Acoustics	Liquid	27
1,2-Dibromo-1,1-difluoroethane	—	+	21.31	Acoustics	Liquid	27
1,2-Dibromo-1,1,2,2-tetrafluoroethane	—	+	27.58	Acoustics	Liquid	20
1,1,2-Tribromotrifluoroethane	—	–	39.4	n.m.r.	Liquid	13
1-Chloro-2-fluorotetrabromoethane	61.8	+	—	n.m.r.	Liquid	28
1-Fluoro-2-bromotetrachloroethane	55.2	–	—	n.m.r.	Liquid	28
1-Fluoro-2,2-dibromotrichloroethane	55.2	–	—	n.m.r.	Liquid	28
n-Butane	—	+	14.21	Acoustics	Liquid	29
2,3-Dimethylbutane	—	+	11.70	Acoustics	Liquid	16
2-Methylbutane	—	–	19.64	Acoustics	Liquid	16
Isobutyl chloride	—	–	15.88	Acoustics	Liquid	16
Isobutyl bromide	—	–	19.64	Acoustics	Liquid	18
Isobutyl iodide	—	–	22.54	Acoustics	Liquid	17
2-Chloro-2-methylbutane	—	–	16.30	Acoustics	Liquid	30
2-Bromo-2-methylbutane	—	–	23.40	Acoustics	Liquid	30
2-Iodo-2-methylbutane	—	–	35.11	Acoustics	Liquid	30
1,2-Dichloro-2-methylpropane	—	+	18.81	Acoustics	Liquid	17
1,2-Dibromo-2-methylpropane	—	+	22.99	Acoustics	Liquid	17
meso-2,3-Dichlorobutane	—	+	21.31	Acoustics	Liquid	30
meso-2,3-Dibromobutane	—	–	26.75	Acoustics	Liquid	30
2,3-Dibromo-2,3-dimethylbutane	—	+	25.0–31.35	Acoustics	Liquid	31
2,3,3-Tribromo-2-methyl butane	56.30	+	58.9	n.m.r.	Soln. in CS_2	34
2,2,3,3-Tetrachlorobutane	56.59	+	63.9	n.m.r.	Soln. in acetone	34
2,2,3,3-Tetrabromobutane	66.75	–	56.5	n.m.r.	Soln. in C_6H_5Cl–acetone	34

The effect of increasing the number of bromine atoms on the C(2)—C(3) fragment can be seen in 3,3-dibromo-2,2-dimethylbutane, 2,3-tribromo-2-methylbutane and 2,2,3,3-tetrabromobutane. A small increase is observed in $\Delta G^{\ddagger}$ and $\Delta H^{\ddagger}$ in comparing the first pair of these. A much larger difference is observed between the last pair and is attributed to the difference between bromine–methyl and bromine–bromine interactions in the transition states. The influence of methyl–halogen interactions in the transition state is evident

Table 5.5 Substituted propanes, butanes and pentanes

Molecule	*Activation barrier* $\Delta H^{\ddagger}_{12}$/kJ mol^{-1}	*Method*	*Reference*
1,2-Dichloropropane	19.64	Acoustics	15
1,2-Dibromopropane	20.48	Acoustics	15
1-Chloro-2-bromopropane	22.99	Acoustics	36
s-Butyl chloride	18.39	Acoustics	17
s-Butyl bromide	20.06	Acoustics	17
s-Butyl iodide	22.15	Acoustics	17
DL-2,3-Dichlorobutane	20.90	Acoustics	30
DL-2,3-Dibromobutane	15.46	Acoustics	30
2-Chloropentane	18.39	Acoustics	36
2-Bromopentane	20.06	Acoustics	36
2-Methylpentane	16.30	Acoustics	30
3-Chloropentane	14.21	Acoustics	36
3-Bromopentane	15.04	Acoustics	36
3-Methylpentane	17.13	Acoustics	16
1,2-Dibromopentane	14.63	Acoustics	36
1,4-Dibromopentane	19.64	Acoustics	36
2-Bromohexane	17.55	Acoustics	36
3-Bromohexane	15.04	Acoustics	36
1,2-Dibromobutane	14.21	Acoustics	36
n-Pentane	16.30	Acoustics	29
n-Hexane	10.86	Acoustics	29
1,2-Dibromo-1,1,2-trifluoroethane	19.22	Acoustics	27
3-Chloro-2,2,3-trimethylbutane	47.6	n.m.r.	35
3-Bromo-2,2,3-trimethylbutane	55.1	n.m.r.	35
3,3-Dichloro-2,2-dimethylbutane	43.89	n.m.r.	35
3,3-Dibromo-2,2-dimethylbutane	47.65	n.m.r.	35
3, Chloro-3-bromo-2,2-dimethylbutane	41.38	n.m.r.	35
3-Chloro-2,2,3-trimethylpentane	57.26	n.m.r.	35
3-Chloropentamethylpentane	49.32	n.m.r.	35

in the comparison of 3,3-dichloro-2,2-dimethylbutane and 2,2,3,3-tetrachlorobutane. The effect of size of alkyl substituent on the barrier height may be seen by comparing 3-chloro-2,2,3-trimethylbutane, 3-chloro-2,2,3-trimethylpentane and 3-chloropentamethylpentane. Although increase of group size does produce an effective increase in the activation parameters, the magnitude of the change is surprisingly small. This is all the more striking because substitution of a bromine atom for a t-butyl group does increase the free energy of activation. Generally, a bromine atom is considered to be comparable in size to a methyl group. A plausible explanation of the relatively small $\Delta G^{\ddagger}$ for 3-chloropentamethylpentane may be the absence of spherical symmetry in the t-butyl group and the existence of a 'cog-wheel' type of arrangement in the transition state. Rather severe methyl–methyl repulsions in the staggered state of 3-chloropentamethylpentane may also serve to decrease the activation barrier.

All the systems described above may be represented by a simple *gauche–trans* notation for the stable conformers[37] of some primary, secondary and tertiary halogeno-alkanes. A modified form of this simple notation may be used for the description of the rotational conformations of long-chain primary monohalogeno-alkanes[33], but is not adequate[38–40] for secondary alkyl halides.

The rotational conformations in these molecules can be generated by projected drawings of the planar carbon backbone forms (fully extended). This approach is illustrated for 3-chloropentane in Figure 5.3. In this figure,

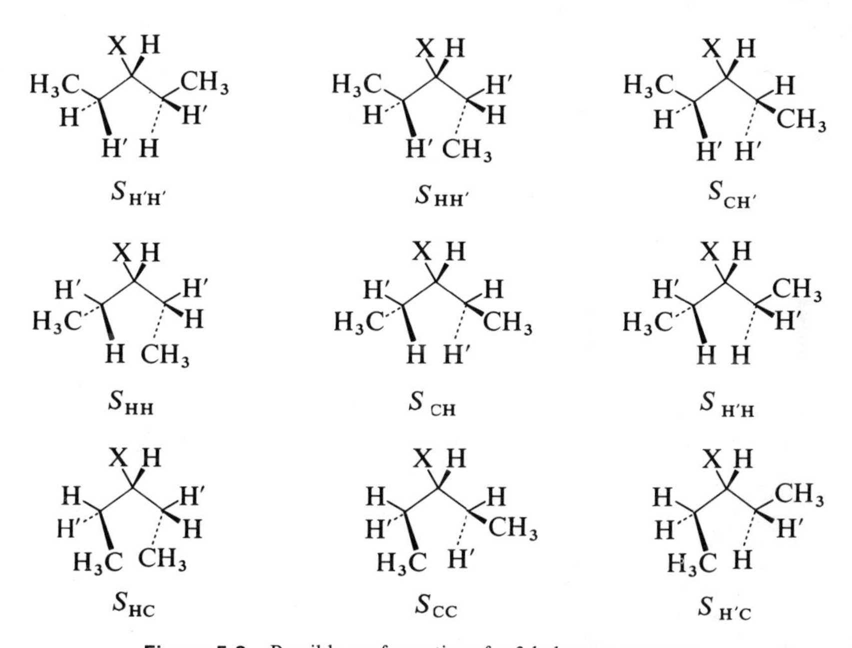

Figure 5.3 Possible conformations for 3-halogeno-pentanes

all the different isomers formed by rotation about the α- and α′-C—C bonds are considered. In order to describe the various isomers using secondary *S* notation it is necessary to denote a hydrogen atom on the α- and α′-carbon atom by hydrogen prime (H′). The different isomers can then be described by the notation S_{AB} where A and B are the atoms or groups attached to the α- and α′-carbons atom and in *trans* or *gauche* positions relative to the halogen atom.

In the first conformer illustrated in Figure 5.3, A and B are both H′; therefore, the isomer is the $S_{H'H'}$ form. In the 3-halogeno alkanes, it has been found that the S_{HH} form does not exist due to steric hindrance. In the 2-halogeno-alkanes the substituents on one of the α-carbon atoms are all hydrogen atoms; therefore, only the forms $S_{CH'}$, $S_{HH'}$ and $S_{H'H'}$ exist. Thus, in a molecule such as (±)-2,3-dichlorobutane, Figure 5.4, the isomeric equilibrium is of the kind:

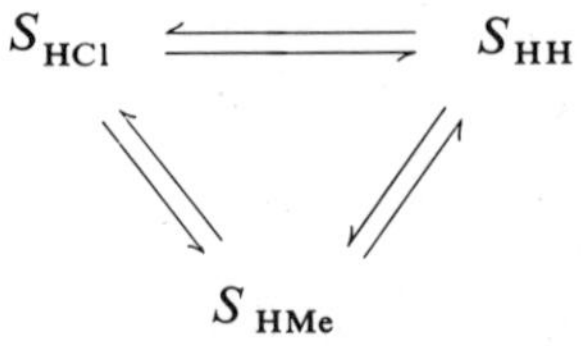

$$\begin{array}{ccc} S_{\mathrm{HCl}} & \rightleftharpoons & S_{\mathrm{HH}} \\ & \nwarrow\!\!\searrow \quad \nearrow\!\!\swarrow & \\ & S_{\mathrm{HMe}} & \end{array}$$

whereas in 3-chlorohexane the situation is more complex. Although in this molecule only three or four conformers have been detected spectroscopically (i.e. with populations in excess of *c.* 5%), the ultrasonic technique is sensitive to populations of *c.* 0.5% and the observed dispersion behaviour may correspond to a more complex reaction scheme[36]. Such a scheme is outlined below:

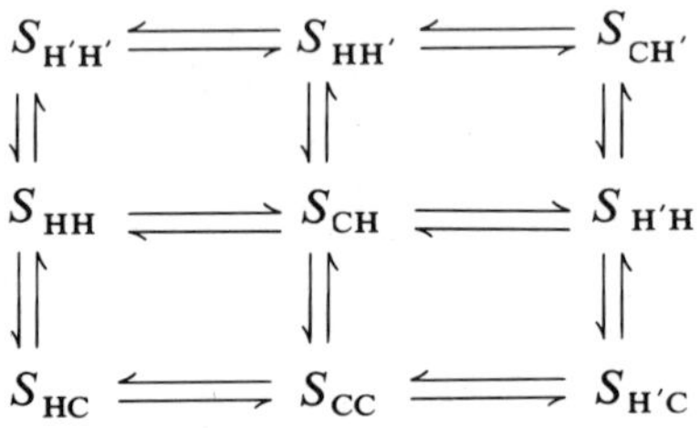

$$\begin{array}{ccccc} S_{\mathrm{H'H'}} & \rightleftharpoons & S_{\mathrm{HH'}} & \rightleftharpoons & S_{\mathrm{CH'}} \\ \downarrow\uparrow & & \downarrow\uparrow & & \downarrow\uparrow \\ S_{\mathrm{HH}} & \rightleftharpoons & S_{\mathrm{CH}} & \rightleftharpoons & S_{\mathrm{H'H}} \\ \downarrow\uparrow & & \downarrow\uparrow & & \downarrow\uparrow \\ S_{\mathrm{HC}} & \rightleftharpoons & S_{\mathrm{CC}} & \rightleftharpoons & S_{\mathrm{H'C}} \end{array}$$

In multi-step conformational equilibria of the type shown above, a spectrum of relaxational times and frequencies should be found, with the number of relaxation times corresponding to the number of independent steps[41]. Because of the coupling between states, the relaxation times will not be the same as those found by treating each step as an isolated two-state equilibrium. It has been found in practice that the observed relaxation data correspond approximately to a single relaxation process, which on analysis yields the

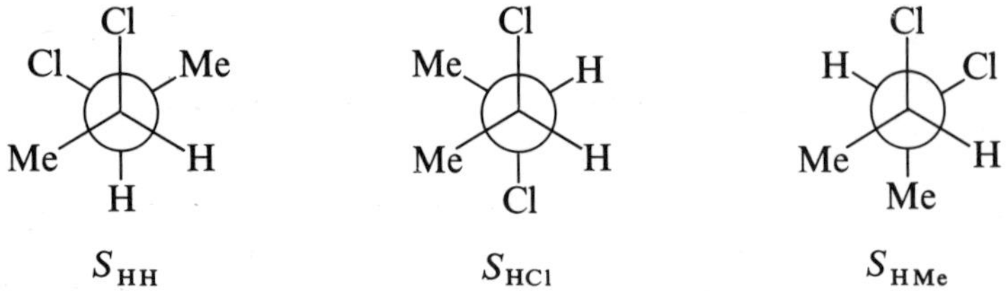

S_{HH} S_{HCl} S_{HMe}

Figure 5.4 (±) 2,3-Dichlorobutane

barriers listed in Tables 5.4 and 5.5. To a first approximation these can be regarded as an average value for the potential barrier hindering rotation in the molecule about bonds adjacent to the carbon–halogen link.

In conclusion of this sub-section, we must remind the reader that although the problem of the origin of barriers around single bonds continues to have a great appeal to quantum chemists, it still remains a subject of great controversy. It will be apparent that no simple explanation will serve to explain the experimentally observed data.

5.3.2 Aldehydes, ketones and similar systems (sp^3–sp^2)

Rotational isomerism about sp^2–sp^3 carbon–carbon single bonds has been much investigated using n.m.r. techniques[42]; but has been little investigated using acoustic relaxation. Measurements have been reported on propionaldehyde and n-butyraldehyde and these indicate that relaxation may be detected in these systems on cooling to low temperature[43].

5.3.3 Unsaturated molecules (sp^2–sp^2 L atoms)

The α,β-unsaturated aldehydes were amongst the first molecules to be investigated acoustically[43]. Spectroscopic studies of molecules containing the basic structure shown in Figure 5.5, show two planar rotational isomeric

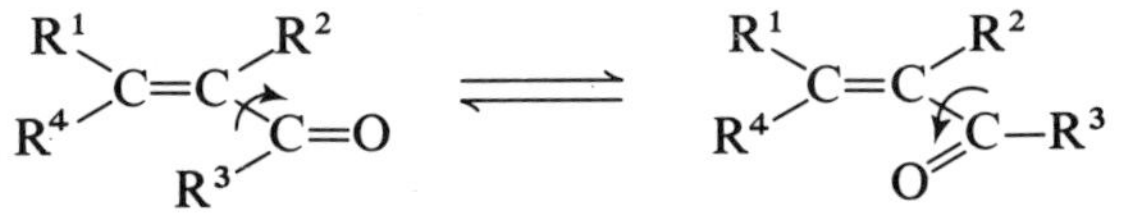

(1) *S-trans* (2) *S-cis*

$R^4 = R^1 = R^2 = R^3 = H$, Acrolein
$R^4 = R^1 = R^3 = H$, $R^2 = CH_3$, α-Methyl acrolein
$R^1 = CH_3$, $R^4 = R^2 = R^3 = H$, Crotonaldehyde
$R^1 = R^2 = CH_3$, $R^4 = R^3 = H$, α,β-Dimethyl acrolein
$R^1 = C_2H_5$, $R^2 = C_3H_7$, $R^4 = R^3 = H$, α-Ethyl-β-propyl acrolein
$R^1 = Ph$, $R^4 = R^2 = R^3 = H$, Cinnamaldehyde
$R^1 = Ph$, $R^2 = CH_3$, $R^4 = R^3 = H$, α-Methyl cinnamaldehyde
$R^1 = Ph$, $R^2 = C_6H_{13}$, $R^4 = R^3 = H$, α-n-Hexyl cinnamaldehyde
$R^1 = C_6H_{11}$, $R^2 = H$, $R^3 = H$, $R^4 = CH_3$, Citral
$R^1 = C_4H_3O$, $R^2 = R^4 = R^3 = H$, β-2-Furyl acrolein
$R^1 = C_4H_3O$, $R^2 = R^4 = H$, $R^3 = CH_3$, β-2-Furylidene acetone
$R^1 = C_4H_3O$, $R^2 = R^4 = H$, $R^3 = C_6H_6$, β-2-Furylacrylophenone

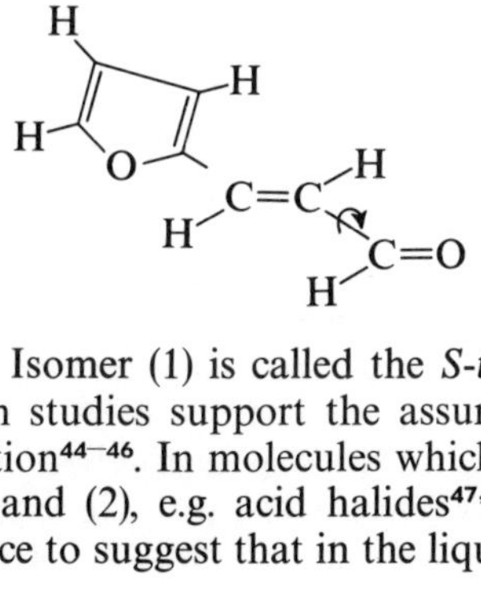

Figure 5.5 Rotational isomerism in α,β unsaturated carbonyl compounds

forms. Isomer (1) is called the *S-trans* form and a number of infrared and Raman studies support the assumption that this form has a planar conformation[44–46]. In molecules which have conformational isomers analogous to (1) and (2), e.g. acid halides[47, 48] and substituted butadienes[49], there is evidence to suggest that in the liquid-phase isomer (2) is non-planar. On the other hand, the microwave spectrum of furan-2-carbaldehyde[50], acrylic acid[51] and acryloyl fluoride[52] show that both isomers (1) and (2) are planar in the vapour phase. The shape of the potential energy function changes from twofold symmetry to threefold depending on whether (2) is planar or *gauche*. In the case of the threefold potential function, two of the energy states will be of equal energy. A wide range of molecules with the basic structure shown in Figure 5.5 have been examined[43, 53, 54]. The lack of detectable relaxation in certain of these molecules may be attributed to activation energies either greater than 40 kJ mol^{-1} or less than 10 kJ mol^{-1}.

The effects of conjugation on the activation energy for internal rotation can be appreciated when the series acrolein, crotonaldehyde and cinnamaldehyde is considered, Table 5.6. The increase in the activation barrier may be

Table 5.6 Energy parameters for unsaturated aldehydes and ketones

Molecule	$\Delta H^{\ddagger}_{12}$/kJ mol^{-1}	$-\Delta H^{\circ}$/kJ mol^{-1}	*Reference*
Acrolein	20.73	8.61	43
Crotonaldehyde	23.03	8.06	43
Cinnamaldehyde	23.49	6.27	53
Methyl acrolein	22.19	12.83	53
Furyacrolein	21.31	5.01	53
α-Methylcinnamaldehyde	21.52	5.4	53
α,β-Dimethylacrolein	23.19	5.4	53
2-Ethyl-3-propylacrolein	21.61	5.4	53
Citral	18.64	—	53
3(2-Furyl)acrylophenone	25.45	5.4	53
2-Furylideneacetone	29.63	5.8	53
α-n-Hexylcinnamaldehyde	19.72	6.6	53
Furan-2-carbaldehyde	45.98	5.4	56
Thiophene-2-carbaldehyde	43.05	5.4	56

attributed to increasing electron-donating properties of the substituent on the *trans* position of the β-carbon atom ($C_6H_5 > CH_3 > H$). A similar trend may be seen in the corresponding series α-methyl acrolein, α,β-dimethyl acrolein and α-methyl cinnamaldehyde. In this system, electrostatic attraction[55] between the α-methyl and the carbonyl groups may lower or raise the energy of the stable conformation state.

The effect of conjugation can be illustrated further by consideration of furan and thiophene-2-carbaldehyde[56]. The differences in observed barrier heights can be attributed mainly to the increased double-bond character of the C—C bond in the —C=C(X)—CHO group as a result of conjugation with the heterocyclic rings[57, 58].

In order to achieve some quantitative measure of the effects of conjugation in these molecules, an attempt has been made to relate acoustic parameters with the electron density on the carbonyl group as estimated from the magnitude of the ^{13}C—H coupling constant ($J(^{13}$C—H)). The ^{13}C coupling constant has been shown to be a good measure of the degree of s character[59–62] and so the extent of conjugation on the relevant carbon atom. In Figure 5.6 are plotted values of $J(^{13}$C—H) against the acoustic determinations of barriers for the more-stable to less-stable isomerisation. A reasonable correlation does exist between $J(^{13}C—H)$ and the barrier heights, a good indication that conjugation does play a definite role in determining the rotational barriers in these molecules.

Acoustic relaxation has also been detected in α,β-unsaturated aldehydes

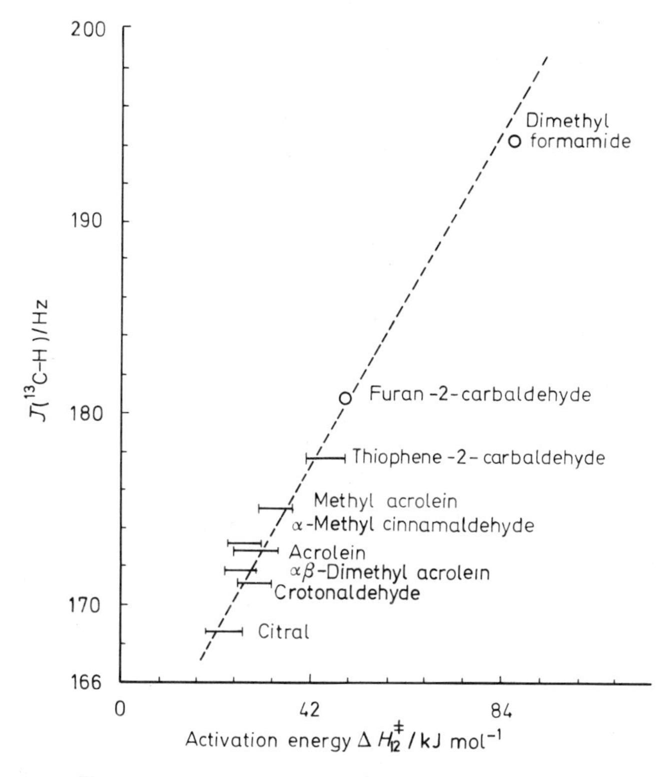

Figure 5.6 Correlation of $J(^{13}C—H)$ with barrier heights

substituted with a furan ring in the β position. It has been suggested that in this case the conformational equilibrium being perturbed is that involving rotation of the furan group[56] with respect to the α,β-unsaturated skeleton.

5.3.4 Vinyl ethers and related systems (sp²—heteroatom)

Relaxation has been observed in vinyl ethers[43] and attributed to rotational isomerism about the vinyl–oxygen linkage in these systems[63]. As with the α,β-unsaturated aldehydes, some uncertainty exists as to the precise geometries of the stable conformations. It may be proposed that the vinyl ethers should adopt both *cis* and *trans* structures. The energy difference in solution between the *trans* and the more stable *cis* form has been estimated spectroscopically as 2.5 kJ mol^{-1} [64–67] in methyl vinyl ether. The precise structure of the *cis* form still remains uncertain and it may or may not possess a completely planar structure. In ethyl vinyl ether the more stable structure has a *cis* form, adopting a 'sickle-shaped' configuration. An energy difference of 2.8 kJ mol^{-1} has been obtained from dipole moment and Kerr constant measurements[69].

Data on a number of vinyl ethers are presented in Table 5.7. N.M.R. studies[70] indicate that the less-stable isomer has the *gauche* conformation.

It is apparent from consideration of methyl vinyl ether and methyl vinyl sulphide (Table 5.7), that the effects of delocalisation of electron density may

Table 5.7 Energetics of internal rotation in vinyl ethers

Molcule	$\Delta H^{\ddagger}$/kJ mol^{-1}	*Reference*
Methyl vinyl ether	16.8	43
Ethyl vinyl ether	15.5	63, 68
2-Chloroethyl vinyl ether	13.4	63, 68
n-Butyl vinyl ether	24.4	63, 68
Methyl vinyl ketone	16.2	63, 68
3-Methyl but-3-ene-2-one	15.2	63, 68
Methyl vinyl sulphide	2.7	63, 68
2,3-Dibromopropene	16.8	63, 68

be subtle[63, 69, 70]. d-Orbital contribution may lead to stabilisation of structures not favoured in the case of methyl vinyl ether.

Dispersion behaviour has also been detected in *o*-chloroanisole and this may be attributed to the rotational isomeric process corresponding to that considered[55] above. The *o*-, *m*- and *p*-chloroanisoles have been studied spectroscopically and shown[71] to exhibit rotational isomerism. Estimates of the total potential function from torsional calculations indicate that the barriers to internal rotation are low. This is in agreement with the observation of relaxation only at low temperatures.

5.3.5 Esters

Esters were amongst the first molecules to be studied[72] using ultrasonic techniques. The acoustic relaxation in esters has been attributed to a perturbation of the equilibrium between the planar *cis* and *trans* forms of the molecule,

Figure 5.7 Rotational isomerism in esters

Figure 5.7. The chief difficulty in studying the relaxation behaviour in esters is that the characteristic frequencies lie in the range 100 kHz–10 MHz where the experimental accuracy has been low until recently. Some of the difficulties encountered have been overcome by working well above the normal liquid boiling point and maintaining the liquid state by application

Table 5.8 Energy parameters for rotational isomerisation of esters

Liquid	*State*	$\Delta H^{\ddagger}$/kJ mol^{-1}	$\Delta S^{\ddagger}$/JK^{-1} mol^{-1}	ΔH°/kJ mol^{-1}	ΔS°/JK^{-1} mol^{-1}	*References*
Methyl formate	Pure	32 ± 2	-36 ± 7	8–12	—	84
	Pure	32	-8	2	4	79
	6% in xylene	29 ± 4	-25 ± 12	8–12	—	84
Ethyl formate	Pure	33 ± 2	-37 ± 5	8–12	—	84
	Pure	26	-40	2	8	79
	6% in xylene	33 ± 3	-40 ± 8	8–12	—	84
	6% in nitrobenzene	31 ± 4	-33 ± 10	—	—	84
n-Propyl formate	Pure	28	-28	15	7	79
i-Propyl formate	7% in xylene	24 ± 2	-18 ± 5	15 ± 2	6 ± 3	73
Methyl acetate	Pure	25	-10	18	0	79
Ethyl acetate	Pure	18	-36	19	-4	79
	Pure	24	—	7	—	74
Methyl propionate	Pure	20	-42	21	7	79
Ethyl propionate	Pure	5	-77	24	6	79
Ethyl n-butyrate	Pure	2	—	—	—	79

of pressure[73]. The characteristic frequencies are then brought into the accurate experimental range (>5 MHz). This procedure is justified since it has been found[74] that the application of pressure does not alter the relaxation frequency of ethyl acetate at a given temperature. This observation also implies that the rotational isomerism does not involve a volume change.

Relaxation behaviour has been observed in a wide variety of esters with characteristic frequencies which lie between 2 and 5 MHz at room temperature[75–79]. An extensive study[79] of a number of esters has shown that the energy of the more stable isomer lies *c.* 37 kJ mol^{-1} below the transition state. The energy of the higher-energy conformation, in which the alkyl groups are *cis* to each other[80, 83] rises with increasing number of carbon atoms on both alkyl groups (Table 5.8). The results suggest that the energy differences arise from steric repulsion of the alkyl groups in the higher-energy conformations. This is supported by the work of Burundukov and Yakovlev on a series of n-alkyl formates[81, 82].

It is interesting to note that a second relaxation has been detected[73] well above the main relaxation region in isopropyl formate. This has been ascribed to rotation about the other C—O bond of the alcohol residue. The relaxation

Table 5.9 Two-state enthalpy parameters for di-esters[85] $ROOC(CH_2)_n$ COOR (in kJ mol)

Ester	*n*	ΔH° kJ mol^{-1}	$\Delta H^\ddagger_{21}$ kJ mol^{-1}	$\Delta H^\ddagger_{12}$ kJ mol^{-1}
Dimethyl oxalate	0	15.88	16.72	32.60
Dimethyl malonate	1	17.97	9.61	27.5
Dimethyl succinate	2	21.73	6.27	28.0
Dimethyl glutarate	3	22.15	4.59	26.75
Dimethyl adipate	4	7.10	3.76	10.86
Diethyl oxalate	0	2.09	25.08	27.1
Diethyl malonate	1	20.48	9.61	30.0
Diethyl succinate	2	21.31	6.27	27.6
Diethyl adipate	4	7.94	4.59	12.5

was found to occur at frequencies too high for the energy parameters to be evaluated.

A number of di-esters have been studied[85]. The results, Table 5.9, show that the enthalpy differences in dimethyl and diethyl malonates and succinates are very close to those found for the corresponding simple esters—methyl acetate and methyl propionate, respectively. In the case of the adipates it is possible to bring the O and C=O groups into a position where dipole–dipole interactions tend to stabilise a 'cyclic conformation', in which the internal rotation of the ester group is considerably hindered.

Relaxation has been observed also in methyl[86], ethyl, phenyl and isoamyl salicylate[87–89], and it has been suggested that this behaviour may be attributed to either intramolecular hydrogen bonding[89] or internal rotation[87, 88].

5.3.6 Amines

Measurements of the ultrasonic absorption in triethylamine[90] and tri-n-butyl amine[91] have revealed a relaxation process which has been attributed to rotational isomerisation. Triethylamine may be considered to exist in three possible conformations, Figure 5.8. This assignment is supported by the fact that no relaxation is observed in trimethylamine, where all the conformations formed by internal rotation about the C—N bond have the same energy. The intramolecular nature of the relaxation process has been

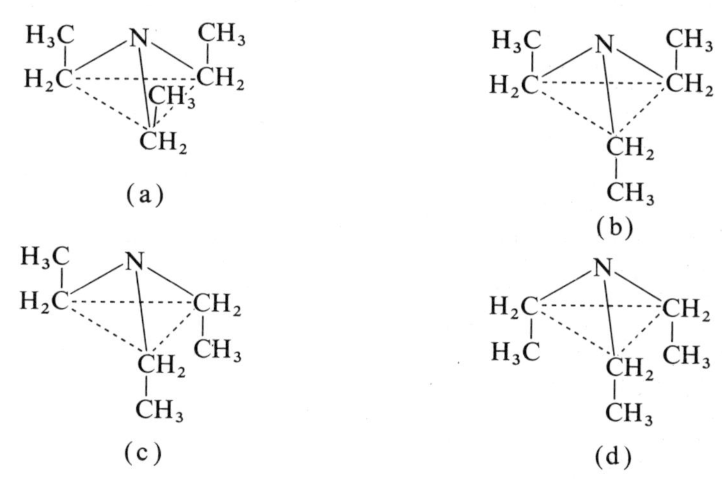

Figure 5.8 Possible conformations for trialkylamines

Table 5.10 Potential energy parameters for amines

Molecule	$-\Delta H^\circ$/ kJ mol^{-1}	$-\Delta S^\circ$/ J mol^{-1} K^{-1}	ΔH_{12}/ kJ mol^{-1}	*Reference*
Triethylamine	14.2	19.6	28.4	90
Triethylamine	24.6	54.3	45.1	96
Tripropylamine	6.3	4.6	18.8	101
Tributylamine	3.7	3.7	17.9	91
Tripentylamine	5.0	3.7	20.0	101
Trihexylamine	4.6	5.4	17.9	101
Tri-isopentylamine	4.6	4.6	12.5	101
Triallylamine	5.8	8.7	24.6	101
Tri(methylallyl)amine	5.4	7.9	17.9	101
3-*N*-Diethylaminopropiononitrile	5.0	2.9	18.8	101
3-*N*-Diethylaminopropylamine	5.4	7.5	26.3	101
4-*N*-Diethylaminobutan-2-one	5.8	8.36	27.2	101
N,*N*-Diethylcyclohexylamine	5.8	7.10	15.4	101
3-*N*-Diethylaminoethylamine	6.6	7.9	24.6	101
1-Diethylamino-2-chloropropane	4.6	1.67	29.2	101
3-*N*-Dipropylaminopropiononitrile	5.0	2.9	24.6	101

confirmed by the observation that the relaxation frequency is independent of pressure up to 3000 atm[92]. The latter data show also that the assumption $(\Delta V^0/V)(C_p/\theta) \ll \Delta H^0$ is applicable in these cases. Data for a number of amines are presented in Table 5.10. The choice of conformers responsible for the relaxation process is difficult since alternative physico-chemical evidence is sparse. From a steric point of view[90], the most stable conformers are a and b. In addition, these forms will be stabilised by the electrostatic attraction between the polarisable alkyl groups and the lone-pair electrons on the nitrogen. Thermochemical[93] and spectroscopic studies[94] indicate that triethylamine is a weaker base than trimethylamine and also primary and secondary amines. This is explained by postulating that at least one of the alkyl groups is folded back so that it projects towards the lone pair of electrons on the nitrogen atom. Thus the most likely conformers to take part in the relaxation process would be a and b.

Further study of molecular models has led to the reasoning that the entropy of a is greater than b. In the experimental results, the sign of the entropy difference indicates that the stable isomer changes from a to b on going from triethyl- to trihexyl-amine. In the higher tri-n-alkylamines parachor calculations[95] also suggest that the more stable form is the conformer b. It has been proposed[96] that both the entropy difference between the isomers and the entropy of activation are temperature-dependent as a result of coupling of the 'umbrella' motion with the rotation process[97].

It is worth noting that the barrier heights obtained here are of the same order of magnitude as those found for rotation about C—C bonds in substituted ethanes. The values for rotation about higher-order C—N bonds as in acid amides (from n.m.r. spectroscopy[98–100]) are much greater, being in the range 60–120 kJ mol^{-1}.

5.4 CYCLIC COMPOUNDS

No single molecule has played such a central role in the development of stereochemistry as cyclohexane. It is, therefore, somewhat surprising that the kinetic aspects of cyclohexane stereochemistry have not yet been completely elucidated. Certain of the energetics for the dynamic changes have been evaluated using n.m.r.[102], but these studies do not provide a complete picture of the potential function governing the conformational interconversion process.

5.4.1 Substituted cyclohexanes

To set the scene for further discussion of acoustic relaxation in cyclic molecules, the theoretical description of the cyclohexane ring inversion will be considered.

Semi-empirical strain calculations[103] indicate that planar cyclohexane, Figure 5.9, has an energy approximately 120 kJ mol^{-1} above the stable chair conformation. It is generally agreed that the energetically most economic pathway for the inversion process crosses a half-chair or 'cyclohexene'-like transition state, and that the motion continues to the metastable boat conformation. The boat, *c.* 24 kJ mol^{-1} above the chair, can alleviate part of the strain by distorting itself to a twist boat and this 'pseudo-rotation' can continue around the ring with great ease. The boat will eventually receive enough energy to cross the half-chair barrier again, thus returning either to the chair from which the whole sequence of events started or crossing over to the inverted chair. In cyclohexane itself the overall rate constant for the chair–chair interconversion, k_{cc}, is therefore just one-half the rate constant of the chair–boat process, k_{cb}.

The monosubstituted cyclohexanes will have differences not only in the ground-state energies of chair forms, but the presence of the substituent will lead to non-equivalent boats, twist boats and half-chairs.

Activation entropies may be discussed using calculations[104–106] based on statistical mechanical arguments. If one accepts the assumption that the interconversion of chair form and twist boat proceeds via the half-chair form, the theoretical entropy of activation for the cyclohexane chair-to-boat

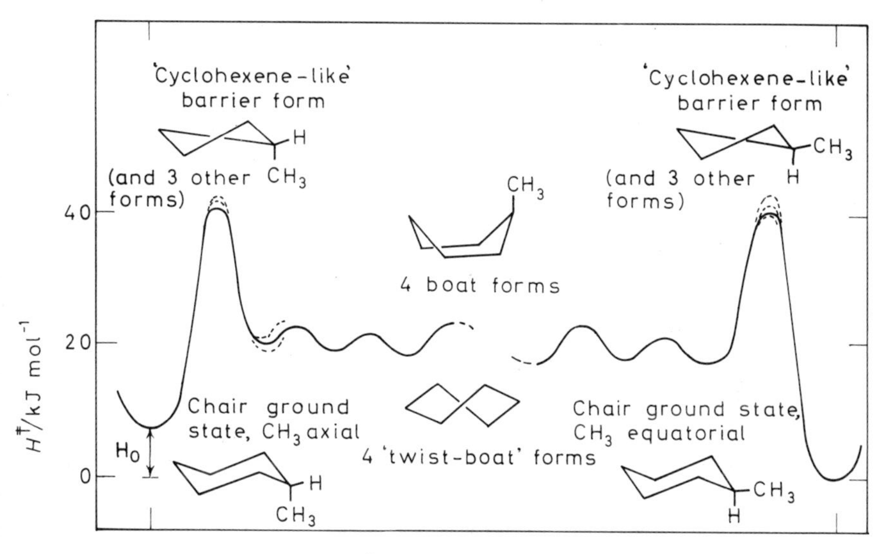

Figure 5.9 Conformational changes in methyl cyclohexane

conversion is bound within the limits $+15.1 \leqslant \Delta S^{\ddagger}_{cb} \leqslant 36$ J mol^{-1} K^{-1}. For substituted cyclohexanes, of course, this range has to be modified according to the appropriate symmetry properties. Specifically, for the mono- and 1,1-di-substituted cyclohexanes one obtains $5.9 \leqslant \Delta S^{\ddagger}_{cb} \leqslant 27$ J mol^{-1} K^{-1}. Acoustic relaxation has been observed in a number of substituted cyclohexane molecules, Table 5.11. The boat is less stable than the chair form due to interactions between the atoms at the top of the boat. In cyclohexane itself the enthalpy difference between the two forms is *c.* 23 kJ mol^{-1} while the energy barrier is *c.* 17 kJ mol^{-1}[111]. As a result, the interconversion is rapid and no ultrasonic relaxation has been observed in liquid cyclohexane at frequencies up to 100 MHz[77, 112]. However, detailed studies have been made on methyl- and ethyl-cyclohexane[113–115] and similar substituted cyclohexanes[112, 116]. The values obtained for chloro- and bromo-cyclohexane agree well with n.m.r.[117] and infrared estimates[118] (Table 5.11).

Extensive studies of *cis*-1,2-, *trans*-1,3- or *cis*-1,4-dimethylcyclohexane have failed to indicate the presence of relaxation. Each of these structures has one axial and one equatorial methyl group, and on ring inversion this yields again one axial and one equatorial substituent. The equilibrium is therefore between conformers of equal energy and is acoustically inactive. In the case of the *trans*-1,2-, *cis*-1,3- and *trans*-1,4-dimethylcyclohexanes, ring inversion

Table 5.11 Energy parameters for ring inversion in cyclic systems

Molecule	$-\Delta H°/$ kJ mol^{-1}	$-\Delta S°/$ J mol^{-1} K^{-1}	$\Delta H_{12}/$ kJ mol^{-1}	*Method*	*Reference*
Methylcyclohexane	12.2	13.0	43	Acoustics	107
	16.9	16.0	46	Acoustics	108
Chlorocyclohexane	—	—	50	Acoustics	108
Bromocyclohexane	—	—	50	Acoustics	108
Cyclohexane	—	—	43.2	n.m.r.	109
Perfluorocyclohexane	—	—	44.4	n.m.r.	110

converts two equatorial methyl groups into two axial groups, and the resulting energy difference leads to an acoustically active equilibrium. Relaxations have been reported in both *trans*-1,2- and *trans*-1,3- but not in the *cis*-1,3-dimethylcyclohexane.

5.4.2 Substituted cyclohexenes

Relaxation has been reported in cyclohexene[77, 115] but has been refuted by Subrahmanyam and Piercy[119]. The earlier observations may have been in error due to incorrect estimation of loss corrections at low frequencies.

Relaxations have been reported[120] in 3- and 4-substituted cyclohexenes, these molecules possessing non-isodynamic equilibrium states. The two possible forms of 2-methylcyclohexene are structurally, and presumably energetically, very similar which might explain the failure to observe a relaxation process in this system. The activation enthalpies, Table 5.12, show only slight variations with change of substituent.

According to the reaction scheme of Anet and Haq[121], the transition state during ring inversion in cyclohexene is a boat form, Figure 5.10, and when one half-chair form is converted into the other, the molecule passes through two such identical boat forms. Ignoring vibrational terms, the entropies of activation for the interconversion process in 3- and 4-substituted cyclohexenes should have values between R ln2 and R ln4 in excess of the half-chair form; the actual figure depends on the number of transition boat forms that the molecules proceed through during the ring inversion process. However, the negative entropies of activation found in the acoustic studies are incompatible with this description, and suggest[120] that there are changes in the low-frequency vibrational modes.

5.4.3 Substituted 1,3-dioxanes

In 1,3-dioxanes, two of the carbon atoms have been replaced by oxygen atoms in the basic cyclohexane structure (Figure 5.10). In 1,3-dioxane, the

Table 5.12 Energetics of conformational change in cyclohexene[120]

Molecule	ΔH_{21}/kJ mol^{-1}	ΔS_{21}/J mol^{-1} K^{-1}
3-Methylcyclohexene	13.8	—
4-Methylcyclohexene	14.2	−17.2
4-Vinylcyclohexene	19.2	—
4-Bromocyclohexene	21.7	−20.3

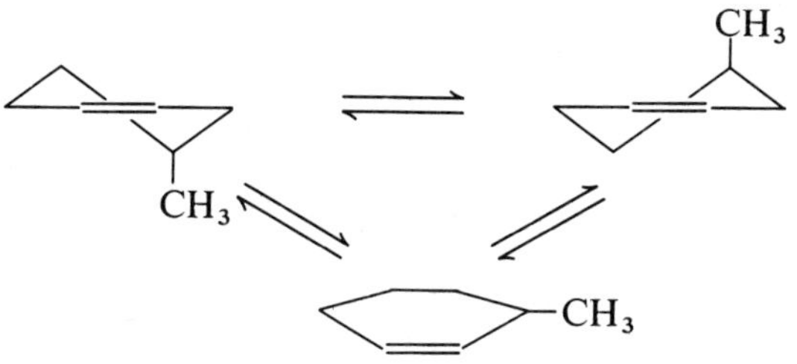

2-Methyl-cyclohexene

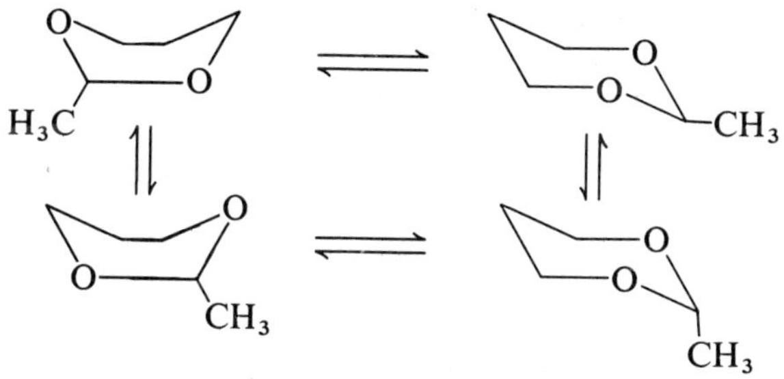

2-Methyl-1,3-dioxane

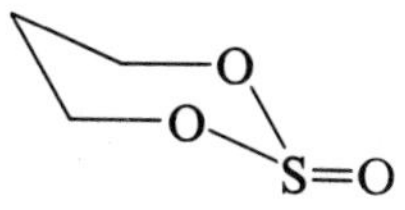

Cyclic sulphites

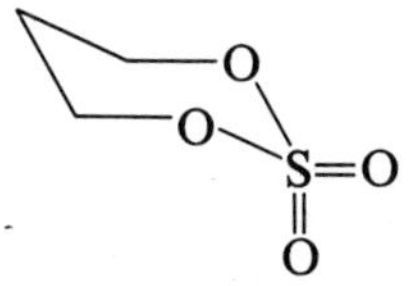

Cyclic sulphates

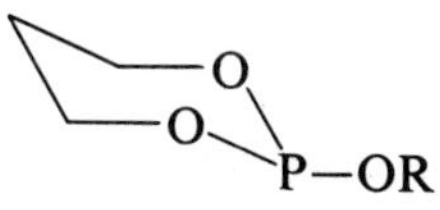

Phosphorinanes

Figure 5.10 Stable conformations and transition states for ring inversion in some cyclic systems

C—O bond is 0.01 nm shorter than the C—C bond in cyclohexane, so that there will be an increase in the interactions between axial substituents at the 2 position and the two axial protons at positions 4 and 6. The chair form in 1,3-dioxane was first suggested by Otto[123] on the basis of dipole moment measurements, and this has been supported by theoretical calculations[124].

Acoustic studies have been reported on a number of mono-, di- and tri-substituted 1,3-dioxanes[125]. The data obtained by analysis of the observed relaxation in terms of a single process are presented in Table 5.13. All the 2-substituted dioxanes showed evidence of relaxation. However, none of the disubstituted molecules were found[125, 126] to undergo dispersion in the temperature and frequency range examined. The absence of relaxation in these latter compounds indicates that the relaxation in ethyl-, n-propyl-, iso-propyl- and n-butyl-dioxane may not be attributed to rotational isomerism of the side chains.

On ring inversion the free-energy differences between axial and equatorial structures in the 2-substituted molecules have been shown to have typical values of *c.* 17 kJ mol^{-1}. In the 2,4-disubstituted molecules ring inversion results in axial methyl groups at the 2 and 4 positions, greatly magnifying the steric interactions. The energy differences have been estimated[127] to be greater than 23 kJ mol^{-1}.

Ring inversion occurs in a number of substituted molecules[128, 129]. The activation entropy in these molecules has a negative sign, which has been attributed[126] to a specific reaction path for the inversion process.

Examination of the acoustic activation enthalpies for ring inversion from the higher energy to the lower energy forms, indicates good agreement with the reverse barriers obtained from n.m.r., when correct allowance is made for the enthalpy difference between states. The low value of the barrier in 4-phenyl-1,3-dioxane can be explained in terms of deformation of the chair structure when the phenyl group is axial. This suggestion is supported by n.m.r. studies[130].

Although, in the above systems, the chair–twist boat equilibrium does not appear to contribute to the relaxation, there is good reason to think that it may be the primary cause for dispersion behaviour in certain other systems (Table 5.13). The assignment of the chair–twist boat equilibrium was made on the basis of a solution study of 2,2-dimethyl-1,3-dioxane. In this molecule, the chair–chair interconversion is isodynamic and hence should be acoustically inactive.

In the 2,2-dimethyl derivatives, 2-axial methyl to 4,6-axial ring-proton interactions occur in both of the chair forms and the inherent strain in the ring structure may be minimised by flattening of the ring at the 2 position. Such a process in the mono 2- and 4-substituted 1,3-dioxanes is considered to be a precursor to the formation of the activated structures involved in the ring inversion process. These conclusions, too, have been supported by n.m.r. studies[131]. Ring distortion leads[132, 133] to a lowering of the activation energy.

Ground-state destabilisation by the syn–axial interaction of the axial alkyl group occurs in 5,5-dimethyl-1,3-dioxane and 5-methyl-5-n-propyl-1,3-dioxane. Neither of these molecules showed evidence of acoustic relaxation in the temperature–frequency range examined. Eliel[134] has shown that

Table 5.13 Energetics of conformational change in 1,3-dioxanes

Molecule	$\Delta H^{\ddagger}$/ kJ mol^{-1}	$\Delta S^{\ddagger}$/ J mol^{-1} K^{-1}	E_a/ kJ mol^{-1}	$10^{11}A$/s^{-1}	ΔG°/ kJ mol^{-1}	Method	References
Chair–chair equilibria							
2-Methyl-1,3-dioxane	18.4	−36.4	20.5	1.82*	—	Acoustics	126
					16.7‡	Equilibration	127
2-Ethyl-1,3-dioxane	18.8	−34.7	20.9	2.94*		Acoustics	126
					16.9‡	Equilibration	127
2-n-Propyl-1,3-dioxane	20.9	−21.3	23.0	1.29*		Acoustics	126
2-Isopropyl-1,3-dioxane	20.1	−26.4	22.6	8.01*		Acoustics	126
					17.4‡	Equilibration	127
2-n-Butyl-1,3-dioxane	14.6	−26.4	16.3	5.42*		Acoustics	126
2-Phenyl-1,3-dioxane	15.5	−13.4	18.4	1.01*		Acoustics	
					13.1‡	Equilibration	127
1,3-Dioxane	42.7†	—	—	—		n.m.r.	128, 129
4-Methyl-1,3-dioxane	49.0†	—	—	—		n.m.r.	128, 129
4-Phenyl-1,3-dioxane	41.0†	—	—	—		n.m.r.	128, 129
Chair–twist boat equilibria							
2,2-Dimethyl-1,3-dioxane	19.0	− 8.5	21.2	11.8*		Acoustics	126
	26.4†	—	—	—		n.m.r.	128, 129
2,2,4-Trimethyl-1,3-dioxane	18.9	−12.5	21.0	31.2*		Acoustics	126
2-Ethyl-2-methyl-1,3-dioxane	12.97	−11.5	15.1	0.852*		Acoustics	126
2-Ethyl-2,4-dimethyl-1,3-dioxane	12.1	−37.7	13.8	1.27*		Acoustics	126
2,2-Diethyl-1,3-dioxane	10.5	−43.1	12.1	0.828*		Acoustics	126
2,2-Diethyl-4-methyl-1,3-dioxane	9.2	−46.0	10.5	0.315*		Acoustics	126
1,3-Dioxane-2-spirocyclopentane	19.4	−11.4	20.9	28.1*		Acoustics	126
4-Methyl-1,3-dioxane-2-spirocyclopentane	19.2	− 9.6	21.7	35·3*		Acoustics	126
5,5-Dimethyl-1,3-dioxane-2-spirocyclopentane	11.3	−50·2	13.4	36·2*		Acoustics	126
5,5-Dimethyl-1,3-dioxane-2,2-spirocyclohexane	11.7	−44.4	13.6	0·08*		Acoustics	126
2,2,5,5-Tetramethyl-1,3-dioxane	10.5	−50.2	12.6	0.397*		Acoustics	126
	38.1†	—	—	—		n.m.r.	128, 129
2-Chloromethyl-1,3-dioxane	19.7	−22.2	20.9	7·39*		Acoustics	126
5,5-Dimethyl-2-chloromethyl-1,3-dioxane	22.2	−18.8	24.3	18.1*		Acoustics	126
2-Methyltetrahydropyran	34.3	− 0.28	36.4	177·1*		Acoustics	126
2-Chloromethyltetrahydropyran	30.1	− 2.5	32.6	157.9*		Acoustics	126
2-Methoxy-1,3-dioxane	36.0	—	38.5	—		Acoustics	126
2-Methoxy-4-methyl-1,3-dioxane	17.2	−23.0	19.2	8.67*		Acoustics	126

*Acoustics measures the metastable to stable rate of interconversion.
†N.M.R. measures the stable to metastable rate of interconversion.
‡Equilibrium measurement of the energy difference between two stable conformations.

the free-energy difference between axial and equatorial alkyl groups in the 5 position of 1,3-dioxane is surprisingly small, *c.* 3.8 kJ mol^{-1}. Thus syn–axial interactions between an axial methyl group and the oxygen lone pairs are small compared to the interaction with syn–axial protons.

Relaxations associated with both ring inversion and side group rotation have been identified in three chloromethyl substituted, methylated-1,3-dioxanes[126].

5.4.4 Tetrahydropyrans

It has been suggested[135] that the substitution of one oxygen atom for a carbon atom in cyclohexane causes only minor ring distortion. Tetrahydropyran itself does not relax in the normal frequency range available for ultrasonic study[126]. However, 2-methyltetrahydropyran shows evidence of processes with high activation energy, as does 2-(chloromethyl)tetrahydropyran. The lack of an observed relaxation in tetrahydropyran suggests either that the activation energy is too large for the process to be detected or else, as in the case of cyclohexane, the chair–chair equilibrium is isodynamic and therefore acoustically inactive.

5.4.5 Cyclic sulphites and sulphates

Ultrasonic relaxation has been observed also in trimethylene sulphite and in several methylated derivatives[136–138]. The energy parameters are summarised in Table 5.14.

The assignment of the relaxation to an axial–equatorial S=O equilibrium may not be assumed without question. From analogy with cyclohexane, chair structures are assumed in which the S=O bond is axial. This assignment is based on low-temperature x-ray data[139], n.m.r. and dipole moment

Table 5.14 Energetics of conformational change in cyclic sulphites[138]

Molecule	$\Delta H°$/ kJ mol^{-1}	$\Delta S°$/ J mol^{-1} k^{-1}	$\Delta H^{‡}$/ kJ mol^{-1}	$\Delta S^{‡}$/ J mol^{-1} k^{-1}
Trimethylene sulphite	5.44	−10.46	23.01	−16.74
4-Methyltrimethylene sulphite	5.02	—	20.08	—
5,5-Dimethyl trimethylene sulphite	6.28	6.28	17.99	−29.29
5,5-Diethyl trimethylene sulphite	5.44	—	17.57	—
5-Methyl-5-ethyl trimethylene sulphite	5.86	—	17.99	—
4,6-Dimethyl trimethylene sulphite	5.02	—	24.27	—
4,5,6-Trimethyl trimethylene sulphite	5.02	—	15.06	—

measurements[140–142]. However, these data are unable to exclude[143] the possibility of twist forms of the molecule. It is not possible to investigate the path of the rotational inversion process by ultrasonic relaxation although a relatively large negative entropy of activation, compared with cyclohexane, suggests that one or two of the half-chairs are favoured. On the other hand, the effects of vibrational mode changes may account partially for the negative activation entropies observed in these molecules.

The activation energies for the ring inversion process tend to be somewhat lower than those found in the related 1,3-dioxane and cyclohexane series. The increased flexibility may be attributed to the fact that the sulphur atom, with its possibility of rehybridisation using low-lying d orbitals, modifies the barrier in such a way that a number of possible conformations can be assumed by the ring.

Acoustic studies on asymmetric sulphate ring structures such as 4-methyl-1,3,2-dioxathian-2,2-dione[144] indicate relaxation in this molecule. The process

Table 5.15 Activation parameters for —CH_2Cl group rotation when attached to cyclic systems

Molecule	$\Delta H^{\ddagger}$/ kJ mol^{-1}	E_a/ kJ mol^{-1}	*Reference*
2-(Chloromethyl)-1,3-dioxane	13.0	14.2	126
5,5-Dimethyl-2-(chloromethyl)-1,3-dioxane	13.4	14.6	126
4-Methyl-2-(chloromethyl)-1,3-dioxane	12.1	13.4	126
2-(Chloromethyl)tetrahydropyran	12.6	—	126
4-(Chloromethyl)-1,3-dioxan-2-one	20.9	—	146

may be attributed to a perturbation of the equilibrium between the equatorial and axial chair isomers, Figure 5.10.

The energy barrier for the equatorial axial ring inversion, Table 5.14, is significantly lower than for the analogous 1,3-dioxane and cyclohexane molecules. If the transition state is a half-chair form, then it is possible to describe a number of these forms all of which differ in energy, and so it is difficult to ascribe a precise path to the inversion process. As in other cases considered, the activation entropy does not appear to correlate with the value expected from theoretical considerations which neglect the vibrational contributions.

5.4.6 Phosphorinanes

A wide variety of cyclic phosphorus systems have been examined using acoustics[145, 147]. Relaxation has been observed in three of these. The liquids are viscous and the data have a high error, and no analysis of the data appears to have been attempted to date.

5.4.7 Rotation of chloromethyl side groups attached to ring structures

The relaxation process associated with the rotational isomerism of the chloromethyl group has been identified in a number of studies of cyclic systems[126, 146] (see Table 5.15). The activation parameters associated with this process compare well with those expected by comparison with the substituted ethanes. The side group rotations do appear to be independent of the process of ring inversion in these systems.

5.5 ROTATIONAL ISOMERISM IN NORMAL HYDROCARBONS

Study of the lower normal hydrocarbons acts as a natural introduction to the discussion of macromolecular systems. Low-temperature measurements on n-butane, n-pentane, n-hexane and n-heptane indicate[148] that rotational isomerism between *trans* and *gauche* forms of these molecules leads to acoustic dispersion phenomena. Relaxation has been observed also in 2-methylbutane, 2-methylpentane, 2,3-dimethylbutane and 3-methylpentane[149, 150].

Recent studies of the normal hydrocarbons from pentane to hexadecane in the frequency range 15–270 MHz indicate the presence of a molecular-weight-dependent high-frequency relaxation process[151]. As the molecular weight of the hydrocarbon increases, so the relaxation frequency appears to decrease. This type of behaviour is typical of normal-mode relaxation in polymers. From theoretical considerations it can be shown[152] that certain of the conformational changes in the low-molecular-weight hydrocarbons, considered as being co-operative, are identical in their description to those of the normal modes of long flexible chains. The acoustic activity of the normal modes considered in viscous relaxation may be rationalised when it is appreciated that certain of these modes of motion lead to a change in the end-to-end distance of the molecule and so in the energy. Also, in these short-chain molecules, Gaussian statistics do not apply and the traditional descriptions of normal mode motion will be somewhat different from that for flexible chains[153–155]. This interaction of the co-operative motion with that of the simple internal rotation process leads to complications in the exact analysis of observed absorption and dispersion behaviour.

5.6 ACOUSTIC STUDIES OF MACROMOLECULES

In the discussion of the possible modes of motion of a polymer chain, we can define three possibilities. The first is the case of independent rotational behaviour in flexible side chains. The second involves independent rotation at each of the rotatable units in the backbone and the third is a co-operative in-phase movement of all backbone units in the chain. This last process is analysed in terms of the normal modes of motion of the chain. A number of papers have appeared in the last 5 years outlining the effects of both side-chain and main-chain motion in polymers.

It has been proposed[155] that the excess absorption of polystyrene in carbon tetrachloride may be attributed to an internal chain rotation. From the temperature dependence of the characteristic frequency of the relaxation process, the barrier to the conformational change $\Delta H^{\ddagger}$ was found to be $27\,\mathrm{kJ\,mol^{-1}}$, while the energy difference between the conformations was $4\,\mathrm{kJ\,mol^{-1}}$.

A complicating factor which may affect interpretation of the results arises in the high-frequency 'tail' of the viscoelastic relaxation of the polymer. This may be expected to contribute to the observed absorption in polymer solutions. The expected magnitude of such an effect can be judged by the comparison of viscoelastic and longitudinal behaviour as studied in liquid polydimethylsiloxane[156,157]. In this case, the viscoelastic behaviour is known[156] and good agreement has been found between calculated and experimental values of sound absorption. A systematic study of a number of narrow-molecular-weight polystyrene samples in toluene solution has shown the effects of the viscoelastic relaxation of the polymer on the longitudinal absorption[158]. The previous report of molecular-weight independence of the absorption in polystyrene[155] may be attributed to the fact that the molecular-weight distribution in the two samples studied was not sufficiently narrow to indicate the influence of viscoelastic relaxation. The data obtained in both studies are in agreement for those samples of common molecular weight[155,158]. It is therefore evident that the absorption observed in the MHz region may not be due solely to backbone segmental motion.

Acoustic studies over the frequency range 1–10 MHz have been reported[159–161] on polystyrene in a variety of solvents, poly(methylmethacrylate), poly(vinyl chloride), poly(vinyl pyrrolidone), poly(vinyl acetate), poly (*p*-bromostyrene), poly(vinyl isobutyl ether) and poly(N-vinyl carbazole). However, the measurements were carried out only at one temperature and it is not possible to comment either on the origin of the relaxation or on the associated activation parameters. However, it would be expected that these processes may be correlated with similar observations from dielectric[162] and n.m.r. relaxation measurements[163].

The acoustic absorption of concentrated solutions of polystyrene in decalin might be described[164] by a double relaxation. It is significant that the concentration at which the relaxation ceases to be described by a single process also corresponds to the point at which entanglement occurs in the polymer solution. The effects of tacticity on the acoustic relaxation in polystyrene have been observed[165]. A similar study of 27 samples of poly(methylmethacrylate) prepared by various methods indicated that the magnitude and position of the dispersion were influenced by the tacticity[166].

Two relaxations have been reported in aqueous solutions of unneutralised poly(acrylic acid) and poly(methacrylic acid). These have been attributed to relaxation of hydrogen-bond association and interaction of water with the polymer[167].

It is evident that relaxation in polymers may be dependent on molecular weight, chain flexibility, tacticity of structure and the magnitude of the interaction of solvent with polymer. Indeed, relaxation due to polymer–solvent interaction has been identified[168–170] in poly(ethylene oxide) in water. It has been suggested[171,172] that the relaxation may be partly described by relaxation of the dynamic shear viscosity.

5.7 ACOUSTIC RELAXATION IN SYSTEMS OF BIOLOGICAL INTEREST

Although this review is concerned primarily with the dynamics of conformational change, relaxation phenomena can arise[173] from the presence of a fast chemical equilibrium. A number of reversible dissociation reactions are important in determining the structure of biomolecules and occur with an interchange frequency in the range accessible to observation by acoustic techniques. Outstanding among these are hydrogen-bond formation and acid–base association phenomena.

In the simple equilibria studied so far, a single relaxation may be observed and the forward and reverse rate constants may be defined by a single relaxation time.

$$A \rightleftharpoons B \quad \tau = (k_{12}+k_{21})^{-1}$$

However, in a dissociation equilibrium of the kind

$$NH_4OH \rightleftharpoons NH_4^+ + OH^-$$

the equilibrium constant K is given by

$$K = \sigma^2 c/(1-\sigma)$$

where c is the concentration of the ammonium hydroxide and σ the degree of dissociation. This process is characterised by a single relaxation time

$$\pi = [k_{12}+k_{21}(c_{NH_4^+}+c_{OH^-})]^{-1}$$

where c_i is the equilibrium concentration of the ion i. For dilute solutions:

$$\tau = [k_{12}+2k_{21}\sqrt{(Kc)}]^{-1}$$

Ultrasonic relaxation ascribable to such an equilibrium has been observed in ammonium hydroxide. The relaxation frequency increases from 22 to 55 MHz as the concentration of ammonium hydroxide increases from 0.5 to 2.5 mol dm^{-3}. A plot of τ^{-1} against $2\sqrt{(Kc)}$ yields a straight line whose slope, k_{21}, is 3.4×10^{10} s^{-1} mol^{-1}. This value is close to that predicted for a diffusion-controlled reaction. It is found that $k_{12} = 6\times10^5$ s^{-1}. The value of this dissociation rate constant shows that the reaction proceeds as shown rather than via the alternative route involving dissociation of water.

$$NH_3+H_2O = NH_3+H^+ + OH^- = NH_4^+ + OH^-$$

The volume change, ΔV^0, was found to be -28 cm^3 mol^{-1}, in excellent agreement with the value derived from partial molar volumes

$$\Delta V = V_{NH_4^+} + V_{OH^-} - V_{NH_3} - V_{H_2O} = -28 \text{ cm}^3 \text{ mol}^{-1}$$

Of particular relevance to biological studies are investigations of monomer-dimer equilibria. Studies of benzoic acid in carbon tetrachloride and toluene suggest that the rate constant for the dimerisation reaction is *c.* 10^9 s^{-1} mol l^{-1}, again indicative of a diffusion-controlled process. In the study of dimerisation and hydrogen-bond formation in biological macromolecules, benzoic acid[174], ε-caprolactam[175], and 2-pyrrolidine[175] are of particular interest as model systems.

The interaction of water with amines[176], dioxane[177], diglycine[178], polyethylene glycol[179–181] and polyglutamic acid[182] have been examined using dielectric and ultrasonic techniques. Studies of the addition of urea, guanidine chloride, ammonium chloride and sodium chloride to aqueous solutions of polyethylene glycol indicate the effects of these molecules on the local structure around the hydrophilic groups of the polymer.

Amino acids in solution may serve as model systems for polypeptide chains. The acoustic relaxation in aqueous solutions of glycine, diglycine and triglycine at the isoelectric point (pH 5–6) is independent of concentration and unaffected by the addition of protein denaturing agents and neutral salts. The observed relaxation has been attributed to perturbation of an equilibrium involving solvation of the zwitterion. The relaxation observed at a pH of *c.* 11 may be assigned to a proton transfer equation of the form

$$RH^+ + OH^- = R + H_2O$$

Similar studies[184,185] have been reported on L-leucine, L-aspartic acid, sodium glutamate, L-lysine hydrochloride, L-cysteine, L-cystine, L-tyrosine, 2-aminoethyl mercaptan and serine and threonine.

The helix-coil transformation, involving hydrogen-bond exchange processes, has been suggested as the mechanism contributing to ultrasonic relaxation in poly(L-lysine)[183] and poly(L-glutamic acid)[184–186].

5.8 CONCLUSION

Studies of acoustic relaxation provide useful information on the dynamics of conformational changes which occur within a frequency range from 10^5 to 10^9 Hz. The technique is most reliable in the provision of conformer interchange rates and the activation parameters between two stable forms. The technique can also provide the free-energy difference between stable conformers, although here care is necessary in the application of certain simplifying assumptions.

A particular advantage of the technique is that it can be applied to systems with a relatively small population of the higher-energy conformers. On the other hand, the acoustic observations of themselves yield no primary information on which energy changes are responsible for the observed relaxation. The details of the conformational change involved can be derived only by comparison with other methods of measurement or by deductive reasoning.

Such uses of the technique have been illustrated with reference to a number of aliphatic and cyclic systems.

It is interesting that ultrasonic irradiation is now being used in connection with molecular conformation in a quite different way. It has been noted that irradiation of a solution in which a carbanionic reaction is taking place may lead to modifications in the stereochemistry of the reaction products. This has been demonstrated[187] in the anionic polymerisation of methylmethacrylate, and it has been suggested that the sound energy perturbs the equilibrium association of the propagating carbanion and its gegen ion.

It is evident that the next few years will see a marked increase in the use of acoustic radiation in studies of molecular conformation.

References

1. Eliel, E. L. and Alinger, N. L. (1967–71). *Topics in Stereochemistry,* Vol. 1–6, (New York: John Wiley)
2. Pethrick, R. A. and Wyn-Jones, E. (1969). *Quart. Rev. Chem. Soc.,* **23,** 301

3. Sheppard, N. (1959). *Advan. Spectroscopy,* **1,** 288
4. Mountain, R. D. (1966). *J. Chem. Phys.,* **44,** 832
5. Herzfeld, K. F. and Litovitz, T. A. (1959). *Absorption and Dispersion of Ultrasonic Waves,* (New York: Academic Press)
6. Lamb, J. (1956). *Physical Acoustics,* Vol. 2, Part 1 (W. P. Mason, editor) (New York: Academic Press)
7. Matheson, A. J. (1971). *Molecular Acoustics,* (London: Wiley-Interscience)
8. Lambert, J. D. (1967). *Quart. Rev. Chem. Soc.,* **21,** 67
9. Gordon, R. G., Klemperer, W. and Steinfeld, J. I. (1968). *Ann. Rev. Phys. Chem.,* **19,** 215
10. Eigen, M. and De Maeyer, L. (1963). *Technique of Organic Chemistry,* Vol. 8, Part 2, 895 (S. L. Freiss, E. S. Lewis and A. Weissberger, editors) (New York: New York)
11. Beyer, R. T. and Letcher, S. V. (1969). *Physical Ultrasonics,* (New York, London: Academic Press)
12. Thompson, D. S., Newmark, R. A. and Sederholm, C. H. (1962). *J. Chem. Phys.,* **37,** 411
13. Newmark, R. A. and Sederholm, C. H. (1965). *J. Chem. Phys.,* **43,** 602
14. Alger, T. D., Gutowsky, H. S. and Vold, R. L. (1967). *J. Chem. Phys.,* **47,** 3130
15. Padmanabhan, R. A. (1960). *J. Sci. Ind. Res. (India),* **19B,** 336
16. Chen, J. H. and Petrauskas, A. A. (1959). *J. Chem. Phys.,* **30,** 304
17. Wyn-Jones, E. and Orville-Thomas, W. J. (1968). *Trans. Faraday Soc.,* **64,** 2907
18. Clark, A. E. and Litovitz, T. A. (1960). *J. Acoust. Soc. Amer.,* **32,** 1221
19. Krebs, K. and Lamb, J. (1958). *Proc. Roy. Soc. (London),* **A244,** 558
20. Crook, K. R., Park, P. J. D. and Wyn-Jones, E. (1969). *J. Chem. Soc. A,* 2910
21. Crook, K. R. and Wyn-Jones, E. (1969). *J. Chem. Phys.,* **50,** 3448
22. Crook, K. R., Wyn-Jones, E. and Orville-Thomas, W. J. (1970). *Trans. Faraday Soc.,* **66,** 1597
23. Pethrick, R. A. and Wyn-Jones, E. (1968). *J. Chem. Phys.,* **49,** 5349
24. Ouellette, R. J. and Williams, S. (1971). *J. Amer. Chem. Soc.,* **93,** 466
25. Bagley, E. B., Nelson, T. P., Chen, S. A. and Barlow, J. W. (1971). *Ind. Eng. Chem. Fundam.,* **10,** 27
26. Piercy, J. E. (1966). *J. Chem. Phys.,* **43,** 4066
27. Pethrick, R. A. and Wyn-Jones, E. (1971). *J. Chem. Soc. A,* 54
28. Weigert, F. J., Winstead, M. B., Garrels, J. I. and Roberts, J. D. (1970). *J. Amer. Chem. Soc.,* **92,** 7359
29. Piercy, J. E. and Seshagiri Rao, M. G. (1967). *J. Chem. Phys.,* **46,** 3951,
30. Park, P. J. D. and Wyn-Jones, E. (1969). *J. Chem. Soc., A,* 646
31. Park, P. J. D. and Wyn-Jones, E. (1968). *J. Chem. Soc. A,* 2064
32. Whiteside, G. M., Sevenair, J. P. and Goetz, R. W. (1967). *J. Amer. Chem. Soc.,* **89,** 1135
33. Westheimer, F. H. (1956). *Steric Effects in Organic Chemistry,* 533, (M. S. Newmann, editor), (New York: Wiley)
34. Szasz, G. J. (1955). *J. Chem. Phys.,* **10,** 2449
35. Hawkins, B. L., Bromster, N., Borcic, S. and Roberts, J. D. (1971). *J. Amer. Chem. Soc.,* **93,** 4472
36. Thomas, T. H., Wyn-Jones, E. and Orville-Thomas, W. J. (1969). *Trans. Faraday Soc.,* **65,** 974
37. Mizushima, S., Shimanouchi, T., Nakamura, K., Hayashi, M. and Tsuchiya, S. (1957). *J. Chem. Phys.,* **26,** 970
38. Gates, P. N., Mooney, E. F. and Willis, H. A. (1967). *Spectrochim. Acta,* **23A,** 2043
39. Shipman, J. J., Folt, V. L. and Krimm, S. (1962). *Spectrochim. Acta,* **18,** 1603
40. Caraculacu, A., Stokr, J. and Schneider, B. (1964). *Collect. Czech. Chem. Commun.,* **29,** 278
41. Eigen, M. and Tamm, K. (1962). *Z. Electrochem.,* **66,** 93, 107
42. Karabatos, G. J. and Fenoglio, D. J. (1970). *Topics in Stereochemistry,* Vol. 5, 1967. (E. L. Eliel and N. L. Allinger, editors), (New York, London: Wiley-Interscience)
43. de Groot, M. S. and Lamb, J. (1957). *Proc. Roy. Soc. (London),* **A242,** 36
44. Cherniak, E. A. and Costain, C. C. (1966). *J. Chem. Phys.,* **45,** 104
45. Suzuki, M. and Kozima, K. (1969). *Bull. Chem. Soc. Japan,* **42,** 2183
46. Bowles, A. J., George, W. O. and Maddams, W. F. (1969). *J. Chem. Soc. B,* 810
47. Khan, A. J. and Jonathan, N. (1969). *J. Chem. Phys.,* **50,** 1801
48. Not allocated
49. Hartman, K. O., Carlson, G. L., Witkowski, R. E. and Fateley, W. G. (1968). *Spectrochim. Acta,* **A24,** 157

50. Moenning, F., Dreizler, H. and Rudolph, H. D. (1965). *Z. Naturforsch*, **20,** 1323
51. Bolton, K., Owen, N. and Sheridan, J. (1968). *Nature (London)*, **218,** 266
52. Keirns, J. J. and Curl, R. F. (1968). *J. Chem. Phys.*, **48,** 3773
53. Pethrick, R. A. and Wyn-Jones, E. (1970). *Trans. Faraday Soc.*, **66,** 2483
54. Khabiullaev, P. K., Aliev, S. S. and Parpiev, K. (1969). *Zh. Fiz. Khim.*, **43**(7), 1894
55. Pethrick, R. A. Unpublished data.
56. Pethrick, R. A. and Wyn-Jones, E. (1969). *J. Chem. Soc. A*, 713
57. Fieser, L. and Fieser, M. (1963). *Topics in Organic Chemistry*, 63, (New York: Reinhold)
58. Dahlquist, K. I. and Forsen, S. (1965). *J. Phys. Chem.*, **69,** 4062
59. McFarlane, W. (1969). *Quart. Rev. Chem. Soc.*, **23,** 187
60. Mooney, E. F. and Winson, P. H. (1969). *Ann. Rev. N.M.R. Spectrosc.*, **2,** 197, (London: Academic Press)
61. Grant, D. M. and Litchmann, W. H. (1965). *J. Amer. Chem. Soc.*, **87,** 3994
62. Orville-Thomas, W. J. (1966). *The Structure of Small Molecules*, 87, (Amsterdam: Elsevier)
63. Crook, K. R. (1970). *Ph.D Thesis*, University of Salford
64. Lide, D. R. and Christensen, D. (1961). *J. Chem. Phys.*, **35,** 1374
65. Owen, N. L. and Sheppard, N. (1964). *Trans. Faraday Soc.*, **60,** 636
66. Owen, N. L. and Sheppard, N. (1966). *Spectrochim. Acta*, **22,** 1101
67. Cahill, P. A., Grold, L. P. and Owen, N. L. (1968). *J. Chem. Phys.*, **48,** 1620
68. Crook, K. R. and Wyn-Jones, E. Unpublished data.
69. Aroney, M. J., Le Fevre, R. J., Ritchie, G. L. D. and Saxby, J. D. (1969). *Aust. J. Chem.*, **22,** 1559
70. Hatada, K., Takishita, M. and Yuke, H. (1968). *Tetrahedron Lett.*, 4621
71. Owen, N. L. and Hester, R. E. (1969). *Spectrochim. Acta*, **25A,** 343
72. Biquard, P. (1936). *Ann. Physik*, **6,** 195
73. Piercy, J. E. and Subrahmanyam, S. V. (1965). *J. Chem. Phys.*, **42,** 1475
74. Slie, W. M. and Litovitz, T. A. (1963). *J. Chem. Phys.*, **39,** 1538
75. Pancholy, M. and Mathur, S. S. (1963). *Acoustica*, **13,** 42
76. Tabuchi, D. (1958). *J. Chem. Phys.*, **28,** 1014
77. Kapovich, J. (1954). *J. Chem. Phys.*, **22,** 1767
78. Huddart, D. H. A. (1950). *M.Sc. Thesis*, University of London
79. Bailey, J. and North, A. M. (1968). *Trans. Faraday Soc.*, **64,** 1499
80. Thompson, H. W. and Torkington, P. (1945). *J. Chem. Soc.*, 640
81. Burundukov, K. M. and Yakovlev, V. F. (1968). *Zh. Fiz. Khim.*, **42,** 2149
82. Burundukov, K. M. and Yakovlev, V. F. (1969). *Akust. Zh.*, **15,** 295
83. Curl, R. F. (1959). *J. Chem. Phys.*, **30,** 1529
84. Subrahmanyam, S. V. and Piercy, J. E. (1965). *J. Acoust. Soc. Amer.*, **37,** 340
85. Bailey, J., North, A. M. and Walker, S. (1970). *J. Molec. Struct.*, **6,** 53
86. Tanaka, Y. (1970). *Acoustica*, **23,** 328
87. Chandrasekhara, R. P. and Rao, K. S. (1970). *J. Phys. Soc. Jap.*, **29,** 1652
88. Chandrasekhara, R. P. and Rao, K. S. (1970). *Curr. Sci.*, 556
89. Yasunago, T., Tatsumota, N., Miura, M. and Inoue, H. (1969). *J. Phys. Chem.*, **73,** 477
90. Heasell, E. L. and Lamb, J. (1956). *Proc. Roy. Soc. (London)*, **A237,** 233
91. Krebs, K. and Lamb, J. (1958). *Proc. Roy. Soc. (London)*, **A244,** 558
92. Litovitz, T. A. and Carnevale, E. A. (1958). *J. Acoust. Soc. Amer.*, **30,** 134
93. Brown, H. C. (1956). *J. Chem. Soc.*, 1248
94. Tames, M., Searles, S., Leighly, E. M. and Mohrman, D. W. (1954). *J. Amer. Chem. Soc.*, **76,** 3983
95. Arbuzov, B. A. and Guzhavina, L. M. (1948). *Dokl. Akad. Nauk. SSSR*, **61,** 63
96. Padmanabhan, R. A. and Heasell, E. L. (1960). *Proc. Phys. Soc.*, **76,** 321
97. Barfield, M. and Baldeschwieler, J. D. (1964). *J. Chem. Phys.*, **41,** 2633
98. Mislow, K. and Raban, M. (1967). *Topics in Stereochemistry*, (N. Allinger and E. L. Eliel, editors), Vol. 1, 1, (New York: Interscience)
99. Conti, F. and von Philipsborn, W. (1967). *Helv. Chim. Acta*, **50,** 603
100. Pritchard, F. E. and Orville-Thomas, W. J. (1962). *J. Molec. Spectrasc.*, **6,** 572
101. Williams, E. J., Thomas, T. H., Wyn-Jones, E. and Orville-Thomas, W. J. (1968). *J. Molec. Struct.*, **2,** 307
102. Binsch, G. (1968). *Topics in Stereochemistry*, (E. L. Eliel and N. L. Allinger, editors), Vol. 3, 97, (New York: Interscience)

103. Davidson, N. (1962). *Statistical Mechanics,* Chapt. 4, (New York: McGraw-Hill)
104. Laidler, K. J. and Polanyi, J. C. (1965). *Progr. Reaction Kinetics,* **3,** 1
105. Beckett, C. W., Pitzer, K. S. and Spitzer, R. (1947). *J. Amer. Chem. Soc.* **69,** 2488
106. Allinger, N. L. and Freiberg, L. A. (1960). *J. Amer. Chem. Soc.,* **82,** 2393
107. Piercy, J. E. (1961). *J. Acoust. Soc. Amer.,* **33,** 198
108. Piercy, J. E. and Subrahmanyam, S. V. (1965). *J. Chem. Phys.,* **42,** 1475
109. Harris, R. K. and Sheppard, N. (1967). *J. Molec. Spectroscopy,* **23,** 231
110. Gutowsky, H. S. and Chen, F. M. (1965). *J. Phys. Chem.,* **69,** 3216
111. Eliel, E. L. (1962). *Stereochemistry of Carbon Compounds,* (New York: McGraw-Hill)
112. Heasell, E. L. and Lamb, J. (1956). *Proc. Phys. Soc. B,* **69,** 869
113. Lamb, J. and Sherwood, J. (1955). *Trans. Faraday Soc.,* **51,** 1674
114. Hall, D. N. (1955). *Trans. Faraday Soc.,* **55,** 1319
115. Pedinoff, M. E. (1962). *J. Chem. Phys.,* **36,** 777
116. Piercy, J. E. (1961). *J. Acoust. Soc. Amer.,* **33,** 198
117. Reeves, L. W. and Stromme, K. O. (1960). *Can. J. Chem.,* **38,** 1241
118. Pentin, Y. A., Shatipov, Z., Kotova, G. G., Kamernitskii, A. V. and Akhrem, A. A. (1963). *Zh. Struct. Khim.,* **4,** 194
119. Subrahmanyam, S. V. and Piercy, J. E. (1965). *J. Chem. Phys.,* **42,** 1845
120. Crook, K. R. and Wyn-Jones, E. (1971). *Trans. Faraday Soc.,* **67,** 660
121. Anet, F. A. L. and Haq, M. Z. (1965). *J. Amer. Chem. Soc.,* **87,** 3147
122. Bogatskii, A. V. and Gareovik, N. L. (1968). *Russ. Chem. Rev.,* **37,** (4)
123. Otto, M. M. (1937). *J. Amer. Chem. Soc.,* **59,** 1590
124. Arbuzov, B. A. (1960). *Bull. Soc. Chim. Fr.,* 1311
125. Hamblin, P. C., White, R. F. M., Eccleston, G. and Wyn-Jones, E. (1969). *Can. J. Chem.,* **47,** 2731
126. Eccleston, G. (1970). *Ph.D. Thesis,* University of Salford
127. Nader, F. W. and Eliel, E. L. (1970). *J. Amer. Chem. Soc.,* **92,** 10, 3050
128. Anderson, J. E. and Brand, J. C. D. (1960). *Trans. Faraday Soc.,* **62,** 517
129. Anet, F. A. L. and Brown, A. J. R. (1967). *J. Amer. Chem. Soc.,* **89,** 760
130. Eliel, E. L. (1970). *Accounts Chem. Res.,* **3,** 1
131. Anteunis, M. and Tavernier, D. (1971). *Tetrahedron,* **27,** 1677
132. Freibolin, H., Schmid, H. G., Kabuss, S. and Faisst, W. (1969). *Org. Mag. Res.,* **1,** 67
133. Anderson, J. E. (1965). *Quart. Rev. Chem. Soc.,* **19,** 426
134. Eliel, E. L. and Knoeber, Sr. M. C. (1968). *J. Amer. Chem. Soc.,* **90,** 3444
135. Hassel, O. (1953). *Quart. Rev. Chem. Soc.,* **7,** 221
136. Pethrick, R. A., Wyn-Jones, E., Hamblin, P. C. and White, R. F. M. (1967–1968). *J. Molec. Struct.,* **1,** 333
137. Hamblin, P. C., White, R. F. M. and Wyn-Jones, E. (1968). *Chem. Commun.,* 1058
138. Eccleston, G., Pethrick, R. A., Wyn-Jones, E., Hamblin, P. C. and White, R. F. M. (1970). *Trans. Faraday Soc.,* **66,** 310
139. Altona, C., Geise, H. J. and Romers, C. (1966). *Rec. Trav. Chim.,* **85,** 1197
140. Arbuzov, B. A. (1960). *Bull. Chim. Soc. Fr.,* 1311
141. van Woerden, H. F. (1964). *Thesis,* Leiden
142. Hamblin, P. C. (1968). *Ph.D. Thesis,* University of London
143. Albiktsen, P. (1971). *Acta Chem. Scand.,* **25,** 478
144. Pethrick, R. A., Wyn-Jones, E., Hamblin, P. C. and White, R. F. M. (1969). *J. Chem. Soc. A,* 1638
145. Crook, K. R. (1970). *Ph.D. Thesis,* University of Salford
146. Pethrick, R. A., Wyn-Jones, E. and White, R. F. M. (1969). *J. Chem. Soc. A,* 1852
147. Chambers, G. (1967). *Ph.D. Thesis,* University of London
148. Piercy, J. E. and Seshagiri Rao, M. G. (1967). *J. Chem. Phys.,* **46,** 3951
149. Young, J. M. and Petrauskas, A. A. (1956). *J. Chem. Phys.,* **25,** 1943
150. Chen, J. H. and Petrauskas, A. A. (1959). *J. Chem. Phys.,* **30,** 304
151. Cochran, M. A., Jones, P. B., North, A. M. and Pethrick, R. A. Unpublished data.
152. Jain, D. V. S. and Pethrick, R. A. Unpublished data.
153. Harris, R. A. and Hearst, J. E. (1966). *J. Chem. Phys.,* **44,** 2595
154. Fixman, M. and Stockmayer, W. H. (1970). *Advan. in Phys. Chem.,* 407
155. Hassler, H. and Bauer, H. J. (1969). *Kolloid Z.,* **230,** 194
156. Barlow, A. J., Harrison, G. and Lamb, J. (1964). *Proc. Roy. Soc. (London),* **A282,** 228
157. Hunter, J. L. and Derdul, P. R. (1967). *J. Acoust. Soc. Amer.,* **42,** 1041

158. Cochran, M. A., North, A. M. and Pethrick, R. A. Unpublished data
159. Candau, S. (1964). *Ann. Phys.,* **9,** 271
160. Cerf, R., Zane, R. and Candau, J. (1961). *C. R. Acad. Paris,* **252,** 681, ibid., 252, 2229, (1962). *ibid.,* 1061
161. Cerf, R. (1965). *Proc. 5th Int. Con. Acoust.,* Paper C13, Liege
162. Block, H. and North, A. M. (1970). *Advan. Mol. Relaxation,* **1,** 309
163. McCall, D. W. (1971). *Accounts Chem. Res.,* **4,** 223
164. Nomura, H., Kato, S. and Miyahara, Y. (1971). *Nippon Kagaku Zasshi,* **91,** 1042
165. Nomura, H., Kato, S. and Miyahara, Y. (1968). *Nippon Kagaku Zasshi,* **89,** 149
166. Michels, B. and Zana, R. (1969). *Kolloid Z-Z Polymer.,* **234,** 1008
167. Nomura, H., Kato, S. and Miyahara, Y. (1969). *Nippon Kagaku Zasshi,* **90,** 250
168. Hames, G. G. and Lewis, T. B. (1966). *J. Phys. Chem.,* **70,** 1610
169. Zane, R., Cerf, R. and Candau, S. (1963). *J. Chim. Phys.,* **60,** 869
170. Kessler, L. W., O'Brien, W. D. and Dunn, F. (1970). *J. Phys. Chem.,* **74,** 4096
171. Hawley, S. A. and Dunn, F. (1969). *J. Chem. Phys.,* **50,** 3523
172. Bloomfield, V. and Zimm, B. H. (1966). *J. Chem. Phys.,* **44,** 315
173. Eigen, M. and de Maeyer, L. (1963). *Techniques of Organic Chemistry,* Vol. 8, (S. L. Friess, E. S. Lewis and A. Weissberger, editors), (New York: Interscience)
174. Tamm, K., Kurtze, G. and Kaiser, R. (1954). *Acoustica,* **4,** 380
175. Pethrick, R. A., Grimshaw, D. and Wyn-Jones, E. Unpublished data
176. Bergmann, K., Eigen, M. and de Maeyer, L. (1963). *Ber. Bunsegnes, Physik. Chem.,* **67,** 819
177. Hammes, G. G. and Knouche, W. (1966). *J. Chem. Phys.,* **45,** 4041
178. Andreae, J. H., Edmonds, P. D. and McKellar, J. F. (1965). *Acoustica,* **15,** 74
179. Hammes, G. G. and Lewis, T. B. (1966). *J. Phys. Chem.,* **70,** 1610
180. Burke, J. J., Hammes, G. G. and Lewis, T. B. (1965). *J. Chem. Phys.,* **42,** 3520
181. Hammes, G. G. and Swann, J. C. (1967). *Biochemistry,* **6,** 1591
182. Hammes, G. G. and Schimmel, P. R. (1967). *J. Amer. Chem. Soc.,* **89,** 442
183. Parker, R. C., Slutsky, L. J. and Applegate, K. R. (1968). *J. Phys. Chem.,* **72,** 3177
184. Inoue, H. (1970). *J. Sci. Hiroshima Univ. Ser., A-2,* **34,** 17
185. White, R. D. and Slutsky, L. J. (1971). *J. Phys. Chem.,* **75,** 161
186. Burke, J. J., Hammes, G. G. and Lewis, T. B. (1965). *J. Chem. Phys.,* **42,** 3520
187. Osawa, Z., Kimura, T. and Kasuga, T. (1969). *J. Polymer Sci.,* **A1, 7,** 2007

6
Gas Electron Diffraction

K. KUCHITSU
The University of Tokyo, Japan

6.1 INTRODUCTION

Electron diffraction is one of the most powerful techniques for determining 'molecular structure' in the gas phase (particularly, so-called 'molecular geometry', meaning bond distances, valence angles and dihedral angles). The present review, however, is intended to cover only a very limited range of the recent problems investigated with this method, namely, the combined use of electron-diffraction (abbreviated here as ED) and spectroscopic (SP) methods. This selection is partly because of the limits of space and of the existence of several excellent review articles[1–4] covering other fields of gas electron diffraction, including newly-developing projects such as electron distributions in atoms and molecules, intramolecular multiple scattering, inelastic scattering, high-temperature measurements and intermolecular interactions. Nevertheless, the subject of this Review seems never to have been surveyed in detail, in spite of its increasing importance as a tool for the precise investigation of the structures of free molecules.

In another review complementary to the present article[5], Cyvin and the Author have discussed various definitions of the geometrical parameters of free molecules and how they can be derived from ED and SP experiments. The merits and limitations of these methods for determining molecular geometry[6] have also been discussed. In short, they measure entirely different observables related to molecular geometry: ED provides the probability distribution of internuclear distances in a one-dimensional chart (radial distribution), whereas SP provides rotational constants, which are inversely proportional to the moments of inertia. The combined use of the ED and SP methods can provide, in a suitable case, more accurate geometrical parameters than either method alone. This is particularly important from the standpoint of ED. Since ED measurements are sometimes subject to serious systematic errors[2,7], the aid of accurate SP data, if available, can improve the accuracy of ED analysis significantly.

Historically, a critical comparison of ED and SP data was made as early as the thirties. With only a few exceptions, internuclear distances and angles determined by ED experiments were reported to be compatible with the corresponding parameters derived from SP (microwave, infrared, or Raman) experiments, and the ED and SP methods were thought to complement each other. ED and SP data were sometimes combined to facilitate analysis and derive more reliable structures. For example, five independent parameters of trifluoromethylacetylene were determined by the simultaneous use of ED intensities and the rotational constants of four isotopic species measured by microwave spectroscopy[8]. An advantage of ED–SP coordination was explicitly demonstrated in this pioneer work.

Studies of intramolecular motions by ED–SP combination started in the 1950s with the development of the sector-microphotometer method[9]. Frequencies of normal vibrations and potential constants (harmonic and anharmonic) derived from SP have been used to interpret ED intensities, from which complementary information on internal motions is obtained usually in terms of mean-square vibrational amplitudes of internuclear distances. The details of the theory and a list of papers relevant to the experimental and theoretical studies of mean-square amplitudes and 'shrinkage effects' are given in Ref. 10. Molecules with large-amplitude vibrations (internal rotation, etc.) have also been studied as an extension of this scheme[2]. A brief account of this problem will be given at the end of this chapter (Section 6.4).

The theory of molecular vibrations plays an important part in the precise determination of molecular geometry, since a probability distribution of internuclear distances averaged over thermal vibrations is measured by ED, while in many of the SP experiments information on molecular geometry comes from the rotational constants for the ground vibrational state involving the influence of zero-point vibrations[11]. Therefore, distance and angle parameters (vibrational averages) derived from the ED and SP methods are not quite equivalent and may sometimes differ significantly[12,13]. A scheme has been presented to define and correlate the distance parameters currently in use in the ED (r_g, r_a, r_α) and SP (r_e, r_z, r_s, r_0) methods with one another[5]. The first part of this article describes a number of applications of this principle to experimental studies of molecular geometry. For the sake of clarity, the significance of relevant distance parameters is given in Table 6.1.

Table 6.1 Symbols and definitions of various distance parameters*

Symbol	*Definition*
r_e	Distance between equilibrium nuclear positions
r_z, r_α^0	Distance between average nuclear positions (ground vibrational state)
r_v	Distance between average nuclear positions (excited vibrational state)
r_α	Distance between average nuclear positions (thermal equilibrium)
r_g	Thermal-average value of internuclear distance
r_0	Effective distance derived directly from ground-state rotational constants
r_s	Effective distance derived from isotopic rotational constants by use of Kraitchman's equations (Ref. 64)
r_a	Constant argument in molecular scattering intensity of electron diffraction

*See Ref. 5 for more complete details.

6.2 DETERMINATION OF MOLECULAR GEOMETRY: A GENERAL SCHEME

From the standpoint of the effective use of the ED and SP methods, we may group molecules into the following four categories:

(a) ED is the only possible means for determining precise molecular geometry in the gas phase
(b) SP is much more effective than ED
(c) and (d) The joint use of ED and SP is possible and should be profitable.

6.2.1 Molecules of Category a

Many of the non-linear, non-polar molecules with relatively large moments of inertia belong to this category: for example, inorganic molecules with high symmetry (X_4, planar XY_3 except for BF_3, XY_4, XY_5, XY_6, XY_7, etc.) and non-polar, medium-sized molecules such as neopentane, cyclohexane, etc. At present, no precise rotational constants for these molecules can be determined by SP; they have no pure rotational spectra in the microwave region, while currently available infrared, ultraviolet and Raman spectrometers have not sufficient resolution to observe rotational fine structures. On the other hand, their structures can easily be determined by ED because of their good symmetry (simple radial distribution). For instance, structural parameters for a large number of metal fluorides have recently been studied by ED[1,2].

6.2.2 Molecules of Category b

For molecules in this category (diatomic, linear and bent XY_2, NH_3, CH_4, etc.), high-resolution spectroscopy can provide more precise structures (an order of magnitude or more) than ED, although the latter by itself can determine the structure unambiguously. The symmetry is such that the number of observable rotational constants is equal to that of independent geometrical parameters. (Pyramidal XY_3 molecules other than ammonia should, in

Table 6.2 Molecules of Category b studied by electron diffraction*

Diatomic molecules (16)

N_2(17), NO(17), O_2(17), Cl_2(17,18), I_2(19)

Polyatomic molecules (20–22)
Hydrides–deuterides
CH_4,CD_4(23.24), NH_3,ND_3(25–27), H_2O,D_2O(28,29)

Non-hydrides
CO_2(30), CS_2(31), SO_2(32), BF_3(33)

*References are given in parentheses. Only those molecules for which high-resolution spectroscopy has given unique, precise structures (r_e, r_z, and r_g) are listed. Electron-diffraction studies have been made by the sector–microphotometer method for the purpose of critical examination of the experimental structure, accuracy and anharmonic effects.

principle, also be included. However, rotational constants around the molecular axis (A or C) for such symmetric tops have only rarely been determined accurately.) For a number of other molecules besides the above list (such as HCN, OCS, NF_3), sufficient numbers of isotopic rotational constants extrapolated to equilibrium have been measured, and precise equilibrium

(r_e) structures have been determined[14]. Therefore, these molecules should be classified into this category.

In the above circumstances, the SP structure can be used in ED studies as an excellent standard for calibration and for a critical check of experimental accuracy. This procedure is particularly effective when anharmonic as well as harmonic potential constants for the molecule are known by SP experiments[15], so that the conversion among various geometrical parameters[5], say from r_g to r_z, can be made with sufficient accuracy. A significant discrepancy in some of the structural parameters, if discovered, indicates the presence of a systematic error in the experimental and/or interpretational procedures of ED. Although much less likely, a misassignment in the SP analysis may be discovered as a cause of discrepancy. The molecules listed in Table 6.2 have so far been investigated for this purpose by the ED method[16–33]. Some of the results are described in more detail in the next section.

6.2.3 Molecules of Category c

For molecules in this category, SP supplies a precise set of rotational constants (usually for the ground vibrational state), yet they are not sufficient to determine a unique structure unless a part of the geometrical parameters are assumed, or unless the rotational constants for other isotopic species are combined. On the other hand, ED by itself can determine a unique structure.

Table 6.3 Molecules of Category c studied by electron diffraction*

Inorganic molecules

H_2S_2(34), B_2H_6†, B_2D_6†(6,35), ONF_3(36), PF_3(37), PBr_3(38), OPF_3(39), $OPCl_3$(39), $SPCl_3$(39), SPF_3(40), AsF_3(41,42), $AsCl_3$(42), $SbCl_3$(43), $FClO_3$(44), $XeOF_4$‡(45)

Organic molecules

acetaldehyde†(46,47), acetone(48,49), carbon suboxide(50), cyanogen(51), diacetylene(52), dimethylacetylene(53), ethane(6,54), ethylene(55,56), glyoxal‡(57), methyl vinyl ether†(58), t-butyl chloride‡(59), propane†(60), propene†(61), trifluoropropene†(61), acetyl chloride†(62), acetyl bromide†(62), cyclobutane(63)

*Only those references are given for which rotational constants derived from ED geometrical parameters have been compared critically with SP rotational constants.

†Assumptions about hydrogen parameters are necessary in the ED analysis. For other molecules the ED structures have been determined without assumptions.

‡Least-squares analyses have been made on the ED intensities and the rotational constants measured by high-resolution spectroscopy as independent observations.

Typical molecules of this category, which have recently been studied by ED in this context[34–63], are listed in Table 6.3.

In some cases, the r_s structure can be derived by means of Kraitchman's method from a set of isotopic rotational constants for molecules of this category[64]. This is the best structure attainable by the SP method alone; it is usually precise and consistent to *c*. 10^{-3} Å, although the physical significance does not necessarily compare with this precision[5]. Otherwise, a conventional 'r_0 structure' (derived with or without additional assumptions about geometrical parameters), which approximately reproduces the observed rotational constants, is reported.

Customarily, electron diffractionists compare the distance and angle parameters derived from ED (r_g, r_a or whatever) directly with the corresponding r_s or r_0 parameters. However, the r_g and r_s parameters for the same internuclear distance can differ from each other by more than 0.01 Å[6, 59, 65, 66]; this exceeds the limit of error of the ED parameter.

For a really critical comparison, it seems more straightforward to compare the rotational constants calculated from the ED structure with those determined by SP, since it is more difficult, at present, to convert r_s or r_0 parameters into the r_g parameter with high accuracy than to calculate rotational constants from the r_g parameters. Corrections for vibrational effects on rotational constants[6, 33] are small, or even negligible, unless the molecule in question has very small moments of inertia and/or very large amplitudes of internal motion. Accordingly, an approximate estimate of the quadratic force constants is sufficient for this purpose (see Section 6.2.4).

Nevertheless, a comparison of this type is limited by the accuracy of ED rotational constants; they usually have much larger uncertainties than those

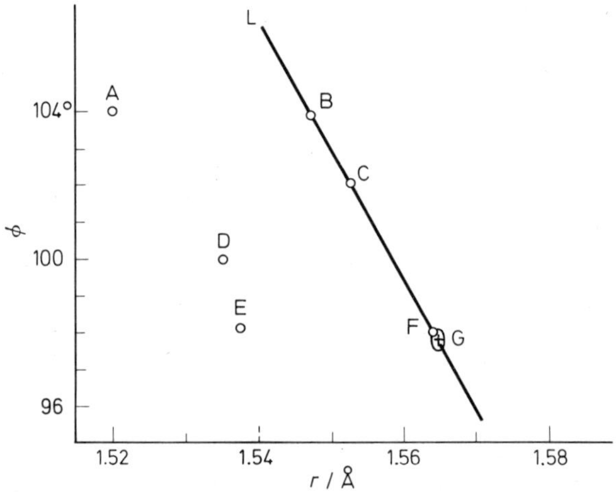

Figure 6.1 Structural parameters for PF_3 (r for the P—F distance and ϕ for the F—P—F angle)[37]. The most probable set and the limit of uncertainty, shown by G, are determined by means of ED. It is consistent with the SP observation since the point G is on the line L representing a nearly linear relation set by the rotational constant B_z determined by microwave spectroscopy. A to F correspond to past experiments
(From Morino *et al.*[37], by courtesy of the American Chemical Society)

of the SP rotational constants, because the random and systematic errors in the ED parameters are 'magnified' in this procedure. As a crude measure, relative uncertainties in the rotational constants (B) and in the distance parameters (r) are related as $|\sigma_B/B| \approx 2|\sigma_r/r|$. A parameter diagram[37], which demonstrates the degree of consistency of ED and SP experiments, is also effective for this comparison (Figure 6.1). SP data can be used as a standard for a calibration of the scale factor, i.e. the overall size of the molecule, or for a check of experimental consistency.

If desired, one may make a conditional least-squares analysis of ED intensities and adjust the variable parameters (particularly, the scale factor) to make the ED rotational constants agree with those of SP [45, 57, 59]. The SP rotational constants may be taken into the analysis as additional observables with appropriately chosen relative weights, by which the constraints can be set either very rigid or loose.

6.2.4 Molecules of Category d

For the rest of the molecules, neither ED nor SP can provide a unique set of geometrical parameters without the use of assumptions, except for the r_s structure which may or may not be derived from isotopic substitution. For this category, a typical pattern of the ED–SP combination is as follows[57]:

(a) Quadratic force constants for the molecule under investigation are determined or estimated from available information on vibrational spectra. Mean-square amplitudes of vibration and small corrections for vibrational effects on the distance parameters and on the rotational constants are calculated by the use of the force constants. For a semi-rigid molecule, even a crude set of these estimates are of considerable help in the ED analysis[67]. The corrections on the rotational constants may be negligible for a molecule of a medium size as compared with experimental uncertainties.

(b) The symmetry of the molecule (linearity, planarity, centre of symmetry, etc.) is determined by vibrational and/or rotational spectroscopy. For instance, a small, negative inertia defect may be interpreted as evidence for a planar structure in equilibrium[68]. Such information is a powerful aid in the ED analysis, since the number of independent parameters can be decreased on account of symmetry.

(c) As mentioned in Section 6.2.3, the rotational constants are used to set firm constraints on adjustable parameters in the ED analysis. Rotational constants for other isotopic species may also be taken into the analysis if sufficiently correct estimates of the isotope effect on geometrical parameters are at hand. (For instance, deuterium-substituted species may be used to improve the accuracy of hydrogen parameters[6, 69].) The use of SP rotational constants in the ED analysis seems to be preferable to the direct transfer of distance and/or angle parameters derived from the SP method, since the rotational constants carry *all* the original information on molecular geometry attainable from the SP experiment.

(d) If the molecule has a large-amplitude motion, the above-mentioned procedure has to be modified. A general theory has recently been worked out[47] for evaluating vibrational corrections to the effective moments of inertia of a molecule containing symmetric internal rotors derived from microwave spectroscopy. In principle, such a treatment can be extended to other large-amplitude motions discussed in Section 6.4.

In this way (and only in this way), the ED method can determine accurately the internuclear distances that differ very slightly from one another (with separations comparable with their mean vibrational amplitudes). Typical examples of the molecules in this category, which have recently been studied in this context, are cyclohexadiene[70], acrolein[57, 69], butadiene[57], vinylacetylene[71], acrylonitrile[65], propynal[66], and isobutene[72]. For instance, the C=O and C≡C bond distances (r_g) in propynal, which are almost equidistant (1.214 and 1.211 Å, respectively), have been determined by this procedure[66].

6.3 APPLICATIONS TO THE STUDIES OF MOLECULAR GEOMETRY

6.3.1 Diatomic molecules

A general theory of the intensity of fast electrons (*c.* 40 keV) scattered from a diatomic molecule is well established[1–4, 16, 73–76]. The so-called 'molecular intensity', $M(s)$, is given in terms of the parameters r_a, l_m and κ, defined in Refs. 16 and 73. They are related to the constants characterising the potential function and represent the effective internuclear distance, the effective root mean square amplitude of vibration and the asymmetry parameter, respectively;

$$sM(s) \propto \exp(-\tfrac{1}{2}l_m^2 s^2) \sin s(r_a - \kappa s^2) \tag{6.1}$$

Experimental molecular intensities for molecules listed in Table 6.2 were analysed by a least-squares method by the use of the asymmetry parameters estimated from anharmonic constants[16, 17]. The r_a and l_m parameters derived from the analysis were reduced to the equilibrium parameters, r_e and l_e, by the use of the theoretical corrections for vibrational effects. The results agreed with the corresponding SP values within experimental standard errors. For the iodine molecule, the variations in the distance and amplitude parameters with temperature[77] were investigated[19].

6.3.2 Hydrogen–deuterium isotope effects and anharmonicity for methane

A critical examination of the ED and SP structures and of the H–D isotope effect on ED intensities was made for methane and deuteromethane[23]. The r_g distances for the C—H bond in CH_4 and for the C—D bond in CD_4 were determined by ED to be $1.106_8(1_2)$ Å and $1.102_7(1_3)$ Å, respectively*, showing an isotope dependence. A schematic explanation of these isotope effects may be made by considering a model potential curve for a one-dimensional anharmonic oscillator exerting a totally symmetric stretching vibration ν_1 (Figure 6.2). A more sophisticated calculation reported in Ref. 24 in terms of the normal coordinates leads to essentially the same conclusion. The vibrational frequency is so high [$\omega_1(CH_4) = 3143\ cm^{-1}$ and $\omega_1(CD_4) = 2224\ cm^{-1}$] that all the systems are practically in the ground vibrational state. Since the zero-point energy and the mean-square amplitude are both proportional to $m_H^{-\frac{1}{2}}$, the ratio of the C—H and C—D r.m.s. amplitudes, l_e(C—H)/l_e(C—D), should be equal to $(m_H/m_D)^{\frac{1}{4}} = 1.19$ according to this model. The observed ratio, 1.14(4), is in reasonable agreement. An approxi-

*The numbers in parentheses represent quoted experimental errors to be attached to the last significant figures. For instance, 1.107(1) represents 1.107 ± 0.001.

mate formula[5, 12, 21] $\langle \Delta r \rangle \approx \frac{3}{2} a_3 \langle \Delta r^2 \rangle$, may be used to estimate the mean displacement of the C—H bond distance, $\langle \Delta r \rangle$, due to the skewness of the potential curve. Since the degree of skewness[16, 78], a_3, of the C—H (C—D)

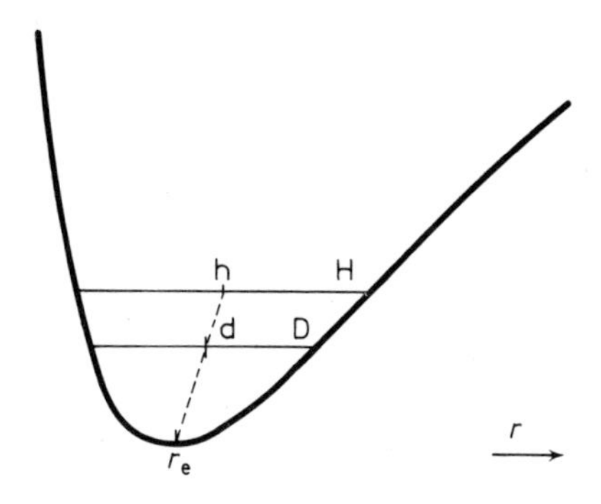

Figure 6.2 A model potential curve of a one-dimensional anharmonic oscillator for CH_4 and CD_4 (exaggerated). Both molecules are essentially in their ground vibrational states at room temperature. The C—H bond in CH_4 (H) has *c.* $\sqrt{2}$ times as much zero-point energy as that of the C—D bond in CD_4 (D), and hence a larger vibrational amplitude. The r_g(C—H) distance (h) is accordingly longer than the r_g(C—D) distance (d). The equilibrium position r_e can be estimated by an extrapolation of the r_g distance to the infinite-mass limit

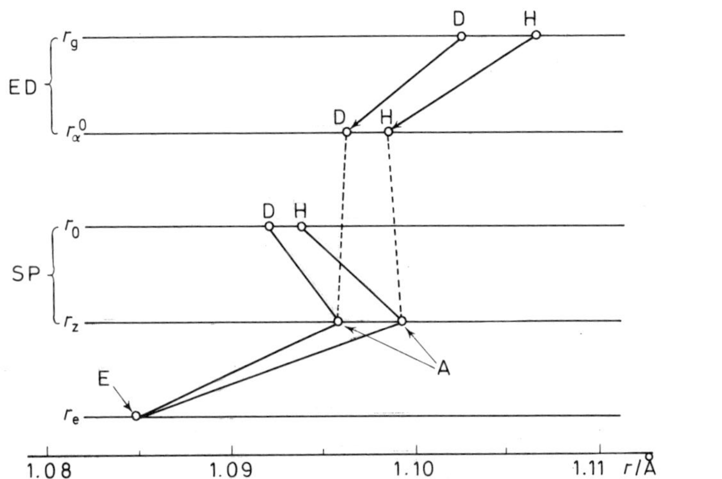

Figure 6.3 The carbon–hydrogen distance parameters of methane. Symbols H and D denote the C—H distance in CH_4 and the C—D distance in CD_4, respectively. Experimental values derived from ED and SP measurements, r_g and r_0 respectively, are different from each other by *c.* 0.01 Å because of the inherent difference in their physical significance. The corrected distances which represent the zero-point average structure (A), r_α^0 and r_z, are in agreement within experimental uncertainties of the order of 0.002 Å. The r_e distance (E) is estimated by a consideration of anharmonicity. See Refs. 23 and 24

bond is estimated to be[23] *c.* 2.6 Å^{-1}, the isotope difference in the r_g distance is expected to be 0.004–0.005 Å. This agrees with the above experimental result. Consequently, the observed isotope effect on the mean-square amplitude is related to the mass dependence of the (harmonic) zero-point energy, while that on the mean bond distance is due to anharmonicity of the bond stretching vibration.

As for the comparison of ED and SP structures, the r_g distances given above (determined by ED) are *c.* 0.01 Å longer than the corresponding SP distances[79, 80], r_0 (C—H) 1.09403(16) Å and r_0 (C—D) = 1.09181(29) Å. After the conversions[5, 81, 82], from r_g to r_α^0 and from r_0 to r_z (see Table 6.1 for definitions), they agree with each other within experimental errors, as shown in Figure 6.3. The r_e (C—H) distance is estimated either by extrapolation from r_g [83] (see Figure 6.2) or by the use of the a_3 parameter given above to be 1.085_0 Å [24]. The effect of anharmonicity of the C—H stretching vibration has also been manifested in the phase parameter κ(C—H) in equation (6.1). The a_3 (C—H) parameter derived from the ED experiment[23], 3.1(5) $Å^{-1}$, is in reasonable agreement with the theoretical estimate, 2.6 $Å^{-1}$.

6.3.3 Isotope effects for other hydrides

Similar effects of hydrogen–deuterium differences in the distances and amplitudes and the effects of anharmonicity have been studied by ED for ammonia[25–27], water[29], methylamine[84], ethylene[6, 55, 56], ethane[6, 54] and diborane[6, 35].

For ethylene and ethane, two independent rotational constants (A_0 and B_0) obtained from infrared and Raman spectroscopy are not sufficient to determine the three independent structural parameters (C—C, C—H and

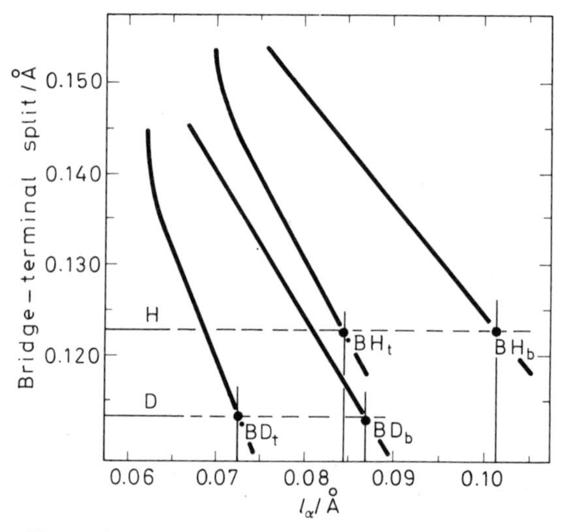

Figure 6.4 Correlation diagram derived from ED experiments for the bridge (b) and terminal (t) B—H distances in B_2H_6 (B_2D_6). (Modification of the original Figure 6 of Ref. 35 according to the discussion in Ref. 6.) Differences in the b and t distances are plotted against their mean amplitudes of vibration. By the use of calculated mean amplitudes (vertical bars) based on SP observations, the b–t splits for B—H and B—D are estimated to be 0.123 and 0.113 Å, respectively
(From Bartell and Carroll[35], by courtesy of the American Institute of Physics)

∠C—C—H) uniquely. (The C_0 constant for ethylene is dependent since the molecule is planar.) The zero-point average rotational constants (B_z^α, $\alpha = A$

and *B*) for the hydride and deuteride were therefore combined, and approximate account was taken of the isotope effect on the r_z parameters[5]. The parameters were determined by a further combination with the r_α^0 parameters obtained from ED, which gave consistent results (with a small adjustment of the ED scale factor in the case of ethane). The accuracy of the SP structures determined depends almost entirely on that of the estimation of the r_z isotope effects. Because of uncertainties from this origin and from the ED experiment, the C—C bond distances in C_2H_6 and C_2D_6 were not determined with sufficient accuracy to confirm a difference (predicted secondary isotope effect[54,85]). A slight secondary isotope effect was also suggested in a recent experimental study of t-butyl chloride and its fully deuterated analogue[59].

In a similar analysis of the B_2H_6 and B_2D_6 structures[6], it was possible to determine the bridge and terminal B—H bond distances separately from the ED intensities[35] by the use of calculated mean amplitudes (Figure 6.4). The r_g(ED) distances (B—B and B—H) are found to be *c.* 0.01 Å longer than the corresponding $r_s(r_0)$ distances determined by infrared spectroscopy[86].

6.3.4 Use of electron-diffraction parameters for spectroscopic assignments

In SP measurements, geometrical parameters determined by ED are often used to calculate preliminary estimates of the rotational constants as a guide for making assignments of spectral lines. For instance, the N—N bond

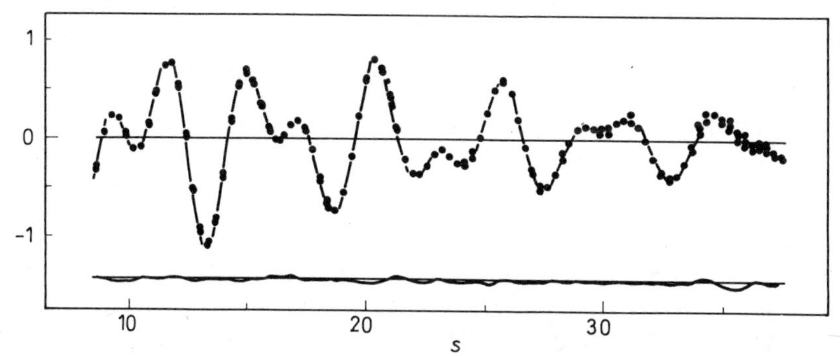

Figure 6.5 Reduced molecular intensity for BF_3 [33]. Values observed from three photographic plates are shown in dots. The solid curve represents the theoretical intensity based on a best-fit structure model, which is very nearly equal to the spectroscopic structure[89]. The lower curve is the difference between the observed and theoretical curves for one of the photographic plates showing the magnitude of random experimental error (From Kuchitsu and Konaka[33], by courtesy of the American Institute of Physics)

distance in hydrazine determined by ED was helpful for infrared spectroscopists to make correct assignments[87].

For boron trifluoride, BF_3, the r_0(B—F) distance was reported from the analysis of the ν_2 infrared band to be[88] 1.295_1 Å, which appeared to be in agreement with the preceding ED parameters, 1.30(2) and 1.31(3) Å. However, a more recent ED measurement[33] turned out to be incompatible with the above r_0 value, since the observed B—F distance ($r_g = 1.313_3(1_0)$ Å) should correspond, according to the above-mentioned conversion scheme to a rotational constant B_0 of 0.3449(6) cm^{-1} (for $^{10}BF_3$ and $^{11}BF_3$), about 2% smaller than that obtained by the infrared analysis (0.3527$_2$ cm^{-1}). This discrepancy was removed by a change in the J assignments[33]. In a more recent infrared measurement with higher resolution[89], B_0 values consistent with the ED value (0.3447$_4$(5) for $^{10}BF_3$ and 0.3450$_5$(5) cm^{-1} for $^{11}BF_3$) were obtained. Figure 6.5 compares the experimental ED molecular intensity with the theoretical curve based on the SP constants.

For phosphorus tribromide, PBr_3, a significant discrepancy was observed in the Br—P—Br angle determined by a recent ED measurement[38] (101.0(4) degrees) with that reported in a previous SP study (106(3) degrees), which gave a set of rotational constants inconsistent with those calculated from the ED geometry. This motivated a remeasurement of the microwave spectrum, which has led to a set of isotopic B_0 constants consistent with the ED geometry[38].

6.3.5 Regularities in the geometry of Group V halides

Group VB halides are one of the series of inorganic molecules that have been studied systematically by the combination of ED and SP methods. The internuclear distances and bond angles in MX_3 and YPX_3, where M = N, P, As and Sb, X = F, Cl, Br and I, and Y = O and S, are listed in Tables 6.4 and 6.5, respectively[90–98].

The following regular trends have been observed in the geometrical parameters (r_g distances and r_α angles) of these molecules[37,39]. Note that some bromide and iodide structures were determined many years ago and have much larger uncertainties.

(a) The M—X bond distances in MX_3 satisfies the empirical rule proposed by Schomaker and Stevenson[99],

$$r(\text{M—X}) = r_M + r_X - c\,|\,x_M - x_X\,|, \tag{6.2}$$

where r_M and x_M are the covalent radius and the electronegativity of the atom M given by Pauling[100], respectively, and c is an empirical parameter. With the parameter c taken as 0.08 Å for all the M—F bonds and 0.04 Å for the rest of the M—X bonds (slightly different from the original prescriptions of Schomaker and Stevenson and of Pauling), the estimated bond distances (except for the N—X in NX_3) agree with the observed r_g values to within 0.03 Å.

(b) The bond angles X—M—X in MX_3 increase from fluorides to iodides and decrease from nitrogen to antimony compounds. Thus, NCl_3 and SbF_3 have the largest and the smallest bond angles in Table 6.4, respectively. Until recently, the F—M—F angles in PF_3 and AsF_3 were erroneously thought to be several degrees larger, but the revised values fit the above regularity[37,38,41,42].

(c) Coordination of a sulphur atom to the lone pair of PX_3 makes the P—X bond distance decrease and the bond angle X—P—X increase by c. 0.03 Å and 2 degrees, respectively, whereas that of an oxygen atom causes larger variations in the same direction, c. 0.05 Å and 3 degrees.

(d) The P—O and P—S bond distances in YPF_3 are shorter than those in

Table 6.4 Internuclear distances and bond angles for Group VB halides*

r(M—X)		M = N	P	As	Sb
X = F	obs.	1.365(2)e 1.371(2)0	1.570(1)g	1.706(2)g 1.710(1)g	1.879(4)0
	calc.†	1.30	1.59	1.69	1.88
Cl	obs.	1.759(2)g	2.040(1)g	2.165(1)g	2.333(3)g
	calc.	1.73	2.05	2.16	2.36
Br	obs.		2.220(3)g	2.329(2)a	2.487(3)‡
	calc.		2.21	2.33	2.51
I	obs.		2.43(4)	2.557(5)g	2.719(2)
	calc.		2.41	2.56	2.72

∠X—M—X		M = N	P	As	Sb
X = F	obs.	102.37(3)e 102.17(3)0	97.8(2)α	96.2(2)α 96.0(3)z	95.0(8)0
Cl	obs.	107.1(5)	100.3(1)α	98.6(4)z	97.0(12)z
Br	obs.		101.0(4)α	99.7(3)	98.2(10)‡
I	obs.		102(2)	100.2(4)	99(1)

References	N	P	As	Sb
F	90	37	41 42	96§
Cl	91	92	42	43§
Br		38§	94§	93‡§
I		93	95§	97§

*Symbols e, g, α, z, and 0 correspond to the definitions of geometrical parameters given in Table 6.1 and Ref. 5.
†Estimated by the modified Schomaker–Stevenson rule. See text.
‡S. Konaka and T. Iijima, unpublished data (1972).
§ED photographs were taken above room temperature. However, the corresponding distance and angle parameters for room temperature are expected to be essentially unchanged within experimental errors.

Table 6.5 Internuclear distances and bond angles in phosphoryl and thiophosphoryl halides*

	r(P—S)	r(P—O)	r(P—X)		
	SPX_3	OPX_3	PX_3	SPX_3	OPX_3
X = F	1.871(4)	1.436(6)	1.570(1)	1.536(3)	1.524(3)
Cl	1.885(5)	1.449(5)	2.040(1)	2.011(3)	1.993(3)
Br	1.89(6)	1.41(7)	2.220(3)	2.13(3)	2.06(3)

	∠X—P—X			*References*		
	PX_3	SPX_3	OPX_3	PX_3	SPX_3	OPX_3
X = F	97.8(2)	100.3(3)	101.3(2)	37	40	39
Cl	100.3(1)	101.8(2)	103.3(2)	92	39	39
Br	101.0(4)	106(3)	108(3)	38	98	98

*Distance and angle parameters are in r_g and r_z structures, respectively, except for those of $SPBr_3$ and $OPBr_3$, which are unidentified. Parameters for these molecules were measured by the visual ED method and have larger uncertainties.

$YPCl_3$ by *c.* 0.01 Å. All the P—O and P—S distances are appreciably smaller than the corresponding sums of the double-bond covalent radii[100], 1.62 and 1.94 Å, respectively.

(e) The F–P–F angles in YPF_3 are smaller than the Cl—P—Cl angles in $YPCl_3$ by *c.* 2 degrees.

The above trends may be accounted for qualitatively by the valence-shell electron-pair repulsion (VSEPR) theory[101]: Gillespie assumes a basic rule that bonding electron pairs or lone pairs behave as if they repel one another on the surface of an atom and maximise their distance apart. In addition, he assumes the following principles: (a) a lone pair takes up more room than a bonding pair, (b) a bonding electron pair takes up less room with increasing electronegativity of the ligand and (c) electron pairs of a multiple bond take up more room than the electron pair of a single bond.

The present series of inorganic molecules seems to be one of the systems to which this VSEPR model can be applied successfully. More quantitative understanding of the geometry of these molecules must await future advances in quantum theory. Attempts in this direction are being made by the use of semi-empirical MO theories. See Refs. 102–104 and the references cited therein.

6.3.6 Carbon–carbon bond distances in conjugated aliphatic molecules

6.3.6.1 Environment effects and their interpretations

In the early thirties, when quantum theory and experimental techniques for determining the geometry of free molecules were established, it was observed that C—C bond distances in organic molecules decrease with bond multiplicity[100]. A few years later, Herzberg *et al.* showed that a C—C single bond in a conjugated system, such as butadiene or diacetylene, is shorter than that in an unconjugated system such as ethane[105]. (This effect is here called the primary environment effect.) A large number of accurate experimental data accumulated in the 1950s, particularly those supplied by microwave spectroscopy, were brought into discussion in the *Epistologue on Carbon Bonds* convened and edited by Dewar[106]. Excellent review papers written by Lide[107], Stoicheff[108], Bastiansen and Traetteberg[109] and a number of other discussions summarise the experimental side of these studies up to 1961. Traetteberg[110] and several other authors have carried on experimental studies related to this problem.

A simple empirical rule representing this environment effect was discovered[111, 112]. As displayed by Stoicheff[108] in Figure 6.6, this effect can be represented by linear functions of the number (n) of bonds adjacent to the C—C bond in question:

$$r(\text{C—C}) = 1.299 + 0.040n \text{ Å} \quad (n = 2\text{–}6) \text{ single bond} \tag{6.3}$$

$$r(\text{C=C}) = 1.226 + 0.028n \text{ Å} \quad (n = 2\text{–}4) \text{ double bond} \tag{6.4}$$

$$r(\text{C≡C}) = 1.207 \text{ Å} \quad \text{triple bond} \tag{6.5}$$

With few exceptions (observed mostly in strained systems), this rule seemed to be applicable to the SP (r_s or r_0) C—C distances, which were reported before 1962 and were believed to be accurate to 0.005 Å.

Various hypotheses have been presented to explain this environment effect in terms of valence theory[113–117]. The following concepts are generally

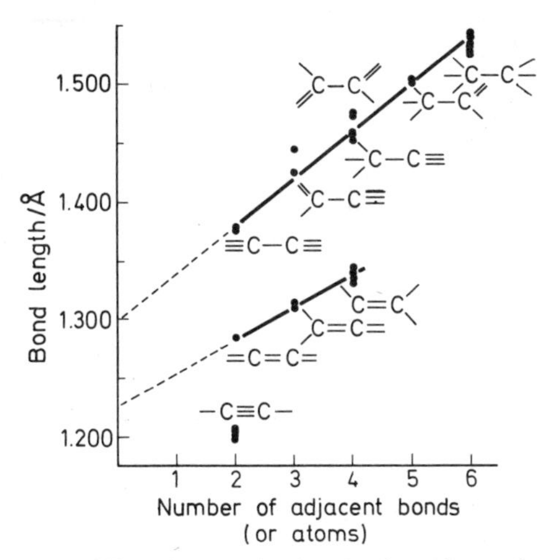

Figure 6.6 Diagram showing the dependence of carbon–carbon bond distances on the number of adjacent bonds or atoms.
(From Stoicheff[108] by courtesy of Pergamon Press)

considered to be the basic factors affecting (either alternatively or co-operatively) the C—C bond distances. However, these oversimplified, intuitive models necessarily have limits of applicability[115, 116].

(a) *Hybridisation* – The bond distances and angles are sometimes thought to be determined primarily by the sizes and directions of the carbon bonding orbitals, which depend on the state of hybridisation[107, 117]. From this stand-point, the Stoicheff rule may be paraphrased in terms of an empirical set of the 'carbon covalent radii'[107, 109], which are characteristic of the hybridisation state. These covalent radii vary roughly linearly with n, which refers to a single carbon atom this time,

$$\rho(\mathrm{sp}^n) \sim \rho_{s0} + \rho_1 n \text{ Å} \quad \text{(single-bond radius)} \tag{6.6}$$

$$\rho(\mathrm{sp}^n) \sim \rho_{d0} + \rho_2 n \text{ Å} \quad \text{(double-bond radius)} \tag{6.7}$$

with $\rho_1 \approx 0.040$ Å and $\rho_2 \approx 0.028$ Å, and therefore, a sufficient condition for the Stoicheff relation is satisfied.

(b) *Electron delocalisation* – According to this scheme, the bond distances in conjugated systems are primarily determined by the distribution of π electrons. This concept is closely related to such familiar terms of valence theory as 'resonance', 'partial multiple-bond character', 'fractional π bond order', 'conjugation', and 'hyperconjugation'[113]. (According to Mulliken[116], the effects of electron delocalisation and hybridisation on the observed bond lengthenings with n are of the same order of magnitude, whereas other causes usually make smaller contributions.)

The effect of electron delocalisation on bond distances is usually displayed in a smooth empirical curve relating observed bond distances (r) with π-bond orders (p) (or overlap populations) calculated by molecular-orbital theory[118].

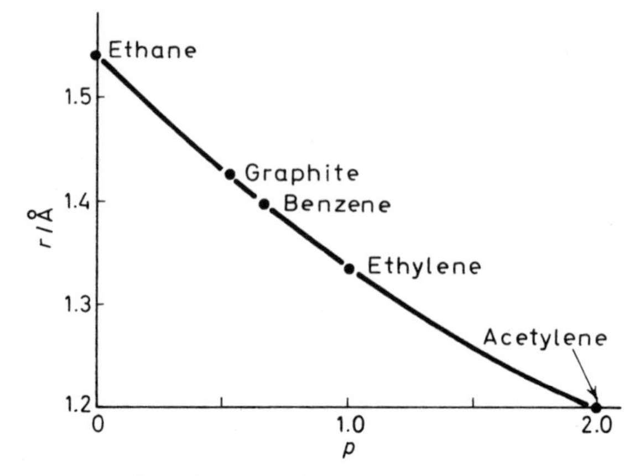

Figure 6.7 Relationship between C—C bond distances (r) and π-bond orders p
(cf. Figure 9.14, Ref. 113)
(From Coulson[113], by courtesy of Clarendon Press)

This p–r relation (Figure 6.7) has been used successfully for predicting C—C bond distances in conjugated hydrocarbons (e.g. those in condensed aromatic molecules) semiquantitatively, or even quantitatively[119–122].

In this framework, the Stoicheff rule results from a regular (nearly linear) increase in the π-bond order with n. Theoretical backgrounds for the p–r curve have been provided in terms of the Hückel or Pariser–Parr–Pople theory, by which the slope of the curve is related to the parameters representing molecular integrals[123–125].

(c) *Steric effect* – An alternative interpretation of the primary environment effect, along with other properties of hydrocarbons, has been presented[102,126] on the basis of a simple assumption of intramolecular van der Waals interactions among non-bonded atom pairs in a molecule. In this scheme, a C—C bond distance decreases with the decrease in n on account of the relief of these steric forces, and bond distances and angles can be predicted by minimising the total energy of the system, in other words, by taking the balance of valence forces (bond stretching and bending) with the non-bonded interactions estimated from an empirical set of interatomic potential functions[127].

This steric model has much utility in various areas of stereochemistry, e.g. quadratic (Urey–Bradley) force field[128,129], anharmonic force field[130], intramolecular strain[127], and conformational analyses[131,132], but it has been subject to controversies as to the physical significance and the limit of applicability[102,114,115].

(d) *Electronegativity and bond ionicity* – This concept has frequently been invoked to estimate the lengths of polar bonds[99,100,114]. However, many of the theories are yet far from quantitative. As one of the problems related to this effect, a systematic study of 'the secondary environment effect' has recently been carried out[133]. When a hydrocarbon is substituted by a heteroatom, the change introduced into the electron distribution influences the distances and angles of neighbouring bonds. For example, the C—C single bonds in butadiene ($H_2C{=}CH{-}CH{=}CH_2$) and glyoxal ($O{=}CH{-}CH{=}O$), which have equal n ($=4$), have significantly different distances[57]. This problem is discussed in more detail in Section 6.3.6.3.

In concluding this section, the following remarks should be made:

(i) Geometrical parameters, like other molecular properties, should ultimately be explained by molecular quantum theory. In days to come, each of the model theories mentioned above will be given its proper place in a more fundamental theory, possibly with close relationships with one another. Rapid progress in *ab initio* and semi-empirical calculations of geometrical parameters[134–136] is being made aiming at this goal.

(ii) For a further investigation of the environment effect, more plentiful measurements of geometrical parameters are necessary. Accuracy is of the utmost importance, since subtle 'chemical shifts' in geometrical parameters may often be obscured by experimental and interpretational uncertainties.

(iii) It is desirable to specify the exact significance of the parameters being measured and to evaluate correct orders of magnitude of experimental and interpretational uncertainties in the parameters so as to permit a really critical comparison among the parameters in different molecules, and among those measured by different methods.

The remarks (ii) and (iii) were made by many of the authors of the above-mentioned Epistologue[107–109, 114, 115]. Particular emphasis was laid there on the necessity of reducing parameters obtained from the ED and SP methods to a common standard. However, since this reduction was thought to be impracticable at the time, these authors based their discussions on either the SP or the ED parameters in order to be 'consistent'[107–109]. With the scheme of the ED–SP combination summarised in the present article, it seems pertinent to enquire whether this scheme can offer more significant and accurate parameter values. The following sections are a brief account of recent studies in this direction.

6.3.6.2 *Stoicheff's diagram in r_g parameters*

A new version of the Stoicheff diagram has been prepared in Figure 6.8. This diagram differs from Stoicheff's original[108] (Figure 6.6) in the following aspects:

(a) Only the C—C bonds in straight-chain hydrocarbons are included in this diagram, whereas the original chart referred to several oxygen, nitrogen and halogen-substituted hydrocarbons with no explicit account of the secondary environment effect. This effect is considered in Section 6.3.6.3.

(b) The C—C distances have been determined (Table 6.6) by a combined use of ED and SP measurements, whereas Stoicheff's main conclusions were based on SP data only.

(c) All the distance parameters have been converted to r_g, whereas the original diagram was based on r_s and r_0 parameters. The differences between r_g and r_e for the C—C distance parameters listed here are estimated to be essentially constant (of the order of 0.010 Å) to within the uncertainties in the distance parameters, since they are mainly dependent on the degree of an-

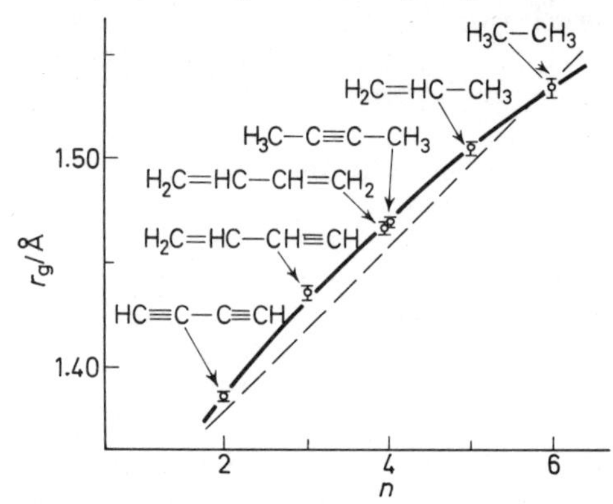

Figure 6.8 Dependence of C—C single bond distances (r_g) in straight-chain hydrocarbons on the number of adjacent bonds or atoms (n). Vertical bars represent estimated limits of experimental error. An empirical quadratic equation (equation (6.8)) and the linear equation given by Stoicheff (equation (6.3) and Figure 6.6) are shown by solid and broken curves, respectively

harmonicity in the C—C stretching vibration and on the C—C mean-square amplitudes[5, 12, 16, 22]. Therefore, the present diagram may further be converted to an approximate r_e disgram by a uniform downward shift of about 0.01 Å.*

(d) Estimated limits of error have been attached to the entries as vertical bars so as to make the comparison more quantitative, whereas no error estimations can be made in the original diagram, because of the lack of clear physical significance of the r_0 and r_s parameters. (Possibly, the uncertainties in these parameters are about 0.005 Å or more.)

In the present diagram are plotted the r_g(C—C) distances in ethane[6], propene[61], butadiene[57, 69], dimethylacetylene[53], vinylacetylene[71], and diacetylene[52], for which the number n is equal to 6, 5, 4, 4, 3 and 2, respectively. The distances in butadiene and dimethylacetylene ($n = 4$) are nearly equal. (Note, however, that the C—C bond in dimethylbut-2-ene, $(CH_3)_2C{=}C(CH_3)_2$ [72], ($n = 5$) is about 0.01 Å longer than that in propene, $CH_3{-}CH{=}CH_2$, possibly because of steric effect[127].) The points shown in Figure 6.8 fit an empirical parabola,

$$r(n) = 1.285 + 0.0533n - 0.0020n^2 \text{ Å}, \tag{6.8}$$

except for vinylacetylene, which deviates upward by about 0.007 Å.

---

*Note, however, that the $r_g - r_e$ difference may depend on environment (presumably by a few thousandths of 1 Å) if higher-order interactions between bond stretching and angle-bending vibrations have significant effect on r_g bond distances[52]. This problem needs further careful investigation.

6.3.6.3 Secondary environment effect on C–C single bond distances

In a microwave study of propynal (HC≡C—CH=O), Costain and Morton[137] remarked that the replacement of =CH_2 in vinylacetylene (HC≡C—CH=CH_2) by =O tends to lengthen the adjacent C—C single bond (r_s) by 0.02 Å. This contrasts notably with a similar comparison made by Lide[107] for the C—C distances in propene (CH_3—CH=CH_2)[138] (r_s = 1.501(4) Å) and in acetaldehyde (CH_3—CH=O)[139] (r_0 = 1,500$_5$(5) Å), which are indistinguishable from each other. The influence of an electronegative neighbour atom on the C—C bond distance was also investigated by Bastiansen and Traetteberg[109]. However, in spite of their careful examination of numerous experimental data published up to 1961, no systematic trend was observed.

The secondary environment effect of oxygen on the r_g(C—C) distances has been observed in the butadiene—acrolein–glyoxal[57, 69] and vinylacetylene–propynal series[66]. The systematic trend shown in Table 6.6 and Figure 6.9

Table 6.6 Bond distances (r_g) in olefinic and acetylenic hydrocarbons and their heteroatom substitutes*

Molecule	C—C	C=C	C≡C	C=O or C≡N	*Ref.*
CH_3—CH_3†	1.533(2)	—	—	—	6
CH_2=CH_2	—	1.337(2)	—	—	55
CH_2=CH—CH_3‡	1.504(3)	1.341(3)	—	—	61
cis-CH_3—CH=CH—CH_3§	1.508(2)	1.347(3)	—	—	140
trans-CH_3—CH=CH—CH_3§	1.510(2)	1.348(3)	—	—	140
$(CH_3)_2C$=CH_2‡	1.507(2)	1.342(3)	—	—	72
$(CH_3)_2C$=$C(CH_3)_2$	1.512(2)	1.355(4)	—	—	72
CH_2=O	—	—	—	1.209(3)	49, 141
CH_3—CH=O†	1.513(4)	—	—	1.208(4)	47
$(CH_3)_2C$=O	1.518(3)	—	—	1.213(4)	49
CH_2=CH—CH=CH_2‡	1.465(3)	1.345(2)	—	—	57, 69
CH_2=CH—CH=O‡	1.484(4)	1.345(3)	—	1.217(3)	57, 69
O=CH—CH=O‡	1.526(3)	—	—	1.212(2)	57, 69
HC≡CH	—	—	1.212(2)	—	53, 142
CH_3—C≡CH	1.470(4)	—	1.210(4)	—	142
CH_3—C≡C—CH_3	1.468(2)	—	1.214(2)	—	53
HC≡N	—	—	—	1.157(2)	143
CH_3—C≡N	1.466(3)	—	—	1.159(3)	143
CH_2=CH—C≡CH‡	1.434(3)	1.344(4)	1.215(3)	—	71
CH_2=CH—C≡N‡	1.438(3)	1.343(4)	—	1.167(4)	65
O=CH—C≡CH‡	1.453(3)	—	1.211(6)	1.214(5)	66
HC≡C—C≡CH	1.384(2)	—	1.218(2)	—	52
N≡C—C≡N†	1.391(2)	—	—	1.162(2)	51, 52

*Determined by recent electron-diffraction studies (in Å). The uncertainties given in parentheses refer to the last significant figures. Most of them are taken directly from the original literature without discrimination between 'limits of error' and 'standard error'.
†Scale factors have been adjusted with reference to the rotational constants.
‡Determined by a least-squares analysis of ED intensities and rotational constants.
§The r_g values have been estimated from the r_a values.

has also been observed in the r_g(C—C) distances in propene[61] and acetaldehyde[47] and in isobutene[72] and acetone[48, 49]. A similar, but less conspicuous, environment effect of nitrogen has been noticed for the r_g(C—C) distances in the diacetylene–cyanogen[52] and vinylacetylene–acrylonitrile series[65].

No complete theory has yet been presented to account for this effect quantitatively, although a set of naive interpretations are not inconceivable, by which at least the sign of the effect may be predicted. According to the three basic schemes given in Section 6.3.6.1, the following qualitative arguments may be made:

(a) The state of hybridisation of the carbon atom is influenced by the adjacent heteroatom so as to increase the covalent radius of the bonding orbital of the carbon atom to the adjacent carbon atom.

(b) The heteroatom tends to suppress the delocalisation of the π electrons into the C—C bond and decrease the π bond order.

(c) The heteroatom has a larger steric effect than the carbon atom.

An attempt to account for this effect more quantitatively in terms of the π-bond order v. observed bond-length curve has turned out to be far from

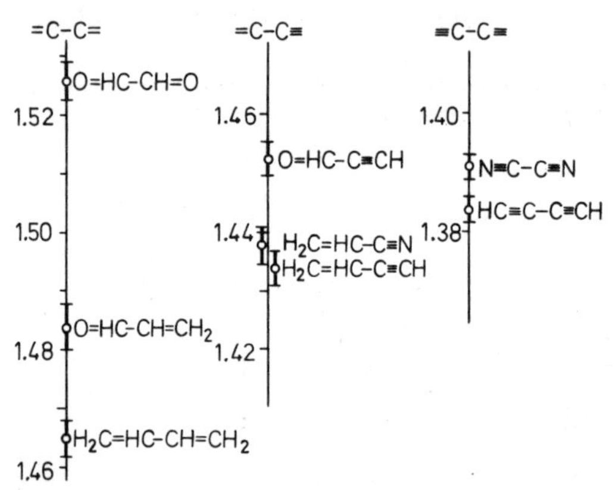

Figure 6.9 C—C Single-bond distances in conjugated aliphatic molecules (r_g) illustrating the effect of heteroatoms. Vertical bars represent limits of experimental error

satisfactory[65]. In a series of analogous molecules, the observed differences in the r_g(C—C) distances are much larger than the estimates by means of empirical or theoretical π-bond orders together with an empirical p–r curve similar to Figure 6.7. The present status may be summarised as follows:

(i) For this comparison, one can use conventional p–r curves proposed by several authors. For instance, the following parameters a and b have been given for the relation:

$$r = a - bp \text{ (Å)} \tag{6.9}$$

where (according to Coulson and Gołebiewski[123]) $a = 1.517$ and $b = 0.18$ for sp^2–sp^2 bonds, or (according to Julg and Pellégatti[122]) $a = 1.520$, 1.48, 1.447 (or 1.46) and $b = 0.19, 0.17, 0.122$ (or 0.13) for sp^2–sp^2, sp^2–sp, and sp–sp bonds, respectively. When the observed r_g(C—C) distances are plotted on these curves, the p parameters for some of the bonds, particularly that of glyoxal, appear to be unrealistic.

(ii) The π-bond orders calculated by means of semi-empirical MO theories are sensitive to the parametrisation of molecular integrals, but in a series of molecules under study, their relative differences, and hence, the local derivatives (dr/dp), seem to be less sensitive.

(iii) The dr/dp slopes for the sp^2–sp^2 and sp^2–sp systems are much larger than those given in equation (6.9). For glyoxal and butadiene, for instance, the p(C—C) values derived by the Pariser–Parr–Pople method are 0.132 and 0.224, respectively[57], and therefore, the slope is $(1.526-1.465)$ Å$/(-0.092) = -0.66$, in contrast to the values given above, -0.18 and -0.17. The slope seems to be essentially unchanged even if the change in the carbon bond radius as a result of the disparity in hybridisation is taken into account[57].

(iv) The above discrepancy may not be too surprising, however, since the p–r curve given in equation (6.9) was originally proposed for pure hydrocarbons, whereas the present problem is concerned with the effect of heteroatom substitution. Nevertheless, a semi-quantitative argument on this effect[65] in the theoretical schemes of Amos[124] and Julg[125] cannot account for the observed trend of the heteroatom effect. Thus, the present problem remains open to a more advanced theoretical approach.

It is interesting to note in this connection that the replacement of $>C=CH_2$ by $>C=O$ also has strong influence on the structure of adjacent three-membered rings (e.g. methylenecyclopropane[144] and cyclopropanone[145]) and on the barrier height of internal rotation of adjacent methyl groups (e.g. propene and isobutene *v.* acetaldehyde and acetone), both studied by microwave spectroscopy[14, 146–148].

6.3.6.4 *Multiple-bond distances*

Since a C—C single bond in a conjugated system is shortened by conjugation as stated above, the corresponding double to triple bonds in the system are expected to increase. The change in multiple-bond distances by conjugation has been discussed by Mulliken[149] from a theoretical point of view. He argued that the multiple bond should be lengthened only slightly since the π-electrons delocalised into adjacent bonds should return to the original bonds because of mutual repulsion*. At the time when Lide wrote his review[107] in the above-mentioned Epistologue, the most accurate measurements available indicated that any lengthening of multiple bonds by conjugation were less than 0.01 Å and probably not over 0.005 Å. Lengthening of this magnitude was also predicted by Bak and Hansen–Nygaard[150] from the shortening of adjacent single bonds in their consideration of the effects of hybridisation and electron delocalisation. They showed that a number of experimental data indicating such lengthenings in multiple bonds were in rough agreement with what they expected and emphasised that further accurate experimental data were useful to make the structure comparison less ambiguous.

Some of the r_g distances listed in Table 6.6 exhibit the expected effect of conjugation of a few thousandths of an angstrom: for example, the differences

---

*Note that one of the more recent calculations of the π-bond orders of acetylene and diacetylene by Julg and Pellégatti[122], together with their estimates of the slope $\Delta r/\Delta p$, seems to suggest an increase of more than 0.01 Å in the C—C distance in the latter molecule[52].

between the bond distances in conjugated and unconjugated systems, r_g(C≡C, diacetylene) − r_g(C≡C, acetylene) ≈ 0.006 Å and r_g(C=C, butadiene) − r_g(C=C, ethylene) ≈ 0.008 Å, exceed the limits of experimental error. Similar significant differences have also been observed for the C≡N and C=O distances, e.g. r_g(C≡N, acrylonitrile) − r_g(C≡N, hydrogen cyanide) ≈ 0.010 Å and r_g(C=O, acrolein) − r_g(C=O, formaldehyde) ≈ 0.008 Å.

A negative correlation between the C=O bond distances in a series of carbonyl compounds and their bond-stretching force constants (or the frequencies of C=O stretching vibrations) has been discussed by a number of authors[62]. A similar trend has here been observed in the relation between the r_g(C≡N) distances and the force constants K(C≡N). For example, the triple bonds in acrylonitrile (CH_2=CH—C≡N, r_g = 1.167 Å and K = 16.94 mdyn $Å^{-1}$ and in cyanogen (N≡C—C≡N, r_g = 1.162 Å and K = 17.26 mdyn $Å^{-1}$)[51] are weaker and longer than that in hydrogen cyanide (H—C≡N, r_g = 1.157 Å and K = 18.71 mdyn $Å^{-1}$ [152]) on account of conjugation.

The C=C bond in dimethylbut-2-ene[72] and *cis*- and *trans*-but-2-enes[140] are significantly (*c.* 0.02 and 0.01 Å, respectively) longer than those in ethylene and propene. Both π-electron delocalisation and the steric effect seem to take part in this effect, but no quantitative account has yet been made.

6.3.6.5 Carbon valence angles

The following empirical rules have been derived in regard to the observed carbon valence angles in conjugated aliphatic systems (Table 6.7):

Table 6.7 Skeleton valence angles (r_z-structure) in olefinic and acetylenic hydrocarbons and with heteroatom substituents (in degrees)

Molecule	∠C—C=C	∠C—C=O	*Ref.*
CH_2=CH—CH_3	124.8(3)	—	61
cis-CH_3—CH=CH—CH_3	125.4(4)†	—	140
	126.7‡	—	153. 154
trans-CH_3—CH=CH—CH_3	123.8(2)	—	140
$(CH_3)_2C$=CH_2	122.1(3)	—	72
	122.0_5‡	—	155
$(CH_3)_2C$=$C(CH_3)_2$	123.7(6)	—	72
CH_3—CH=O	—	124.3(5)	47
$(CH_3)_2C$=O	—	122.0(1)	48, 49
CH_2=CH—CH=CH_2	123.3(5)	—	57
CH_2=CH—CH=O	120.3(7)	123.3(7)	69
O=CH—CH=O	—	121.2(2)	57
CH_2=CH—C≡CH§	123.1(5)	—	71
CH_2=CH—C≡N‖	121.7(5)	—	65
O=CH—C≡CH¶	—	124.2(2)	66

*See footnote * of Table 6.6.
†‡r_a and r_0 structures, respectively.
§¶The C—C≡C angles are 178.6(3) and 177.9(12), respectively, in the direction *trans* to the double bond.
‖The C—C≡N angle is 178.2(10) *trans* to the double bond.

(a) There is no conspicuous 'primary environment effect', i.e. the C—C=CH_2 angle is essentially equal to the corresponding C—C=O angle to within

1 degree (propene–acetaldehyde, isobutene–acetone, butadiene–acrolein, acrolein–glyoxal, vinylacetylene–propynal).

(b) The C(1)—C(2)=CH_2 (or C(1)—C(2)=O) angle in a conjugated system decreases by *c.* 2 degrees when the CH_2 (or CH) group attached to the C(1) atom is replaced by an oxygen (or nitrogen) atom (butadiene–acrolein, acrolein-glyoxal, vinylacetylene–acrylonitrile), (secondary environment effect).

(c) The single bond and triple bond in a conjugated system are not collinear. The triple bond is bent in the direction *trans* to the double bond by *c.* 2 degrees (vinylacetylene, acrylonitrile, and propynal).

(d) A regular relationship observed for methyl-substituted ethylenes may be accounted for by a naive steric model that the *gem* methyl groups and the *cis* methyl groups repel each other by 2 and 1 degrees, respectively. According to this model, the C—C=C (or C—C=O) angles are predicted to be *cis*-but-2-ene ≈ 125 degree > propene ≈ *trans*-but-2-ene ≈ acetaldehyde ~ 124 degrees > dimethyl-p-but-2-ene ~ 123 degrees > isobutene ≈ acetone ~ 122 degrees. The observed relative orders are correctly reproduced, and the predicted values are within 1 degree of the observed values.

6.3.6.6 C–H *bond distances and* C–C–H *angles*

It is generally more difficult to determine C—H bond distances and C—C—H valence angles accurately than to determine distances and angles among heavier atoms, even when ED and SP data are combined in the analysis. From the experimental side, the hydrogen atom is the weakest scatterer of electrons and the rotational constants are the least sensitive to hydrogen positions. From the theoretical side, bond distances and angles involving hydrogen have the largest vibrational effects and the differences among the structural parameters under different definitions are the largest.

In spite of the above difficulties, the following empirical estimates of the r_g(C—H) distances can be made from the data listed in Table 6.8:

r_g(C—H, aldehyde)	–C(=O)—H	1.13 Å
r_g(C—H, methyl)	—C(—H)H_2	1.11—1.12 Å
r_g(C—H, vinyl)	—C(—H)=CH_2	1.10—1.11 Å
r_g(C—H, ethynyl)	—C≡C—H	1.08—1.09 Å

The difference in the aldehyde and vinyl C—H distances is in the direction of the secondary environment effect of oxygen mentioned in Section 6.3.6.3. Very little is known about other environment effects on C—H distances and valence angles involving hydrogen (Tables 6.8 and 6.9) because of the lack of experimental accuracy, in spite of their great importance in valence theory,

Table 6.8 C—H bond distances (r_g) in olefinic and acetylenic hydrocarbons and heteroatom substituents* (in Å)

Molecule	C—H$_m$	C—H$_v$	C—H$_e$	C—H$_a$	*Ref.*
CH_3—CH_3	1.111(2)	—	—	—	6
CH_2=CH_2	—	1.103(2)	—	—	6, 55
CH_2=CH—CH_3	1.119(8)	1.111(8)	—	—	61
$(CH_3)_2C$=CH_2	1.115(8)	(1.103)	—	—	72
$(CH_3)_2C$=$C(CH_3)_2$	1.122(5)	—	—	—	72
CH_2=O	—	—	—	1.123	49, 141
CH_3—CH=O	1.105(6)	—	—	1.127(9)	47
$(CH_3)_2C$=O	1.101(3)	—	—	—	48, 49
CH_2=CH—CH=CH_2	—	1.108(4)	—	—	57, 69
CH_2=CH—CH=O	—	1.100(10)	—	1.13(3)	57, 69
O=CH—CH=O	—	—	—	1.132(8)	57, 69
HC≡CH	—	—	1.078(2)	—	142
CH_3—C≡C—CH_3	1.116(6)	—	—	—	53
HC≡N†	—	—	1.084	—	143
CH_2=CH—C≡CH	—	1.106(10)	1.09(2)	—	71
CH_2=CH—C≡N	—	1.114(7)	—	—	65
O=CH—C≡CH	—	—	1.086(5)	1.131(4)	66
HC≡C—C≡CH	—	—	1.094(10)	—	52

*See footnote * of Table 6.6. m = methyl, v = vinyl (average), e = ethynyl, and a = aldehyde.
†Estimated from spectroscopic r_e distance.

Table 6.9 Bond angles involving hydrogen in olefinic and acetylenic hydrocarbons and heteroatom substituents* (in degrees)

Molecule	C—CH_2—H	C—C (=CH_2)—H	C—C (=O)—H	C=C—H	O=C—H	*Ref.*
CH_3—CH_3	111.5(3)	—	—	—	—	6
CH_2=CH_2	—	—	—	121.9(4)	—	6, 55
CH_2=CH—CH_3	110.7(9)	114.7	—	120.5(8)	—	61
$(CH_3)_2C$=CH_2	112.0(11)	—	—	121.3(25)	—	72
$(CH_3)_2C$=$C(CH_3)_2$	113.9(20)	—	—	—	—	72
CH_2=O	—	—	—	—	121.7(3)	141, 156
CH_3—CH=O	110.1(9)	—	115.3(3)	—	120.4	47
$(CH_3)_2C$=O	110.6(4)	—	—	—	—	49
CH_2=CH—CH=CH_2	—	114.9	—	121.8(12)	—	57
CH_2=CH—CH=O	—	116.8	114.4(27)	121.3	122.3(27)	69
O=CH—CH=O	—	—	112.2	—	126.6(17)	57
CH_3—C≡C—CH_3	110.7(4)	—	—	—	—	53
CH_2=CH—C≡CH	—	116.3	—	120.6(15)	—	71
CH_2=CH—C≡N	—	118.6	—	119.7(21)	—	65
O=CH—C≡CH	—	—	113.7	—	122.1(8)	66

*See footnote * of Table 6.6. The angles are in the r_z structure of the r_α structure, which is thought to be nearly equivalent to the r_z structure. Values listed without uncertainties are derived from other independent angles by assuming planarity. The tilt of the methyl group is averaged.

particularly in relation to the $^{13}C—H$ coupling constants in n.m.r. spectra[157,158] and to the quadratic force constants[128].

6.3.7 Strained molecules

One of the problems of great current interest is the structure of highly strained molecules, and a great many studies have been made both in the gas phase (electron diffraction, microwave spectroscopy, etc.) and in the crystal phase (x-ray and neutron diffraction, etc.). Again, this seems to be a subject to which the ED plus SP analysis can contribute effectively. So far, however, very few studies have been made in this context. Critical checks and improvements of 'model force fields' will be made in the future in the light of precise structures determined in this way.

6.3.7.1 *Aliphatic molecules*

The background of the problem and recent ED studies have been reviewed in detail in relation to various steric models[2,102,127]. The innermost C—C bond in tri-t-butylmethane[159], 1.611(5) Å, is the longest reported for an acyclic hydrocarbon, much longer than the C—C bonds in neopentane[160], 1.541(2), *sym*-tetramethylethane[127,161], 1.540(2), and hexamethylethane[127,161], 1.548(2) Å. The r_g(C—Cl) distance 1.828(5) Å in t-butyl chloride and its fully deuterated analogue, studied recently by a careful ED plus SP analysis[59], is appreciably different from the r_s(C—Cl) distance for the same compound[162], 1.803(2) Å, and is longer than any C—Cl distance determined precisely.

Regular variations in the C—C distances and in the carbon valence angles in methyl-substituted ethylenes[72] are pointed out in Section 6.3.6.2 and 6.3.6.4. In this connection, the barrier heights of methyl internal rotation (V_3) in substituted propenes measured by microwave spectroscopy show remarkable regularity[14,146–148]. Substitutions of one of the vinyl hydrogen atoms in the *secondary*, *cis*, and *trans* positions of propene by methyl, fluorine, etc. change the V_3 values for unsubstituted propene (1.98 kcal mol^{-1}) to 2.2–2.7, 0.6–1.4 and 1.9–2.2 kcal mol^{-1}, respectively. These changes are interpreted in terms of steric interactions[14,148,154].

6.3.7.2 *Ring molecules*

The r_g(C—C) bond distances in cycloalkanes, C_3H_6 to C_6H_{12}, are 1.512(3)[163], 1.557(3)[63], 1.546(2)[164] and 1.540(2) Å [63], respectively. The longest bond in C_4 and the shortest in C_3 are explained as the effects of intramolecular strain[165] and bent bonds[166], respectively (Figure 6.10).

Recent microwave studies have given r_s-structures of a number of cyclic, heterocyclic and aromatic compounds[14,146,147]. For instance, three-membered rings with heteroatoms (N, O, S, etc.) have widely different C—C bond distances (1.47 to 1.59 Å) and apparent bond angles (say 48.4 degrees, ∠CSC in $(CH_2)_2S$ [167] to 64.6 degrees, ∠C(CO)C in $(CH_2)_2CO$ [145]). On the

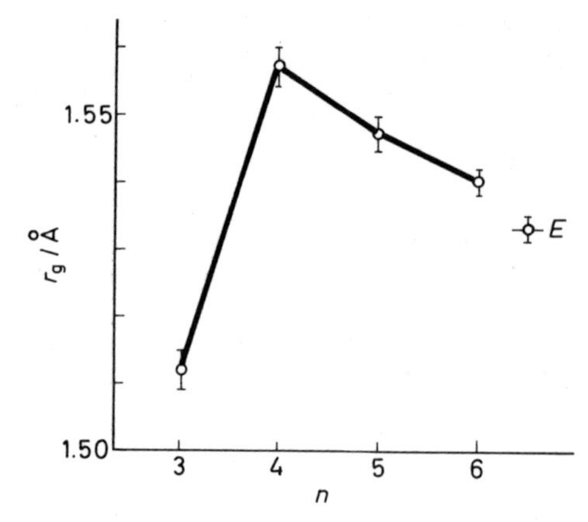

Figure 6.10 C—C bond distances (r_g) in cycloalkanes, C_nH_{2n} (n = 3–6). The r_g (C—C) distance in ethane is shown by the symbol *E*. Vertical bars represent estimated limits of experimental error. See text for references.

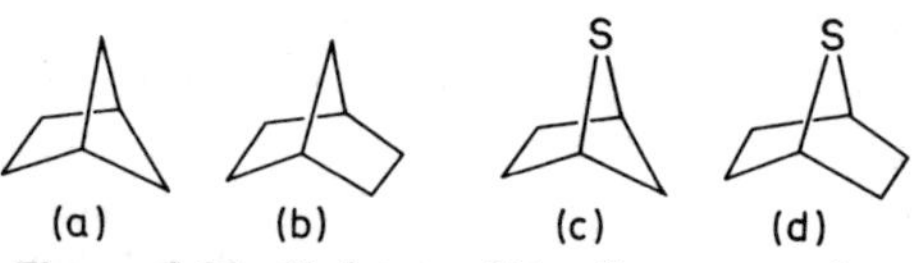

Figure 6.11 Skeletons of bicyclic compounds.
(a) bicyclo[2.1.1]hexane, (b) bicyclo[2.2.1]heptane,
(b) 5-thiabicyclo[2.1.1]hexane, and
(d) 7-thiabicyclo[2.2.1]heptane

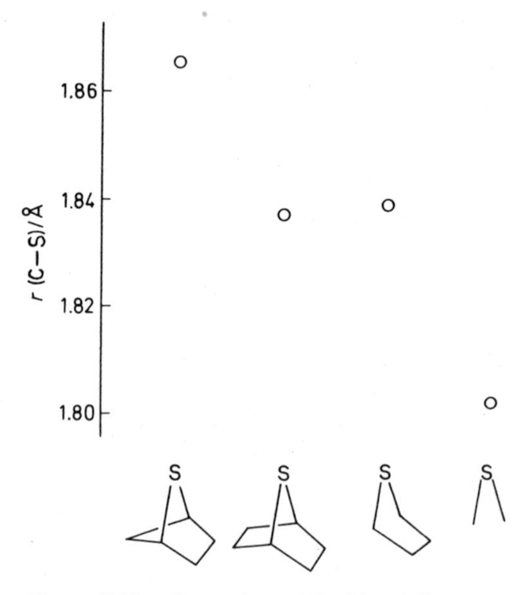

Figure 6.12 Comparison of C—S bond distances

other hand, similar systems as well as more complicated bicyclic compounds have recently been investigated by ED[1,2]. For example, systematic changes in valence angles and dihedral angles have been observed when double bonds or nitrogen atoms are introduced into bicyclo [2.2.1] and bicyclo [2.2.2] hydrocarbons[67,168,169]. The effect of replacement of CH_2 with S is illustrated in Figure 6.11, where regular increases in r_g(C—S) bond distances and sulphur valence angles with the degree of intramolecular strain are demonstrated[170], as shown in Figures 6.12 and 6.13. The differences of the order of

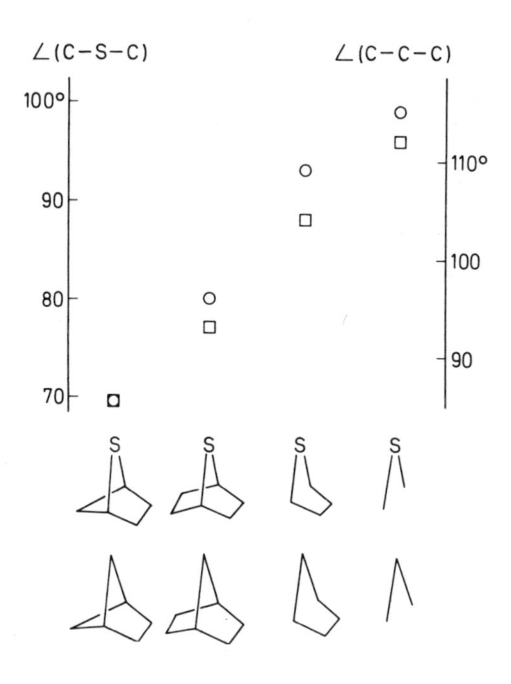

Figure 6.13 Comparison of carbon- and sulphur-valence angles

13–15 degrees between the C—CH_2—C angles in hydrocarbons and the C—S—C angles in the corresponding sulphides are in contrast to the difference of 17.3 degrees between the H—CH_2—H in methane and the H—S—H in hydrogen sulphide[171]. According to a recent microwave study of 7-thiabicyclo[2.2.1]heptane[172], the structure determined by the ED experiment can reproduce the observed rotational constants if the difference in the C(2)—C(3) and C(1)—C(2) bond distances, which were assumed to be equal in the ED analysis[170], is set equal to 0.03 Å. By means of this ED–SP combination, the skeleton structure of this molecule has been determined. Similar differences in the C(2)—C(3) and C(1)—C(2) distances have also been observed (with appreciable uncertainties, however) in bicyclo[2.2.1]heptane (norbornane)[67] and bicyclo[2.2.2]octane[173].

It is hoped that these SP and ED studies, which are now being made nearly independently, will be related with each other more closely in the near future.

6.4 LARGE-AMPLITUDE MOTIONS

The dynamics of 'non-rigid molecules' is one of the most active fields in current structural chemistry. The following nine categories are generally known as 'large-amplitude motions', to which the ordinary theory of normal vibrations[174] based on the assumption of infinitesimal harmonic oscillations is not applicable. Each of these motions is characterised by a specific potential function (e.g. a double minimum with a small hump) controlled by a delicate balance of various intramolecular forces. The most precise information on the potential parameters comes from the measurements of differences among the vibrational and rotational energy levels (sometimes very closely spaced), the rotational constants for each vibrational state, and relative populations by means of microwave[14,146,147,175] and/or far infrared and Raman spectroscopy. An intricate quantum-mechanical treatment is required, even when the large-amplitude motion in question is assumed to be completely separable from other vibrations (say skeletal deformations), since (a) the coordinates are not rectilinear but curvilinear, (b) the effective mass depends on the displacement of this motion and (c) coupling of angular momenta due to this motion and overall rotation is present. The problem is much more complicated if this motion is not separable from other vibrations.

Electron diffraction provides quantitative or semi-quantitative information on the potential function, $V(q)$, in regard to a 'slowly-varying coordinate'[176,177], q, through the probability distribution function, $P(r)$, of an internuclear distance r[178], which varies with q. Usually, the problem is treated in the following classical approximation developed by Karle, Morino and their co-workers[179–181]:

$$P(r) = \int [2\pi l_f^2(q)]^{-\frac{1}{2}} \exp\{-[r(q)-r_e]^2/2l_f^2(q)\} w(q) \mathrm{d}q \tag{6.10}$$

where

$$w(q) = \exp[-V(q)/kT] \Big/ \int \exp[-V(q)/kT] \mathrm{d}q \tag{6.11}$$

and $r(q)$ and $l_f^2(q)$ are the instantaneous internuclear distance and the mean-square amplitude of the 'semi-rigid framework', respectively, calculated at the coordinate q. The temperature dependence of $P(r)$ due to the Boltzmann weight function $w(q)$ and the frame amplitude l_f^2 has also been studied experimentally[182,183]. Since primary sources of information on $V(q)$ from ED are the first and second moments of $P(r)$, what is usually practicable is to estimate only a small number of adjustable parameters in $V(q)$ rather than to determine a fine detail of the potential uniquely. Therefore, a complementary use of accurate observations of rotational and vibrational spectra is indispensable, and there is much to be explored in future studies. In general, reliability of the analysis depends on the quality of the measurements of ED intensity and the correct estimation of frame amplitudes. Since a contribution to ED intensity from an atom pair depending on such a large-amplitude motion damps out rapidly with the scattering angle, an intensity analysis of small-angle scattering is often effective[184–186]. Since a complete survey of the recent ED studies related to this subject has been made by Hilderbrandt and Bonham[2], only an outline will be presented in the following sections.

6.4.1 Quasi-linearity

The presence of an intermediate between linear and permanently bent molecules was discussed in detail for disiloxane[177]. Theories have recently been developed in relation to the ultraviolet spectra of a number of triatomic molecules in excited electronic states[187]. Electron diffraction[50, 188, 189] and far-infrared studies[190] of carbon suboxide indicated that this molecule is 'quasi-linear' in regard to the lowest C=C=C bending mode (ν_7) and possibly has a small hump in the linear position.

6.4.2 Wagging and inversion

The wagging motions of the NH_2 group, or inversions, in amines[191] and amides[192], as well as the CH_2 wagging motions in ketene and diazomethane[193], have been studied extensively by spectroscopy[14, 175]. However, ED intensity is insensitive to this motion because the scattering power of hydrogen is so small and because the relevant signal fades out rapidly with the scattering angle.

6.4.3 Puckering of four- and five-membered rings

The potential in regard to the ring-puckering motion in cyclobutane determined by ED[63, 194, 195] is consistent with that derived from recent infrared and Raman measurements[196]. The potential has a minimum at a dihedral angle of *c.* 35 degrees and a barrier of 518 cm^{-1} at the ring plane[196]. According to SP studies, the replacement of one of the CH_2 groups by $C{=}CH_2$, O, C=O, S, Se and Si lowers the barrier height to 7.6, 15.3, 160, 274, 383 and 442 cm^{-1}, respectively[197–202]. Cyclopentene has a potential minimum at a dihedral angle (C_s symmetry) of 23.3 degrees and a barrier of 232 cm^{-1} [203].

6.4.4 Pseudo-rotation in five-membered rings

Pseudo-rotation in cyclopentane has been studied in detail by ED[164] besides infrared and Raman spectroscopy[204, 205] and thermochemical measurements[206, 207]. The molecule is believed to have essentially free pseudo-rotation (no potential difference between the C_2 and C_s symmetry conformations and anywhere in between) with an equilibrium out-of-plane displacement, q_e, of 0.44 Å [164] and a barrier of 1.8×10^3 cm^{-1} at the planar conformation (D_{5h}) [205].

A similar pseudo-rotation has been observed for tetrahydrofuran[208–210]. The equilibrium conformation (thermal-average displacement $q_g \approx 0.38$ Å [210]), is *c.* 27, 57 and 1220 cm^{-1} lower than the C_2, C_s and C_{2v} forms, respectively[208]. The q_e values for the above molecules determined by ED are slightly, but not significantly, smaller than the effective q values derived from SP measurements[63, 164, 211]. Replacement of the oxygen atom in tetrahydrofuran by S, SiH_2, C=O, Se, and GeH_2 stabilises the C_2 form, and the C—C bond opposite to the heteroatom X is twisted by *c.* 20 degrees from the C—X—C plane[212, 213]. The energy differences between the C_s and C_2 forms are 773, 1362, 1687, 1882 and 2065 cm^{-1}, respectively, according to SP measurements[213].

6.4.5 Pseudo-rotation in inorganic fluorides

Large-amplitude three-dimensional vibrations analogous to the above pseudo-rotation have been observed for ReF_7 [214] and IF_7 [215]. According to the ED studies of Bartell and his co-workers, their structures deviate significantly from D_{5h} symmetry and are presumed to be 'dynamically non-rigid'. Their interpretation of the experimental data is that the five-membered ring formed by the equatorial fluorine atoms experiences essentially free pseudo-rotation (e″ mode) which is coupled with the e'_1 axial-fluorine bending mode through a large cubic potential term. In a schematic expression, bond–bond repulsions thrust the equatorial fluorine atoms out of plane, as much as 9 and 7.5 degrees for ReF_7 and IF_7, respectively, and this out-of-plane displacement, in turn, induces an axial bend of the order of 8 and 4.5 degrees, respectively.

By the technique of deflection of molecular beams in an inhomogeneous electric field, Klemperer *et al.*[216] observed polar components of ReF_7, which increased intensity as the temperature was lowered. A parallel but much smaller effect was observed for IF_7. Their observations show that these molecules do not have a rigid polar structure, and are compatible with the non-rigid distorted geometry mentioned above.

An analogous motion has been observed for XeF_6 [217, 218]. In contrast to XeF_2 and XeF_4 (linear[219] and square planar[220], respectively), XeF_6 has a non-rigid O_h structure with the t_{1u} bending mode undergoing very large amplitudes of vibration. This motion is coupled in phase with the t_{2g} bending through a large cubic potential term. A model assuming repulsive interactions[101, 102] among the xenon lone pair and Xe—F bonding electrons works only qualitatively, and further experimental and theoretical studies are needed for a more complete interpretation.

The equatorial and axial P—F bonds in PF_5 (D_{3h} symmetry) are non-equivalent (1.534(4) and 1.577(5) Å, respectively) as determined by ED [221], whereas all the fluorine nuclei are equivalent in the n.m.r. spectrum above $-150\,°C$, indicating rapid exchange of the axial and equatorial bonds[222]. This exchange is presumed to take place through bending motions similar to pseudo-rotation[222, 223].

6.4.6 Torsion in bicyclic molecules

Bicyclo[2.2.2]octane (Figure 6.14) is a quasi-D_{3h} molecule with an r.m.s. amplitude of torsion around the C(1)—C(4) axis of 7.2 degrees and a small potential hump of *c.* 35 cm^{-1} at the D_{3h} position[173]. This observation has been confirmed by microwave[224] and x-ray[225] studies of a number of its derivative compounds. Similar potentials have also been observed for 1-aza[226] and 1,4-diaza compounds[227]. An empirical model assuming a threefold torsional

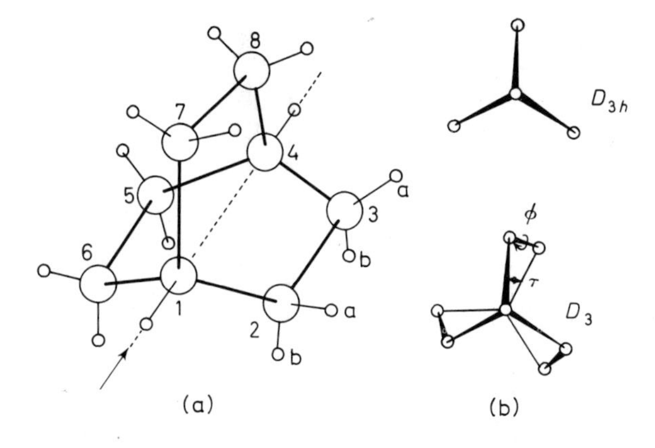

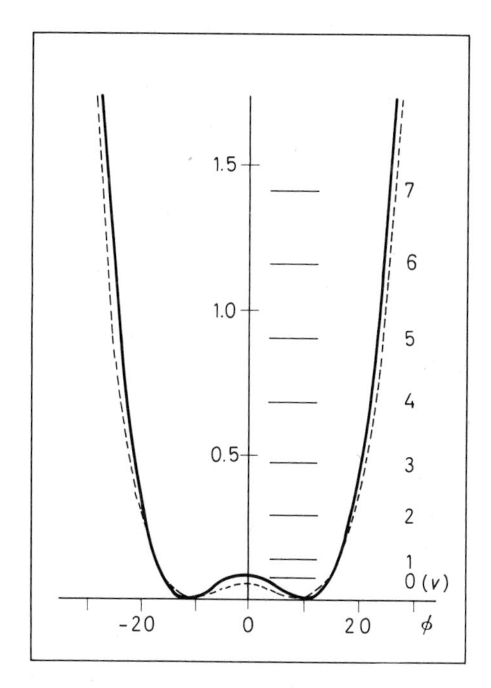

Figure 6.14 (a) Torsional motion in bicyclo [2.2.2]octane around the C(1)–C(4) axis[173].
(b) potential functions in terms of the torsional angle ϕ. Solid curve: estimated from the ED experiment, broken curve: calculated from a model function
(From Yokozeki *et al.*[173], by courtesy of the Chemical Society of Japan)

barrier around C—C bonds in addition to intramolecular strain due to bending displacements and non-bonded interactions[127] reproduces such a potential function quite satisfactorily[172, 224–227] (Figure 6.14).

6.4.7 Internal rotation

Internal rotation and rotational isomerism have been studied by means of SP and ED for a large number of molecules and parameters characterising

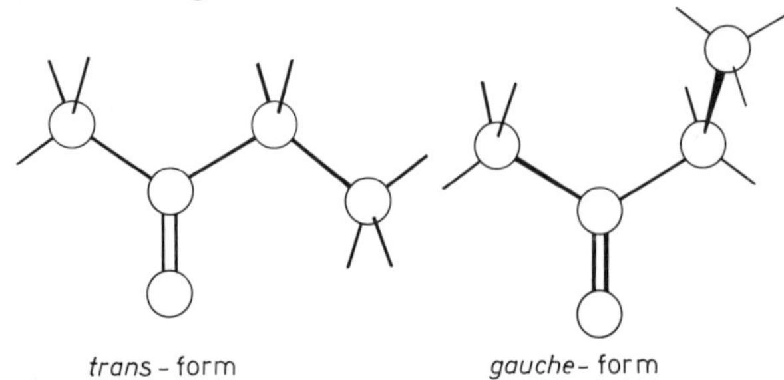

Figure 6.15 Rotational isomerism in methyl ethyl ketone[185] (From Abe *et al.*[185], by courtesy of Elsevier Publishing Co.)

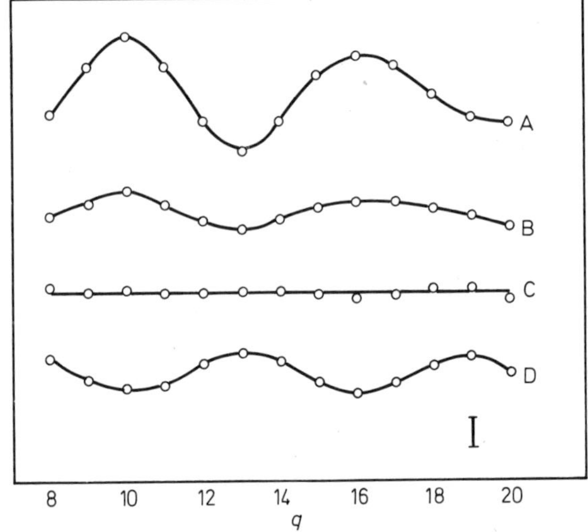

Figure 6.16 Background functions for different compositions of the *trans* and *gauche* conformers of methyl ethyl ketone[185]. The functions are adjusted in such a way that for an ideal experimental condition and a correct model the function is constant. The fractions of the *trans* conformation are assumed to be 80, 90, 95 and 100% for A, B, C and D, respectively. The vertical bar represents the estimated limit of experimental uncertainty. From this analysis, the fraction of the *trans* conformer is estimated to be 95±3%
(From Abe *et al.*[185], by courtesy of Elsevier Publishing Co.)

the potential function (equilibrium conformations, populations, and torsional amplitudes around the stable positions) have been obtained[148, 228, 228a]. The merits and demerits of the ED method for this purpose were discussed by Wilson[228] and Lowe[148] in comparison with other experimental methods. In general, it is easier to apply the ED method to a molecule with good symmetry and with relatively heavy atoms or groups attached to both ends of the axis of internal rotation (e.g. hexachloroethane[179, 180]). For example, the use of small-angle intensities (a background analysis[184]) for estimating the relative population of coexisting conformers in methyl ethyl ketone is shown in Figures 6.15 and 6.16[185].

6.4.8 Vibronic interaction

Polyatomic linear and non-linear molecules in degenerate electronic states show Renner–Teller and Jahn–Teller effects, respectively, because of the interaction of electronic and vibrational angular momenta[229]. For the former effect, no ED study has ever been made. An ED study[230] of VCl_4 searching for the latter effect indicated a distortion of about 0.05 Å from T_d symmetry in the direction of the e bending mode (ν_2), but the evidence was not regarded as conclusive because of the lack of information on the vibrational levels for ν_2. A recent infrared analysis[231] of ReF_6 presented evidence for a Jahn–Teller displacement of the fluorine atoms in the t_{2g} (ν_5) direction with a potential minimum of $-74\ cm^{-1}$ at 0.046 Å from the O_h configuration, whereas ED data contained no evidence for this distortion[232].

6.4.9 Weak intra- and inter-molecular bonding

Large-amplitude motions also take place in such weak bonds as intramolecular hydrogen bonding (oxalic acid[233], acetylacetone[234], etc.) and intermolecular hydrogen bonding (formic acid dimer[235], acetic acid dimer[236], and polymeric hydrogen fluoride[237]), where examples are taken from recent ED studies. Bondings in complexes of group-III elements (e.g. dimers of trimethyl aluminium[238] and aluminium trichloride[239]) and in metallocenes[240, 241] have also been studied recently by ED. The van der Waals molecules (Ar_2, Xe_2, etc.) studied by Audit[242] in a supersonic jet are one of the most interesting systems in this category, for which a joint approach by ED and SP techniques seems promising.

Acknowledgements

The author wishes to thank Professor Robert K. Bohn of the University of Connecticut and Dr. Tsutomu Fukuyama of the University of Tokyo for reading the manuscript and making helpful comments.

References

1. Bauer, S. H. (1970). *Physical Chemistry, An Advanced Treatise,* Vol. 4, Chapter 14. (D. Henderson, editor) (New York: Academic Press)
2. Hilderbrandt, R. J. and Bonham, R. A. (1971). *Ann. Rev. Phys. Chem.,* **22,** 279
3. Karle, J. (1972). *Determination of Organic Structures by Physical Methods,* V. (F. C. Nachod and J. J. Zuckerman, editors) (New York: Academic Press)
4. Bartell, L. S. (1972). *Physical Methods in Chemistry,* 4th Edition. (A. Weissberger and B. W. Rossiter, editors) (New York: Interscience)

5. Kuchitsu, K. and Cyvin, S. J. (1972). *Molecular Vibrations and Structure Studies,* Chapter 12 (S. J. Cyvin, editor). (Amsterdam: Elsevier)
6. Kuchitsu, K. (1968). *J. Chem. Phys.,* **49,** 4456
7. Kuchitsu, K. (1972). Reference 5, Chapter 30
8. Shoolery, J. N., Shulman, R. G., Sheehan, W. F., Jr., Schomaker, V. and Yost, D. M. (1951). *J. Chem. Phys.,* **19,** 1364
9. Karle, I. L. and Karle, J. (1949). *J. Chem. Phys.,* **17,** 1052; **18,** 957, 963
10. Cyvin, S. J. (1968). *Molecular Vibrations and Mean Square Amplitudes.* (Oslo: Universitetsforlaget and Amsterdam: Elsevier)
11. Lide, D. R., Jr. (1964). *Ann. Rev. Phys. Chem.,* **15,** 225, and the references cited therein
12. Bartell, L. S. (1955). *J. Chem. Phys.,* **23,** 1219
13. Bartell, L. S. and Bonham, R. A. (1959). *J. Chem. Phys.,* **31,** 400
14. Morino, Y. and Hirota, E. (1969). *Ann. Rev. Phys. Chem.,* **20,** 139
15. Morino, Y. (1969). *Pure Appl. Chem.,* **18,** 323
16. Kuchitsu, K. (1967). *Bull. Chem. Soc. Jap.,* **40,** 498
17. Bartell, L. S. and Kuchitsu, K. (1962). *J. Phys. Soc. Jap.,* **17,** Supplement B-II, 20
18. Shibata, S. (1963). *J. Phys. Chem.,* **67,** 2256
19. Ukaji, T. and Kuchitsu, K. (1966). *Bull. Chem. Soc. Jap.,* **39,** 2153
20. Iijima, T. and Morino, Y. (1963). *Bull. Chem. Soc. Jap.,* **36,** 412
21. Bartell, L. S. (1963). *J. Chem. Phys.,* **38,** 1827
22. Kuchitsu, K. (1967). *Bull. Chem. Soc. Jap.,* **40,** 505
23. Bartell, L. S., Kuchitsu, K. and deNeui, R. J. (1961). *J. Chem. Phys.,* **35,** 1211
24. Kuchitsu, K. and Bartell, L. S. (1962). *J. Chem. Phys.,* **36,** 2470
25. Bastiansen, O. and Beagley, B. (1964). *Acta Chem. Scand.,* **18,** 2077
26. Kuchitsu, K., Guillory, J. P. and Bartell, L. S. (1968). *J. Chem. Phys.,* **49,** 2488
27. Morino, Y., Kuchitsu, K. and Yamamoto, S. (1968). *Spectrochim. Acta,* **24A,** 335
28. Kuchitsu, K. and Bartell, L. S. (1962). *J. Chem. Phys.,* **36,** 2460
29. Shibata, S. and Bartell, L. S. (1965). *J. Chem. Phys.,* **42,** 1147; *J. Molec. Struct.* (in the press)
30. Murata, Y., Kuchitsu, K. and Kimura, M. (1970). *Jap. J. Appl. Phys.,* **9,** 591
31. Morino, Y. and Iijima, T. (1962). *Bull. Chem. Soc. Jap.,* **35,** 1661
32. Haase, J. and Winnewisser, M. (1968). *Z. Naturforsch.,* **23a,** 61
33. Kuchitsu, K. and Konaka, S. (1966). *J. Chem. Phys.,* **45,** 4342
34. Winnewisser, M. and Haase, J. (1968). *Z. Naturforsch.,* **23a,** 56
35. Bartell, L. S. and Carroll, B. L. (1965). *J. Chem. Phys.,* **42,** 1135
36. Plato, V., Hartford, W. D. and Hedberg, K. (1970). *J. Chem. Phys.,* **53,** 3488
37. Morino, Y., Kuchitsu, K. and Moritani, T. (1969). *Inorg. Chem.,* **8,** 867
38. Kuchitsu, K., Shibata, T., Yokozeki, A. and Matsumura, C. (1971). *Inorg. Chem.,* **10,** 2584
39. Moritani, T., Kuchitsu, K. and Morino, Y. (1971). *Inorg. Chem.,* **10,** 344
40. Karakida, K., Fukuyama, T. and Kuchitsu, K. (to be published)
41. Clippard, F. B., Jr. and Bartell, L. S. (1970). *Inorg. Chem.,* **9,** 805
42. Konaka, S. (1970). *Bull. Chem. Soc. Jap.,* **43,** 3107
43. Konaka, S. and Kimura, M. (1972). *Bull. Chem. Soc. Jap.* (to be published)
44. Clark, A. H., Beagley, B., Cruickshank, D. W. J. and Hewitt, T. G. (1970). *J. Chem. Soc. A,* 872
45. Jacob, E. J., Thompson, H. B. and Bartell, L. S. (1971). *J. Molec. Struct.,* **8,** 383
46. Iijima, T. and Kimura, M. (1969). *Bull. Chem. Soc. Jap.,* **42,** 2159
47. Iijima, T. and Tsuchiya, S. (1972). *J. Molec. Spectrosc.,* (in the press)
48. Iijima, T. (1970). *Bull. Chem. Soc. Jap.,* **43,** 1049
49. Iijima, T. (to be published)
50. Tanimoto, M., Kuchitsu, K. and Morino, Y. (1970). *Bull. Chem. Soc. Jap.,* **43,** 2776
51. Morino, Y., Kuchitsu, K., Hori, Y. and Tanimoto, M. (1968). *Bull. Chem. Soc. Jap.,* **41,** 2349
52. Tanimoto, M., Kuchitsu, K. and Morino, Y. (1971). *Bull. Chem. Soc. Jap.,* **44,** 386
53. Tanimoto, M., Kuchitsu, K. and Morino, Y. (1969). *Bull. Chem. Soc. Jap.,* **42,** 2519
54. Bartell, L. S. and Higginbotham, H. K. (1965). *J. Chem. Phys.,* **42,** 851
55. Kuchitsu, K. (1966). *J. Chem. Phys.,* **44,** 906
56. Bartell, L. S., Roth, E. A., Hollowell, C. D., Kuchitsu, K. and Young, J. E., Jr. (1965). *J. Chem. Phys.,* **42,** 2683

57. Kuchitsu, K., Fukuyama, T. and Morino, Y. (1968). *J. Molec. Struct.*, **1,** 463
58. Owen, N. L. and Seip, H. M. (1970). *Chem. Phys. Lett.*, **5,** 162
59. Hilderbrandt, R. L. and Wieser, J. D. (1971). *J. Chem. Phys.*, **55,** 4648; (1972). *J. Chem. Phys.*, **56,** 1143
60. Iijima, T. (1972). *Bull. Chem. Soc. Jap.*, **45,** (in the press)
61. Tokue, I., Fukuyama, T. and Kuchitsu, K. (1972). *Bull. Chem. Soc. Jap.* (to be published)
62. Tsuchiya, S. and Kimura, M. (1972). *Bull. Chem. Soc. Jap.*, **45,** 736
63. Kambara, H., Kuchitsu, K. and Morino, Y. (1972). *Bull. Chem. Soc. Jap.*, (to be published)
64. Kraitchman, J. (1953). *Amer. J. Phys.*, **21,** 17; Costain, C. C. (1958). *J. Chem. Phys.*, **29,** 864; Costain, C. C. (1966). *Trans. Amer. Crystallogr. Assoc.*, **2,** 157
65. Fukuyama, T. and Kuchitsu, K. (1970). *J. Molec. Struct.*, **5,** 131
66. Sugié, M., Fukuyama, T. and Kuchitsu, K. (1972). *J. Molec. Struct.*, to be published
67. Yokozeki, A. and Kuchitsu, K. (1971). *Bull. Chem. Soc. Jap.*, **44,** 2356
68. Oka, T. and Morino, Y. (1961). *J. Molec. Spectrosc.*, **6,** 472; (1962). *ibid.*, **8,** 9; (1963). *ibid.*, **11,** 349; Kuchitsu, K., Oka, T. and Morino, Y. (1965). *J. Molec. Spectrosc.*, **15,** 51
69. Kuchitsu, K., Fukuyama, T. and Morino, Y. (1969). *J. Molec. Struct.*, **4,** 41
70. Dallinga, G. and Toneman, L. H. (1967). *J. Molec. Struct.*, **1,** 11
71. Fukuyama, T., Kuchitsu, K. and Morino, Y. (1969). *Bull. Chem. Soc. Jap.*, **42,** 379
72. Tokue, I., Fukuyama, T. and Kuchitsu, K. (1972). *Bull. Chem. Soc. Jap.*, (to be published)
73. Kuchitsu, K. and Bartell, L. S. (1961). *J. Chem. Phys.*, **35,** 1945
74. Bonham, R. A. and Peacher, J. L. (1963). *J. Chem. Phys.*, **38,** 2319
75. Almenningen, A., Bastiansen, O., Haaland, A. and Seip, H. M. (1965). *Angew. Chem. Internat. Ed.*, **4,** 819
76. Bonham, R. A. and Su, L. S. (1966). *J. Chem. Phys.*, **45,** 2827
77. Ng, E. W., Su, L. S. and Bonham, R. A. (1969). *J. Chem. Phys.*, **50,** 2038
78. Kuchitsu, K. and Morino, Y. (1965). *Bull. Chem. Soc. Jap.*, **38,** 805, 814
79. Thomas, M. A. and Welsh, H. L. (1960). *Can. J. Phys.*, **38,** 1291
80. Olafson, R. A., Thomas, M. A. and Welsh, H. L. (1961). *Can. J. Phys.*, **39,** 419
81. Morino, Y., Kuchitsu, K. and Oka, T. (1962). *J. Chem. Phys.*, **36,** 1108
82. Laurie, V. W. and Herschbach, D. R. (1962). *J. Chem. Phys.*, **37,** 1687
83. Oka, T. and Morino, Y. (1962). *J. Molec. Spectrosc.*, **8,** 302
84. Higginbotham, H. K. and Bartell, L. S. (1965). *J. Chem. Phys.*, **42,** 1131
85. Bartell, L. S. (1962). *J. Chem. Phys.*, **36,** 3495
86. Lafferty, W. J., Maki, A. G. and Coyle, T. D. (1970). *J. Molec. Spectrosc.*, **33,** 345
87. Morino, Y., Iijima, T. and Murata, Y. (1960). *Bull. Chem. Soc. Jap.*, **33,** 46
88. Nielsen, A. H. (1954). *J. Chem. Phys.*, **22,** 659
89. Ginn, S. G. W., Kenny, J. K. and Overend, J. (1968). *J. Chem. Phys.*, **48,** 1571
90. Ootake, M., Matsumura, C. and Morino, Y. (1968). *J. Molec. Spectrosc.*, **28,** 325
91. Bürgi, H. B., Stedman, D. and Bartell, L. S. (1971). *J. Molec, Struct.*, **10,** 31
92. Hedberg, K. and Iwasaki, M. (1962). *J. Chem. Phys.*, **36,** 589
93. Swingle, S. M. (1950). Cited in Sutton, L. E., editor (1958). *Tables of Interatomic Distances and Configuration in Molecules and Ions.* (London: Chemical Society)
94. Barnhart, D. M., cited by Hedberg, K. (1966). *Trans. Amer. Crystallogr. Assoc.*, **2,** 79
95. Morino, Y., Ukaji, T. and Ito, T. (1966). *Bull. Chem. Soc. Jap.*, **39,** 71
96. Ukaji, T. and Uchimura, H. (1969). unplished data; Matsumura, C. and Takeo, H. (to to published)
97. Almenningen, A. and Bjorvatten, T. (1963). *Acta. Chem. Scand.*, **17,** 2573
98. Secrist, J. H. and Brockway, L. O. (1944). *J. Amer. Chem. Soc.*, **66,** 1941
99. Schomaker, V. and Stevenson, D. P. (1941). *J. Amer. Chem. Soc.*, **63,** 37
100. Pauling, L. (1960). *The Nature of the Chemical Bond,* Third Edition. (New York: Cornell University Press)
101. Gillespie, R. J. (1963). *J. Chem. Educ.*, **40,** 295; (1967). *Angew. Chem.*, **79,** 885; *ibid., Internat. Edn.*, **6,** 819; (1970). *J. Chem. Educ.*, **47,** 18
102. Bartell, L. S. (1968). *J. Chem. Educ.*, **45,** 754
103. Bartell, L. S., Su, L. S. and Yow, H. (1970). *Inorg. Chem.*, **9,** 1903
104. Musher, J. I. (1969). *Angew. Chem.*, **81,** 68; *ibid., Internat. Edit.*, **8,** 54
105. Herzberg, G., Patat, F. and Verleger, H. (1937). *J. Phys. Chem.*, **41,** 123
106. Dewar, M. J. S. (1962). *Tetrahedron,* **17,** 123

107. Lide, D. R., Jr. (1962). *Tetrahedron,* **17,** 125
108. Stoicheff, B. P. (1962). *Tetrahedron,* **17,** 135
109. Bastiansen, O. and Traetteberg, M. (1962). *ibid.,* **17,** 147
110. Haugen, W. and Traetteberg, M. (1967). *Selected Topics in Structural Chemistry,* p. 113. (Oslo: Universitetsforlaget); Traetteberg, M. (1968). *Acta. Chem. Scand.,* **22,** 628,2294, 2305; (1969). *Doktorarbeid,* University of Trondheim
111. Herzberg, G. and Stoicheff, B. P. (1955). *Nature (London),* **175,** 79
112. Costain, C. C. and Stoicheff, B. P. (1959). *J. Chem. Phys.,* **30,** 777
113. Coulson, C. A. (1952). *Valence.* (Oxford: Clarendon Press)
114. Bartell, L. S. (1962). *Tetrahedron,* **17,** 177
115. Wilson, E. B., Jr. (1962). *Tetrahedron,* **17,** 191
116. Mulliken, R. S. (1962). *Tetrahedron,* **17,** 247
117. Dewar, M. J. S. and Schmeising, H. N. (1959). *Tetrahedron,* **5,** 166
118. Coulson, C. A. (1939). *Proc. Roy. Soc. (London),* **A169,** 413
119. Coulson, C. A., Daudel, R. and Robertson, J. M. (1951). *Proc. Roy. Soc. (London),* **A207,** 306; Cruickshank, D. W. J. (1962). *Tetrahedron,* **17,** 155
120. Coulson, C. A. (1951). *Proc. Roy. Soc. (London),* **A207,** 91
121. Pauling, L. (1952). *J. Phys. Chem.,* **56,** 361
122. Julg, A. and Pellégatti, A. (1964). *Theoret. Chim. Acta,* **2,** 202, 396; Pellégatti, A. (1967). *Thesis,* Marseille
123. Coulson, C. A. and Golebiewski, A. (1961). *Proc. Phys. Soc. (London),* **78,** 1310
124. Amos, A. T. (1967). *Theoret. Chim. Acta,* **8,** 91
125. Julg, A. (1968). *J. Chim. Phys.,* **65,** 541
126. Bartell. L. S. (1960). *J. Chem. Phys.,* **32,** 827
127. Jacob, E. J., Thompson, H. B. and Bartell, L. S. (1967). *J. Chem. Phys.,* **47,** 3736
128. Shimanouchi, T. (1970). *Physical Chemistry, An Advanced Treatise,* Vol. 4, Chapter 6, p. 233. (New York: Academic Press)
129. Bartell, L. S. and Kuchitsu, K. (1962). *J. Chem. Phys.,* **37,** 691
130. Kuchitsu, K. and Morino, Y. (1966). *Spectrochim. Acta,* **22,** 33
131. Bartell, L. S. and Kohl, D. A. (1963). *J. Chem. Phys.,* **39,** 3097
132. e.g., Scott, R. A. and Scheraga, H. A. (1966). *J. Chem. Phys.,* **44,** 3054
133. Morino, Y., Kuchitsu, K., Fukuyama, T. and Tanimoto, M. (1969). *Acta Crystallogr.,* **A25,** S127
134. Allen, L. C. (1970). *Ann. Rev. Phys. Chem.,* **20,** 315
135. Pople, J. A. and Beveridge, D. L. (1970). *Approximate Molecular Orbital Theory.* (New York: McGraw-Hill)
136. Davidson, R. B., Jorgensen, W. L. and Allen, L. C. (1970). *J. Amer. Chem. Soc.,* **92,** 749
137. Costain, C. C. and Morton, J. R. (1959). *J. Chem. Phys.,* **31,** 389
138. Lide, D. R., Jr. and Christensen, D. (1961). *J. Chem. Phys.,* **35,** 1374
139. Kilb, R. W., Lin, C. C. and Wilson, E. B., Jr. (1957). *J. Chem. Phys.,* **26,** 1695
140. Almenningen, A., Anfinsen, I. M. and Haaland, A. (1970). *Acta Chem. Scand.,* **24,** 43
141. Kato, C., Konaka, S., Iijima, T. and Kimura, M. (1969). *Bull. Chem. Soc. Jap.,* **42,** 2148
142. Tanimoto, M., Karakida, K., Fukuyama, T. and Kuchitsu, K. (to be published)
143. Karakida, K., Fukuyama, T. and Kuchitsu, K. (to be published)
144. Laurie, V. W. and Stigliani, W. M. (1970). *J. Amer. Chem. Soc.,* **92,** 1485
145. Pochan, J. M., Baldwin, J. E. and Flygare, W. H. (1969). *J. Amer. Chem. Soc.,* **91,** 1896
146. Flygare, W. H. (1967). *Ann. Rev. Phys. Chem.,* **18,** 325
147. Rudolph, H. D. (1970). *Ann. Rev. Phys. Chem.,* **21,** 73
148. Lowe, J. P. (1968). *Progr. Phys. Org. Chem.,* **6,** 1
149. Mulliken, R. S. (1959). *Tetrahedron,* **6,** 68
150. Bak, B. and Hansen-Nygaard, L. (1960). *J. Chem. Phys.,* **33,** 418
151. Rosenberg, A. and Devlin, J. P. (1965). *Spectrochim. Acta,* **21,** 1613
152. Nakagawa, T. and Morino, Y. (1969). *Bull. Chem. Soc. Jap.,* **42,** 2212
153. Sarachman, T. N. (1968). *J. Chem. Phys.,* **49,** 3146
154. Kondo, S., Sukurai, Y., Hirota, E. and Morino, Y. (1970). *J. Molec. Spectrosc.,* **34,** 231
155. Scharpen, L. H. and Laurie, V. W. (1963). *J. Chem. Phys.,* **39,** 1732
156. Takagi, K. and Oka, T. (1963). *J. Phys. Soc. Jap.,* **18,** 1174
157. McFarlane, W. (1969). *Quart. Rev. Chem. Soc.,* **23,** 187
158. Nakatsuji, H., Morishima, I., Kato, H. and Yonezawa, T. (1971). *Bull. Chem. Soc. Jap.,* **44,** 2010

159. Burgi, H. B. and Bartell, L. S. (1972). *J. Amer. Chem. Soc.*, **94,** (in the press); Bartell, L. S. and Burgi, H. B. (1972), *ibid.*, **94,** (in the press)
160. Beagley, B., Brown, D. P. and Monaghan, J. J. (1969). *J. Molec. Struct.*, **4,** 233
161. Boates, T. L. (1966). *Thesis, Iowa State University*
162. Lide, D. R., Jr. and Jen, M. (1963). *J. Chem. Phys.*, **38,** 1504
163. Bastiansen, O., Fritsch, F. N. and Hedberg, K. (1964). *Acta Crystallogr.*, **17,** 538
164. Adams, W. J., Geise, H. J. and Bartell, L. S. (1970). *J. Amer. Chem. Soc.*, **92,** 5013
165. Bernett, W. A. (1967). *J. Chem. Educ.*, **44,** 17
166. Bak, B. and Led, J. J. (1969). *J. Molec. Struct.*, **3,** 379
167. Cunningham, G. L., Boyd, A. W., Meyers, R. J., Gwinn, W. D. and LeVan, W. I. (1951). *J. Chem. Phys.*, **19,** 676. A misprint in the original paper ($\angle$CSC in their Table 14 should read $\angle$CCS) has been revised
168. Yokozeki, A. and Kuchitsu, K. (1971). *Bull. Chem. Soc. Jap.*, **44,** 72
169. Yokozeki, A. and Kuchitsu, K. (1971). *Bull. Chem. Soc. Jap.*, **44,** 1783
170. Fukuyama, T., Kuchitsu, K., Tamaru, Y., Yoshida, Z. and Tabushi, I. (1971). *J. Amer. Chem. Soc.*, **93,** 2799
171. Allen, H. C., Jr. and Plyler, E. K. (1956). *J. Chem. Phys.*, **25,** 1132
172. Hirota, E. (1972). Private communication
173. Yokozeki, A., Kuchitsu, K. and Morino, Y. (1970). *Bull. Chem. Soc. Jap.*, **43,** 2017
174. Wilson, E. B., Jr., Decius, J. C. and Cross, P. C. (1955). *Molecular Vibrations.* (New York: McGraw-Hill)
175. Laurie, V. W. (1970). *Accounts Chem. Res.*, **4,** 331
176. Newton, R. R. and Thomas, L. H. (1948). *J. Chem. Phys.*, **16,** 310
177. Thorson, W. R. and Nakagawa, I. (1960). *J. Chem. Phys.*, **33,** 994
178. Debye, P. (1941). *J. Chem. Phys.*, **9,** 55
179. Swick, D. A. and Karle, J. (1954). *J. Chem. Phys.*, **22,** 1242; (1955). *J. Chem. Phys.*, **23,** 1499
180. Morino, Y. and Hirota, E. (1958). *J. Chem. Phys.*, **28,** 185
181. Karle, J. (1966). *J. Chem. Phys.*, **45,** 4149
182. Ryan, R. R. and Hedberg, K. (1969). *J. Chem. Phys.*, **50,** 4986
183. Almenningen, A., Bastiansen, O., Fernholt, L. and Hedberg, K. (1971). *Acta Chem. Scand.*, **25,** 1946
184. Morino, Y. and Kuchitsu, K. (1958). *J. Chem. Phys.*, **28,** 175
185. Abe, M., Kuchitsu, K. and Shimanouchi, T. (1969). *J. Molec. Struct.*, **4,** 245
186. Yokozeki, A. and Kuchitsu, K. (1971). *Bull. Chem. Soc. Jap.*, **44,** 2926
187. Hougen, J. T., Bunker, P. R. and Johns, J. W. C. (1970). *J. Molec. Spectrosc.*, **34,** 136; Redding, R. W. and Hougen, J. T. (1971). *J. Molec. Spectrosc.*, **37,** 366; Brown, F. B. and Charles, N. G. (1971). *J. Chem. Phys.*, **55,** 4481
188. Almenningen, A., Arnesen, S. P., Bastiansen, O., Seip, H. M. and Seip, R. (1968). *Chem. Phys. Lett.*, **1,** 569
189. Clark, A. and Seip, H. M. (1970). *Chem. Phys. Lett.*, **6,** 452
190. Pickett, H. M. and Strauss, H. L. (1969). *J. Chem. Phys.*, **51,** 952; and the references cited therein
191. e.g., Tamagake, K., Tsuboi, M., Takagi, K. and Kojima, T. (1971). *J. Molec. Spectrosc.*, **39,** 454
192. e.g., Costain, C. C. and Dowling, J. M. (1960). *J. Chem. Phys.*, **32,** 158
193. Moore, C. B. and Pimentel, G. C. (1964). *J. Chem. Phys.*, **40,** 1529
194. Dunitz, J. D. and Schomaker, V. (1952). *J. Chem. Phys.*, **20,** 1703
195. Almenningen, A., Bastiansen, O. and Skancke, P. N. (1961). *Acta Chem. Scand.*, **15,** 711
196. Miller, F. A. and Capwell, R. J. (1971). *Spectrochim. Acta*, **27A,** 947; and the references cited therein
197. Scharpen, L. H. and Laurie, V. W. (1968). *J. Chem. Phys.*, **49,** 3041
198. Chan, S. I., Borgers, T. R., Russell, J. W., Strauss, H. L. and Gwinn, W. D. (1966). *J. Chem. Phys.*, **44,** 1103
199. Scharpen, L. H. and Laurie, V. W. (1968). *J. Chem. Phys.*, **49,** 221
200. Harris, D. O., Harrington, H. W., Luntz, A. C. and Gwinn, W. D. (1966). *J. Chem. Phys.*, **44**, 3467
201. Petit, M. G., Gibson, J. S. and Harris, D. O. (1970). *J. Chem. Phys.*, **53,** 3408
202. Pringle, W. C., Jr. (1971). *J. Chem. Phys.*, **54,** 4979
203. Scharpen, L. H. (1968). *J. Chem. Phys.*, **48,** 3552

204. Durig, J. R. and Wertz, D. W. (1968). *J. Chem. Phys.*, **49,** 2118
205. Carreira, L. A., Jiang, G. J., Person, W. B. and Willis, J. N., Jr. (1972). *J. Chem. Phys.*, **56,** 1440
206. Kilpatrick, J. E., Pitzer, K. S. and Spitzer, R. (1947). *J. Amer. Chem. Soc.*, **69,** 2483
207. Pickett, H. M. and Strauss, H. L. (1971). *J. Chem. Phys.*, **55,** 324
208. Engerholm, G. G., Luntz, A. C., Gwinn, W. D. and Harris, D. O. (1969). *J. Chem. Phys.*, **50,** 2446
209. Almenningen, A., Seip, H. M. and Willadsen, T. (1969). *Acta Chem. Scand.*, **23,** 2748
210. Geise, H. J., Adams, W. J. and Bartell, L. S. (1969). *Tetrahedron,* **25,** 3045
211. Bartell, L. S. (1971). Private communication; Hirakawa, A., Tsuboi, M., Kambara, H. and Kuchitsu, K. (1972). to be published
212. Seip, H. M. (1971). *J. Chem. Phys.*, **54,** 440
213. Green, W. H., Harvey, A. B. and Greenhouse, J. A. (1971). *J. Chem. Phys.*, **54,** 850
214. Jacob, E. J. and Bartell, L. S. (1970). *J. Chem. Phys.*, **53,** 2235
215. Adams, W. J., Thompson, H. B. and Bartell, L. S. (1970). *J. Chem. Phys.*, **53,** 4040
216. Kaiser, E. W., Muenter, J. S., Klemperer, W. and Falconer, W. E. (1970). *J. Chem. Phys.*, **53,** 53
217. Gavin, R. M., Jr. and Bartell, L. S. (1968). *J. Chem. Phys.*, **48,** 2460
218. Bartell, L. S. and Gavin, R. M., Jr. (1968). *J. Chem. Phys.*, **48,** 2466
219. Reichman, S. and Schreiner, F. (1969). *J. Chem. Phys.*, **51,** 2355
220. Bohn, R. K., Katada, K., Martinez, J. V. and Bauer, S. H. (1963). *Noble Gas Compounds,* p. 238. (Chicago: University of Chicago Press)
221. Hansen, K. W. and Bartell, L. S. (1965). *Inorg. Chem.*, **4,** 1775
222. Berry, R. S. (1960). *J. Chem. Phys.*, **32,** 933
223. Ugi, I., Marquarding, D., Klusacek, H., Gillespie, P. and Ramirez, F. (1971). *Accounts Chem. Res.*, **4,** 288; Miller, F. A. and Capwell, R. J. (1971). *Spectrochim. Acta,* **27A,** 125
224. Hirota, E. (1971). *J. Molec. Spectrosc.*, **38,** 367
225. Ermer, O. and Dunitz, J. D. (1969). *Helv. Chim. Acta,* **52,** 1861
226. Hirota, E. and Suenaga, S. (1972). *J. Molec. Spectrosc.* (in the press)
227. Yokozeki, A. and Kuchitsu, K. (1971). *Bull. Chem. Soc. Jap.*, **44,** 72
228. Wilson, E. B., Jr. (1959). *Advan. Chem. Phys.*, **2,** 367
228a. Bastiansen, O., Seip, H. M. and Boggs, J. E. (1971). *Perspectives in Structural Chemistry,* Vol. 4, 60 (J. D. Dunitz and J. A. Ibers, editors). (New York: J. Wiley)
229. Herzberg, G. (1966). *Molecular Spectra and Molecular Structure III. Electronic Spectra and Electronic Structure of Polyatomic Molecules,* Chapter 1, Section 2 (Princeton, New Jersey: D. Van Nostrand)
230. Morino, Y. and Uehara, H. (1966). *J. Chem. Phys.*, **45,** 4543
231. Brand, J. C. D., Goodman, G. L. and Weinstock, B. (1971). *J. Molec. Spectrosc.*, **38,** 449
232. Jacob, E. J. and Bartell, L. S. (1970). *J. Chem. Phys.*, **53,** 2231
233. Nahlovska, Z., Nahlovsky, B. and Strand, T. G. (1970). *Acta. Chem. Scand.*, **24,** 2617
234. Lowrey, A. H., George, C., D'Antonio, P. and Karle, J. (1971). *J. Amer. Chem. Soc.*, **93,** 6399
235. Almenningen, A., Bastiansen, O. and Motzfelt, T. (1969). *Acta Chem. Scand.*, **23,** 2848; (1970). *Acta Chem. Scand.*, **24,** 747
236. Derrissen, J. L. (1971). *J. Molec. Struct.*, **7,** 67
237. Janzen, J. and Bartell, L. S. (1969). *J. Chem. Phys.*, **50,** 3611
238. Almenningen, A., Halvorsen, S. and Haaland, A. (1971). *Acta Chem. Scand.*, **25,** 1937
239. Zasorin, E. Z. and Rambidi, N. G. (1967). *Zh. Strukt. Khim.*, **8,** 391, 591
240. Haaland, A. and Nilsson, J. E. (1968). *Acta Chem. Scand.*, **22,** 2653
241. Hedberg, L. and Hedberg, K. (1970). *J. Chem. Phys.*, **53,** 1228
242. Audit, P. (1969). *J. Phys. (Paris),* **30,** 192

7
The Theory of Dielectric Polarisation

A. D. BUCKINGHAM
University Chemical Laboratory, Cambridge

7.1 INTRODUCTION

7.1.1 Terminology

A dielectric is a non-conductor of electricity. The dielectric properties of a material determine its response to an applied electric field $\boldsymbol{E}$. When a dielectric is inserted between the plates of a parallel plate condenser, its capacitance is increased by a factor ε_r called the relative permittivity, or sometimes the dielectric constant. The medium is polarised by the field, that is, it has an electric dipole moment induced in it, and this enhances both the electric displacement $\boldsymbol{D} = \varepsilon_r \varepsilon_0 \boldsymbol{E}$ and the charge that the conducting plates hold when a fixed potential difference exists between them. The quantity ε_0 is the permittivity of a vacuum and in SI units $4\pi\varepsilon_0 = 10^7 c^{-2} = 1.11265 \times 10^{-10}$ C V^{-1} m^{-1} where c is the speed of light in m s^{-1}; in the electrostatic system of units $4\pi\varepsilon_0 = 1$. The induced moment is related to the polarity of the molecules comprising the system and hence to their structure; thus a measurement of the polarisation $\boldsymbol{P} = (\varepsilon_r - 1)\varepsilon_0 \boldsymbol{E}$ induced by a uniform field $\boldsymbol{E}$ yields information about molecular structure. The theory of dielectric polarisation relates $\boldsymbol{P}/\boldsymbol{E}$ to the microscopic or molecular properties of the material. It has been a topic of interest for nearly a century and remains active.

7.1.2 Early theories

A basic difficulty in formulating a general theory of the relative permittivity ε_r is that the effective field acting on a molecule (sometimes called the local field) is not necessarily equal to the applied macroscopic field $\boldsymbol{E}$. In a gas at low pressure the two fields are the same, but in an imperfect gas, or in a liquid or solid, they are different. The macroscopic field $\boldsymbol{E}$ is the *average* field at a point in the material; there may be large fluctuations in the actual field strength, but the mean field in a medium in which $\boldsymbol{E}$ is uniform is equal to minus the gradient of the mean potential. In a parallel plate condenser in which U is the voltage between the plates which are separated by d metres, $E = U/d$ V m^{-1} and the direction of $\boldsymbol{E}$ is from the higher to the lower potential.

The Clausius–Mossotti formula for ε_r [1,2], namely

$$(\varepsilon_r - 1)/(\varepsilon_r + 2) = (N\alpha/3\varepsilon_0 V) \tag{7.1}$$

where N is the number of molecules in the volume V and α is the polarisability of a single molecule, that is the dipole moment induced in an isolated molecule by unit field, was deduced by equating the effective field to the Lorentz field[3]

$$\boldsymbol{E}_{\text{Lorentz}} = \boldsymbol{E} + \frac{1}{3\varepsilon_0}\boldsymbol{P} = \boldsymbol{E} + \frac{\varepsilon_r - 1}{3}\boldsymbol{E} = \frac{\varepsilon_r + 2}{3}\boldsymbol{E} \tag{7.2}$$

If a spherical sample of dielectric is placed into a uniform field $\boldsymbol{E}_0$ which exists in a vacuum due to charges at infinity, the field inside the sphere can easily be shown by classical electrostatics to be uniform and equal to $\boldsymbol{E} = 3\boldsymbol{E}_0/(\varepsilon_r + 2)$ (see for example Stratton[4], p. 205). Hence the external field $\boldsymbol{E}_0$ is equal to $\boldsymbol{E}_{\text{Lorentz}}$ in this case.

The Lorentz field was also used by Debye in his pioneering work on the electric dipole moments of molecules[5, 6]. His equation is

$$\frac{\varepsilon_r - 1}{\varepsilon_r + 2} = \frac{N}{3\varepsilon_0 V}\left(\alpha + \frac{\mu^2}{3kT}\right) \tag{7.3}$$

where μ is the magnitude of the permanent dipole moment of an isolated molecule. The dipole moment enters as its square because its contribution to the polarisation $\boldsymbol{P}$ is proportional to $\overline{\mu_E}$, the mean value of the component of $\boldsymbol{\mu}$ in the direction of the field, and for a classical rotator in weak fields

$$\overline{\mu_E} = \mu \overline{\cos\theta} = (\mu^2/3kT)E_{\text{Lorentz}} \tag{7.4}$$

where θ is the angle between $\boldsymbol{\mu}$ and $\boldsymbol{E}$ and the bar denotes a statistical average.

A major difficulty with the Clausius–Mossotti and Debye equations occurs when the right-hand sides of equations (7.1) and (7.3) reach the value of unity when the density is increased (or the temperature decreased in the case of equation (7.3))—then $\varepsilon_r = \pm\infty$ when the right-hand sides equal $1 \mp \delta$. Various attempts were made to overcome this problem; Debye[7] and Fowler[8] introduced orientational correlation between the dipole moments of neighbouring molecules, and Onsager[9] considered a molecule in a real spherical cavity of radius a in a continuum having the bulk properties of the medium. The field inside the cavity when the infinite medium carries a field $\boldsymbol{E}$ is Onsager's cavity field $\boldsymbol{G}$ and classical electrostatics yields[9]

$$\boldsymbol{G} = \frac{3\varepsilon_r}{2\varepsilon_r + 1}\boldsymbol{E} = \boldsymbol{E} + \frac{\varepsilon_r - 1}{2\varepsilon_r + 1}\boldsymbol{E} \tag{7.5}$$

The cavity field $\boldsymbol{G}$ is the actual field in a real cavity and differs from the Lorentz field which is the sum of $\boldsymbol{E}$ and the field of the charges in the medium outside a hypothetical sphere in the infinite medium. Thus

$$\boldsymbol{E}_{\text{Lorentz}} = \boldsymbol{E} + \frac{1}{3\varepsilon_0}\boldsymbol{P} = \frac{\varepsilon_r + 2}{3}\boldsymbol{E} = \frac{(\varepsilon_r + 2)(2\varepsilon_r + 1)}{9\varepsilon_r}\boldsymbol{G} \tag{7.6}$$

Onsager also introduced the reaction field $\boldsymbol{R}$ as the field in the cavity arising from the polarisation of the medium by the dipole moment of a point polarisable dipole representing a single molecule. The reaction field is proportional to the total dipole moment $\boldsymbol{m}$ of the molecule. Classical electrostatics leads to the result[9]

$$\boldsymbol{R} = \frac{2(\varepsilon_r - 1)}{(2\varepsilon_r + 1)}\frac{\boldsymbol{m}}{4\pi\varepsilon_0 a^3} \tag{7.7}$$

and $\boldsymbol{m}$ is the sum of the moment $\boldsymbol{\mu}$ when $E = 0$ and an induced moment proportional to $\boldsymbol{E}$. Hence

$$\boldsymbol{m} = \boldsymbol{\mu}_0 + \alpha(\boldsymbol{G} + \boldsymbol{R}) \tag{7.8}$$

where $\boldsymbol{\mu}_0$ is the permanent moment of a single molecule in the gaseous phase. From equations (7.7) and (7.8)

$$\boldsymbol{m} = \boldsymbol{\mu} + \frac{3\varepsilon_r}{2\varepsilon_r + 1}\left[1 - \frac{2(\varepsilon_r - 1)}{2\varepsilon_r + 1}\frac{\alpha}{4\pi\varepsilon_0 a^3}\right]^{-1}\alpha\boldsymbol{E} \tag{7.9}$$

$$\boldsymbol{\mu} = \left[1 - \frac{2(\varepsilon_r - 1)}{2\varepsilon_r + 1}\frac{\alpha}{4\pi\varepsilon_0 a^3}\right]^{-1}\boldsymbol{\mu}_0 \tag{7.10}$$

To obtain Onsager's effective field it is convenient to write $\boldsymbol{R}$ as the sum of a term proportional to $\boldsymbol{E}$ and a term proportional to $\boldsymbol{\mu}$. Thus

$$\boldsymbol{R} = \boldsymbol{R}_E + \boldsymbol{R}_\mu \tag{7.11}$$

where

$$\boldsymbol{R}_E = \frac{2(\varepsilon_r - 1)}{2\varepsilon_r + 1}\left[1 - \frac{2(\varepsilon_r - 1)}{2\varepsilon_r + 1}\frac{\alpha}{4\pi\varepsilon_0 a^3}\right]^{-1}\frac{\alpha \boldsymbol{G}}{4\pi\varepsilon_0 a^3} \tag{7.12}$$

$$\boldsymbol{R}_\mu = \frac{2(\varepsilon_r - 1)}{2\varepsilon_r + 1}\frac{\boldsymbol{\mu}}{4\pi\varepsilon_0 a^3} \tag{7.13}$$

The effective field is $\boldsymbol{G} + \boldsymbol{R}_E = \boldsymbol{E}_{\text{Onsager}}$,

$$\boldsymbol{E}_{\text{Onsager}} = \frac{3\varepsilon_r}{2\varepsilon_r + 1}\left[1 - \frac{2(\varepsilon_r - 1)}{(2\varepsilon_r + 1)}\frac{\alpha}{4\pi\varepsilon_0 a^3}\right]^{-1}\boldsymbol{E} \tag{7.14}$$

and for this model

$$(\varepsilon_r - 1)E = \frac{N\overline{\boldsymbol{m}\cdot\boldsymbol{E}}}{\varepsilon_0 VE} = \frac{N}{\varepsilon_0 V}\left[\alpha E_{\text{Onsager}} + \mu\,\overline{\cos\theta}\right] \tag{7.15}$$

Onsager did not equate $\mu\,\overline{\cos\theta}$ to $\mu^2 E_{\text{Onsager}}/3kT$; he noticed that the reaction field $\boldsymbol{R}$ is parallel to $\boldsymbol{m}$ and therefore does not contribute to the torque on $\boldsymbol{m}$. He deduced that $\boldsymbol{G}$ is the field orienting $\boldsymbol{\mu}$ and hence that

$$\mu\,\overline{\cos\theta} = \frac{\mu^2 G}{3kT} = \frac{3\varepsilon_r}{2\varepsilon_r + 1}\frac{\mu^2 E}{3kT} \tag{7.16}$$

From equations (7.15) and (7.16)

$$\frac{(\varepsilon_r - 1)(2\varepsilon_r + 1)}{9\varepsilon_r}\left[1 - \frac{2(\varepsilon_r - 1)\alpha}{(2\varepsilon_r + 1)4\pi\varepsilon_0 a^3}\right] = \frac{N}{3\varepsilon_0 V}\left(\alpha + \frac{\mu\mu_0}{3kT}\right) \tag{7.17}$$

Unlike equation (7.3), equation (7.17) does not misbehave when its right hand side goes through unity.

Onsager supposed further that

$$\frac{\alpha}{4\pi\varepsilon_0 a^3} = \frac{\varepsilon_\infty - 1}{\varepsilon_\infty + 2} \tag{7.18}$$

where ε_∞ is the relative permittivity for optical frequencies extrapolated to very long wavelengths; ε_∞ depends on the 'distortion' polarisation and not on the 'orientation' polarisation proportional to $\mu\,\overline{\cos\theta}$. Actually, ε_∞ is not well defined for a real molecular system and the splitting of the total polarisation is artificial, particularly when the molecular structure is non-rigid; however, in many cases there is little absorption between molecular rotation frequencies and optical frequencies and there is then minimal ambiguity in the appropriate value of ε_∞. However, far-infrared absorption spectra have recently been recorded for a number of liquids and it appears that in general absorption does occur over the entire frequency range from radio to optical frequencies[10].

Using equation (7.18), the permanent moment in equation (7.10) is

$$\boldsymbol{\mu} = \frac{(2\varepsilon_r + 1)(\varepsilon_\infty + 2)}{3(2\varepsilon_r + \varepsilon_\infty)}\boldsymbol{\mu}_0 \tag{7.19}$$

Also, the effective field reduces to

$$\boldsymbol{E}_{\text{Onsager}} = \frac{\varepsilon_r(\varepsilon_\infty + 2)}{(2\varepsilon_r + \varepsilon_\infty)} \boldsymbol{E} \tag{7.20}$$

and equation (7.17) to

$$\frac{(\varepsilon_r - 1)(2\varepsilon_r + \varepsilon_\infty)}{3\varepsilon_r(\varepsilon_\infty + 2)} = \frac{N}{3\varepsilon_0 V}\left(\alpha + \frac{\mu\mu_0}{3kT}\right) \tag{7.21}$$

The left-hand side of equation (7.21) is shown as a function of ε_r and ε_∞ in Figure 7.1. Unlike the Debye equation (7.3), there is no catastrophe in

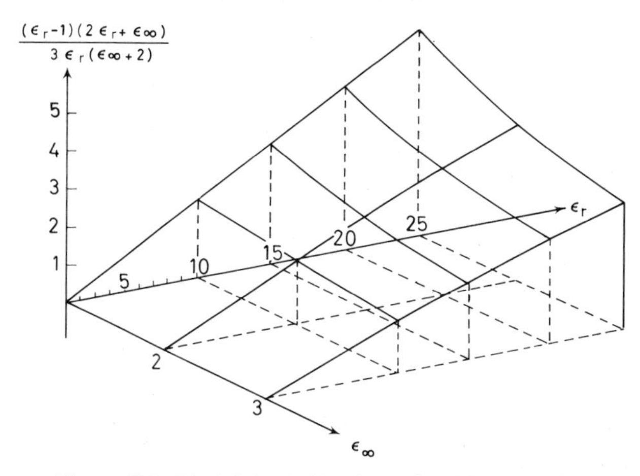

Figure 7.1 The left-hand side of equation 7.21 *v.* ε_r and ε_∞

Onsager's equation. Actually, μ in equation (7.21) is, in Onsager's model, a function of ε_r (see equation (7.19)) but this does not affect the qualitative result.

Onsager introduced the additional assumption that the volume of a single cavity is equal to V/N, that is

$$\frac{4\pi N a^3}{3V} = 1 \tag{7.22}$$

Then equation (7.21) reduces to the well-known equation

$$\frac{(\varepsilon_r - \varepsilon_\infty)(2\varepsilon_r + \varepsilon_\infty)}{\varepsilon_r(\varepsilon_\infty + 2)^2} = \frac{N\mu_0^2}{9\varepsilon_0 VkT} \tag{7.23}$$

Various generalisations have been incorporated into Onsager's model; Wilson[11] included anisotropy in the polarisability and Scholte[12], Böttcher[13] and others[13a] have considered ellipsoids in place of the sphere. Although Onsager's work represented a major advance, it is not a general statistical-mechanical theory of dielectric polarisation for it is based on a simplified model of molecular behaviour in a polar liquid. It is inappropriate to an imperfect gas where the assumption of a constant value of μ is clearly not consistent with the fluctuations that exist when a pair of polarisable dipoles interact[14, 18]. If $(\varepsilon_r - 1)V/(\varepsilon_r + 2)$ is expanded as a power series in V_m^{-1} where

V_m is the molar volume of the gas:

$$\frac{\varepsilon_r - 1}{\varepsilon_r + 2} V_m = A_\varepsilon + \frac{B_\varepsilon}{V_m} + \frac{C_\varepsilon}{V_m^2} + \ldots \tag{7.24}$$

then the observables A_ε, B_ε, C_ε ... are called the first, second, third, ..., *dielectric virial coefficients*[15]; for both the Debye and Onsager models, B_ε is zero whereas in fact B_ε makes a significant contribution to the polarisation and may be either positive or negative[16].

There has been much spirited discussion about the validity of equation (7.16) for $\overline{\cos\theta}$ in Onsager's model. Harris and Alder[17] and Harris[18] developed a statistical-mechanical theory which they applied to Onsager's model; they deduced a formula differing from equation (7.21) only in that μ^2 replaces $\mu\mu_0$, which is equivalent to equating $\overline{\cos\theta}$ to $\mu E_{\text{Onsager}}/3kT$. However Fröhlich[19] and O'Dwyer[20] argued, correctly it seems, that Onsager's equation is the right result for the model. In any case, as shown in Section 7.2, very general statistical-mechanical reasoning, applicable to non-rigid as well as to rigid molecules, indicates that Onsager's expression for $\overline{\cos\theta}$, namely $\mu G/3kT$, is appropriate. (See equation (7.41).)

One further point to note about Onsager's equations is that equation (7.17) does *not* reduce to the Clausius–Mossotti equation (7.1) when $\mu_0 = \mu = 0$, unless the additional assumption of equation (7.18) is incorporated; this point has been discussed by Brown[21].

7.2 GENERAL STATISTICAL-MECHANICAL THEORIES

7.2.1 Kirkwood's approach

The first successful general theory relating the permittivity of a polar fluid to the properties of the molecules comprising the fluid was propounded by Kirkwood[22] in 1939. If the fluid is comprised of rigid, i.e. non-polarisable, dipoles, then the theory is rigorous within the limits of applicability of classical statistical mechanics. If the molecules are non-rigid, then the theory provides a rigorous expression for $\overline{\cos\theta}$ [23, 24] where θ is the angle between the field $\boldsymbol{E}$ in the medium and a molecule-fixed axis such as the direction of the dipole moment in a free gaseous molecule in its ground state.

Kirkwood chose to introduce a macroscopic sample containing N identical molecules occupying a spherical volume V into a region of free space in which fixed external charges produce a uniform field $\boldsymbol{E}_0$ in vacuo. Classical electrostatics requires that the field $\boldsymbol{E}$ in the sphere is uniform and given by (see p. 242)

$$\boldsymbol{E} = [3/(\varepsilon_r + 2)]\boldsymbol{E}_0 \tag{7.25}$$

7.2.1.1 *The shape of the sample*

One may enquire if Kirkwood's choice of a spherical sample of fluid imposes any restriction on the validity of his theory of dielectric polarisation. It does not, provided one is concerned with homogeneous dielectrics, for in any internal region Q of a macroscopic specimen the source of the polarisation $\boldsymbol{P}$ is $\boldsymbol{E}$, the field in that region; the molecules experience a field $\boldsymbol{E}$ and that is why the region acquires a mean dipole moment. The source of $\boldsymbol{E}$, which, for a given $\boldsymbol{E}$, is dependent on the shape of the macroscopic sample, is irrelevant. What external influence other than $\boldsymbol{E}$ could yield the dipole moment of the region Q? Magnetic fields or electric field-gradients might affect $\boldsymbol{P}$, but their slight consequences form part of another subject – non-linear dielectrics. Surface forces might interfere, but their range is limited to regions in the vicinity of the surface and by taking a large enough sample the proportion of molecules that are near the surface of a three-dimensional sample may be made vanishingly small. Thus the ratio $\boldsymbol{P}/\boldsymbol{E}$ is the same in the interior of a large sphere and in a slab between parallel plates provided the thermodynamic state is the same (i.e. the same temperature, density and chemical potential). Thus the relative permittivity ε_r is a genuine equilibrium bulk property. Its independence of shape has recently been explored for a model fluid comprised of rigid polar molecules by Nienhuis and Deutch[25]; the constancy of $\boldsymbol{P}/\boldsymbol{E}$ suggests to them that it must be possible to express ε_r in terms of microscopic properties of the fluid, that is, the properties of a molecule and its near neighbours.

7.2.1.2 *Internal degrees of freedom*

The sphere in an external field $\boldsymbol{E}_0$ is a tractable theoretical model, for $\boldsymbol{E}_0$ unlike $\boldsymbol{E}$, can be fixed when averaging over molecular positions and orientations τ ($\boldsymbol{E}$ is a temperature-dependent quantity which involves averaging over τ when $\boldsymbol{E}_0$ is fixed). It is reasonable to consider the positions and orientations of the molecules in the fluid as classical variables and to treat the internal electronic and vibrational coordinates quantum mechanically. These internal degrees of freedom endow molecules with polarisabilities and other properties ascribable to non-rigidity. In Fröhlich's general theory of dielectric polarisation[26] the electronic and nuclear coordinates are not distinguished; all are treated classically, but this is not appropriate for electrons in molecules since the separation of electronic states is generally large compared to kT. It is preferable to use a type of Born–Oppenheimer approach and to take a quantum-mechanical expectation value over the electronic and vibrational coordinates in the presence of the field for fixed positions and orientations τ of all the molecules. This leads to a total dipole moment $\boldsymbol{M}(\tau, \boldsymbol{E}_0)$ which is a function of the independent variables τ and $\boldsymbol{E}_0$. The final averaging over τ may then be accomplished using classical statistical mechanics, since the translational and rotational motion of molecules in a dense fluid is well described by classical mechanics, the separation of the occupied levels being very small compared to kT.

7.2.1.3 *Statistical averaging*

The instantaneous potential energy $V(\tau, \boldsymbol{E}_0)$ of the sphere in the field may be expanded as a power series in $\boldsymbol{E}_0$:

$$V(\tau, \boldsymbol{E}_0) = V(\tau) - \boldsymbol{M}(\tau) \cdot \boldsymbol{E}_0 - \ldots \tag{7.26}$$

where $V(\tau)$ and $\boldsymbol{M}(\tau)$ are the potential energy and dipole moment of the sphere in the absence of the field when the molecules are in the positions and orientations τ. The omitted terms in equation (7.26) are proportional to $\boldsymbol{E}_0^2$, $\boldsymbol{E}_0^3$, etc. and are associated with the polarisation induced by the action of the field on the internal degrees of freedom.

If $\boldsymbol{e}$ is the unit vector in the direction of $\boldsymbol{E}_0$ and $\boldsymbol{E}$, so that $\boldsymbol{E}_0 = E_0\boldsymbol{e}$, then $\cos\theta = \boldsymbol{a}\cdot\boldsymbol{e}$ where θ is the angle between $\boldsymbol{e}$ and an axis $\boldsymbol{a}$ fixed in a particular molecule (say molecule 1). Classical statistical mechanics requires that the mean value of $\cos\theta$ in the field is

$$\overline{\cos\theta} = \overline{\boldsymbol{a}\cdot\boldsymbol{e}} = \frac{\int(\boldsymbol{a}\cdot\boldsymbol{e})\exp\left[-V_n(\tau,\boldsymbol{E}_0)/kT\right]\mathrm{d}\tau}{\int\exp\left[-V_n(\tau,\boldsymbol{E}_0)/kT\right]\mathrm{d}\tau} \tag{7.27}$$

Equations (7.26) and (7.27) yield, on neglecting terms in higher powers of E_0,

$$\overline{\cos\theta} = \langle\cos\theta\rangle + \left[\langle(\boldsymbol{a}\cdot\boldsymbol{e})(\boldsymbol{M}(\tau)\cdot\boldsymbol{e})\rangle - \langle\boldsymbol{a}\cdot\boldsymbol{e}\rangle\langle\boldsymbol{M}(\tau)\cdot\boldsymbol{e}\rangle\right]E_0/kT$$

where the angular brackets denote a statistical average over all τ in the absence of the field; thus

$$\langle X\rangle = \frac{\int X(\tau)\exp\left[-V(\tau)/kT\right]\mathrm{d}\tau}{\int\exp\left[-V(\tau)/kT\right]\mathrm{d}\tau} \tag{7.28}$$

If the fluid is isotropic in the absence of the field,

$$\langle\cos\theta\rangle = \langle\boldsymbol{M}(\tau)\cdot\boldsymbol{e}\rangle = 0, \quad \langle(\boldsymbol{a}\cdot\boldsymbol{e})(\boldsymbol{M}(\tau)\cdot\boldsymbol{e})\rangle = \tfrac{1}{3}\langle\boldsymbol{M}(\tau)\cdot\boldsymbol{a}\rangle = \tfrac{1}{3}\langle M_a(\tau)\rangle$$

and

$$\overline{\cos\theta} = \langle M_a(\tau)\rangle E_0/3kT \tag{7.29}$$

From equations (7.25) and (7.29)

$$\overline{\cos\theta} = \langle M_a(\tau)\rangle(\varepsilon_r+2)E/9kT \tag{7.30}$$

7.2.1.4 Relationship between the mean macroscopic and microscopic moments

Equation (7.30) is a macroscopic equation and it is appropriate only to a spherical sample. Kirkwood showed in the Appendix to his paper[22] how to relate $\langle M_a(\tau)\rangle$ to $\overline{m_a}$, the mean dipole moment, in the direction of the $\boldsymbol{a}$ axis fixed in molecule 1, of this molecule and its neighbours in a small sphere centred on it. This 'inner sphere' is small in relation to the entire sample but otherwise its radius is arbitrarily large. Its mean moment $\overline{m_a}$ is a true microscopic moment in the sense that the polarisation is concentrated in the vicinity of molecule 1; $\overline{m_a}$ is therefore independent of the shape of the bulk sample. Kirkwood's equation relating the macroscopic moment $\langle M_a(\tau)\rangle$ of the sphere to $\overline{m_a}$ is

$$\langle M_a(\tau)\rangle = \frac{9\varepsilon_r}{(2\varepsilon_r+1)(\varepsilon_r+2)}\overline{m_a} \tag{7.31}$$

and this important result is now carefully derived.

The moment $\langle M_a(\tau)\rangle$ of the sphere is an average over all configurations τ. This averaging is conveniently carried out in two stages, first over all configurations of all molecules except 1 and finally over all positions and orientations of molecule 1. Thus, consider molecule 1 with its centre of mass fixed at the position O_1 in the interior of the sphere (see Figure 7.2). The

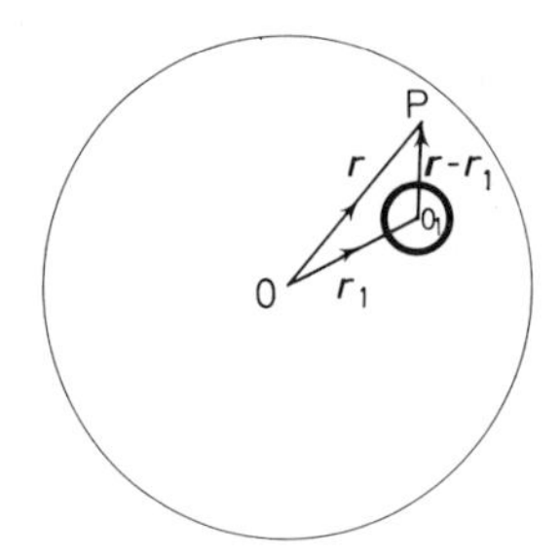

Figure 7.2 The position of the fixed molecule 1 at the centre O_1 of the inner sphere of radius r_0 in the large sphere of radius R. The point P is at an arbitrary point $\boldsymbol{r}$ outside the inner sphere and at $\boldsymbol{r}-\boldsymbol{r}_1$ relative to O_1

orientation of molecule 1 is also fixed, but all other molecules are allowed to take up any configuration in the sphere of radius R. When $E_0 = 0$, the mean dipole moment of the entire sphere in the direction of the $\boldsymbol{a}$ axis of the fixed molecule 1 is

$$\overline{M_a}(\tau_1) = \frac{\int\ldots\int M_a(\tau_1, \tau_2, \ldots, \tau_N)\exp\left[-V(\tau_1, \tau_2, \ldots, \tau_N)/kT\right]\mathrm{d}\tau_2\ldots\mathrm{d}\tau_N}{\int\ldots\int\exp\left[-V(\tau_1, \tau_2, \ldots, \tau_N)/kT\right]\mathrm{d}\tau_2\ldots\mathrm{d}\tau_N} \tag{7.32}$$

The $N-1$ molecules 2 to N are unconstrained except by the fixed molecule 1 and by the imaginary spherical boundary enclosing the volume V. The near neighbours of molecule 1 experience short-range coupling with it, but this diminishes with increasing separation until at macroscopic distances the interactions with molecule 1 may be described in terms of the bulk properties of the fluid. Thus Kirkwood drew a small macroscopic sphere of radius r_0 around the fixed molecule 1 at O_1. The radius r_0 is chosen so that $r_0 \to \infty$ and $r_0/R \to 0$. The region outside the sphere of radius r_0 contributes to $\overline{M_a}(\tau_1)$ as if it were a continuum of relative permittivity ε_r. Classical electrostatics can then be employed to determine the mean potential in any region outside the small sphere. The potential must satisfy Laplace's equation and in Cartesian tensor notation is therefore of the form

$$\phi = \frac{\mu_\alpha^*}{4\pi\varepsilon_0\varepsilon_r}\frac{r_\alpha - r_{1\alpha}}{|\boldsymbol{r}-\boldsymbol{r}_1|^3} + \frac{\Theta_{\alpha\beta}^*(r_\alpha - r_{1\alpha})(r_\beta - r_{1\beta})}{4\pi\varepsilon_0\varepsilon_r|\boldsymbol{r}-\boldsymbol{r}_1|^5} + \ldots$$

$$+ \sum_{n=1}^{\infty} B_{\alpha\beta\gamma\ldots\nu}^{(n)} {}^r\alpha {}^r\beta {}^r\gamma \ldots, {}^r\nu \tag{7.33}$$

in the region between the spherical boundaries, and

$$\phi_0 = \frac{\overline{M_\alpha}(\tau_1)r_\alpha}{4\pi\varepsilon_0 r^3} + \frac{\overline{\Theta_{\alpha\beta}}(\tau_1)r_\alpha r_\beta}{4\pi\varepsilon_0 r^5} + \ldots \tag{7.34}$$

in the exterior region. The usual tensor summation convention is used (i.e. a repeated Greek suffix implies a summation over all three Cartesian components; thus $M_\alpha r_\alpha \equiv M_x r_x + M_y r_y + M_z r_z \equiv \boldsymbol{M}\cdot\boldsymbol{r}$); μ_α^* and $\Theta_{\alpha\beta}^*$ are the so-called *external* dipole and quadrupole moments of the inner sphere of radius r_0 at $\boldsymbol{r}_1$, and $\overline{M_\alpha}(\tau_1)$ and $\overline{\Theta_{\alpha\beta}}(\tau_1)$ are the mean dipole and quadrupole moments of the entire sample when the configuration τ_1 of molecule 1 is fixed, and they generate the mean potential ϕ_0 in equation (7.34) at any point outside the sphere of radius R.

The potentials ϕ and ϕ_0 satisfy the boundary conditions of electrostatics at $r = R$, namely

$$\left.\begin{aligned} \phi &= \phi_0 \text{ at } r = R \\ \varepsilon_r \left(\frac{\partial \phi}{\partial r}\right)_{r=R} &= \left(\frac{\partial \phi_0}{\partial r}\right)_{r=R} \end{aligned}\right\} \tag{7.35}$$

Hence from (7.33), (7.34) and (7.35)

$$\left.\begin{aligned} B_\alpha^{(1)} &= \frac{2(\varepsilon_r - 1)}{3\varepsilon_r} \frac{\overline{M_\alpha}(\tau_1)}{4\pi\varepsilon_0 R^3} \\ \overline{M_\alpha}(\tau_1) &= \frac{3}{\varepsilon_r + 2} \mu_\alpha^* \end{aligned}\right\} \tag{7.36}$$

The mean dipole moment of the large sphere, $\overline{M_\alpha}(\tau_1)$, may also be equated to the sum of that of the inner sphere at $\boldsymbol{r}_1$ and the moment induced in the region between the spheres, if V_0 is the volume of the inner sphere,

$$\overline{M_\alpha}(\tau_1) = \overline{m_\alpha} + \int_{V_0}^{V} P_\alpha \, \mathrm{d}\tau \tag{7.37}$$

where the integral of the polarisation P_α is taken over the volume between the spherical boundaries. The mean moment $\overline{m_\alpha}$ of the inner sphere centred on the fixed molecule 1 is independent of the position and orientation τ_1 except for a negligible fraction of the molecules near the surface of the sample. The polarisation is simply related to the gradient of ϕ, $P_\alpha = -\varepsilon_0(\varepsilon_r - 1)\nabla_\alpha \phi$, so equations (7.33) and (7.37) yield

$$\overline{M_\alpha}(\tau_1) = \overline{m_\alpha} - \varepsilon_0(\varepsilon_r - 1) B_\alpha^{(1)} \frac{4\pi}{3}(R^3 - r_0^3) \tag{7.38}$$

The multipolar fields proportional to $\boldsymbol{\mu}^*, \boldsymbol{\Theta}^*$, etc. make no contribution to $\int P_\alpha \, \mathrm{d}\tau$.

From equations (7.36) and (7.38)

$$\overline{M_\alpha}(\tau_1) = \frac{9\varepsilon_r}{(2\varepsilon_r + 1)(\varepsilon_r + 2)} \overline{m_\alpha} \left[1 - \frac{2(\varepsilon_r - 1)^2}{(2\varepsilon_r + 1)(\varepsilon_r + 2)} \left(\frac{r_0}{R}\right)^3\right]^{-1} \tag{7.39}$$

which is independent of τ_1. The second part of the averaging in $\langle M_\alpha(\tau)\rangle$, that is the average over all positions $\boldsymbol{r}_1$ and orientations of molecule 1, introduces no change so that $\langle M_\alpha(\tau)\rangle = \overline{M_\alpha}(\tau_1)$; however the mean values of moments higher than the dipole are affected by this averaging, since they are dependent on the choice of origin[24]. In the limit $r_0/R \to 0$, equation (7.39) with $\alpha = a$ leads to the Kirkwood equation (7.31). Since

$$\frac{9\varepsilon_r}{(2\varepsilon_r + 1)(\varepsilon_r + 2)} = 1 - \frac{2(\varepsilon_r - 1)^2}{(2\varepsilon_r + 1)(\varepsilon_r + 2)} \tag{7.40}$$

the mean macroscopic moment $\overline{M_\alpha(\tau_1)}$ may be interpreted as arising from the moment $\overline{m_\alpha}$ of the fixed molecule and its neighbours in the inner sphere surrounding it (and this is the microscopic moment that is independent of the shape of the bulk sample) together with a diffuse or long-range part induced uniformly in the entire sphere by surface charges resulting from the moments of the inner sphere. The long-range moment is directly proportional, but opposite in sign, to the mean moment $\overline{\boldsymbol{m}}$ of the inner sphere; it is insignificant in any thin spherical shell surrounding the inner sphere, so $\overline{\boldsymbol{m}}$ is independent of the radius r_0 provided $r_0/R \to 0$.

Combining equations (7.30) and (7.31) gives

$$\overline{\cos\theta} = \frac{3\varepsilon_r}{2\varepsilon_r+1}\,\frac{\overline{m_a}E}{3kT} \tag{7.41}$$

which, unlike equation (7.30), is a general microscopic equation relating $\overline{\cos\theta}$ to the field E in the medium; it is independent of the shape of the bulk sample and is applicable to polarisable as well as to rigid molecules[23]. It shows that if $\overline{m_a}$ is zero then the molecular $\boldsymbol{a}$ axis is not polarised in any direction in space by the field $\boldsymbol{E}$. If the molecules are non-rigid, like ammonia or 1,2-dichloroethane, equation (7.41) is still appropriate provided the large-amplitude internal motion leads to a spacing of the occupied states that is small compared to kT, thereby justifying a classical description. The $\boldsymbol{a}$ axis might be a principal axis of the inertial tensor or it could be associated with one of the bonds.

The microscopic moment $\overline{m_a}$ is the mean moment of an inner sphere, rather than that of some other shape, for otherwise there would be a contribution to $\int P_\alpha\,d\tau$ in equation (7.37) arising from the fields of the external moments $\boldsymbol{\mu}^*, \boldsymbol{\Theta}^*, \ldots$ in equation (7.33), and an associated change in the inner

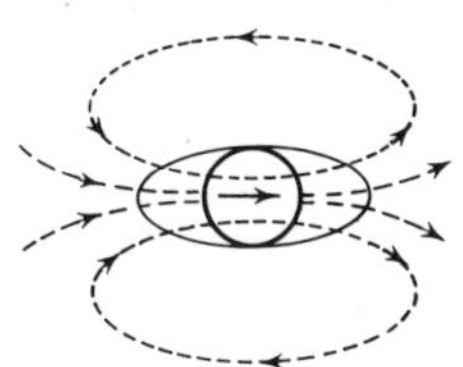

Figure 7.3 The additional moment induced in a prolate ellipsoid by the field surrounding the inner sphere.

moment itself. This is illustrated in Figure 7.3. The total mean moment $\overline{M_a(\tau_1)}$ must be independent of the shape of the inner volume surrounding the fixed molecule 1; however the sphere is convenient because $\overline{m_a}$ is then localised near molecule 1, an increase in the radius r_0 having no affect in the limit of large r_0.

7.2.1.5 The boundary of Kirkwood's inner sphere

In a recent paper, Mandel[27] suggests that just as Onsager's theory[9] suffers from the neglect of short-range interactions, so too does Kirkwood's[22] since there are molecules at the boundary of the inner sphere. In Onsager's theory the inner sphere contains a single molecule and $\overline{m_a} = \mu_a$ where μ_a is the component in the direction of the molecular $\boldsymbol{a}$ axis of the moment $\boldsymbol{\mu}$ in equation (7.10). The short-range interactions with neighbouring molecules were clearly and deliberately neglected by Onsager, and one of the primary objectives of Kirkwood was to remove this limitation. Mandel's point[27] is that the region outside the inner sphere of radius r_0 actually has a molecular nature and it may therefore be erroneous to treat it as if it were a continuum; he concludes that Kirkwood's equation (7.31) is approximate because the short-range interaction energy ΔU between the molecules inside and outside the inner sphere has been neglected. Mandel[27] added that 'it will be difficult to estimate what influence the neglect of ΔU will have on the value of ε_r'. Actually there is no error from this cause in Kirkwood's theory of dielectric polarisation. Equation (7.31) relating $\langle M_a(\tau)\rangle$ to $\overline{m_a}$ *is* 'entirely rigorous within the limits of applicability of classical statistical mechanics'[22]. The surface of the inner sphere has no effect on the molecules of the sample – they cross it without hindrance and there are fluctuations in the number of molecules inside. It was introduced to separate the local or microscopic moment associated with the fixed molecule 1 and its neighbours in short-range interaction with it from the uniform polarisation of the entire sphere by the surface charge resulting from the fixing of molecule 1. What is required is the mean moment of the whole sphere when a single molecule 1 is fixed in position and orientation. The other molecules are constrained only through interaction with molecule 1 and by the need to remain in the entire sample. The mean potential at a *macroscopic* distance from 1 is given by equation (7.33), that is by the laws of classical electrostatics which lead through the boundary conditions (equation (7.35)) and equation (7.37) to Kirkwood's equation (7.31). The fact that molecules near the surface of the imaginary inner sphere of radius r_0 centred on 1 are in short-range interaction with their neighbours across the boundary is no limitation – indeed, these interactions contribute to the permittivity of the fluid as well as to its continuity at this surface.

A limitation of Mandel's criticism[27] can be exposed by considering the effects of a continuum outside the sphere of radius r_0 containing n fixed molecules whose total moment is $\boldsymbol{m}(\tau_1, \ldots, \tau_n)$. *For this model* the moment $\boldsymbol{M}$ of the entire sphere is $[(9\varepsilon_r)/(2\varepsilon_r+1)(\varepsilon_r+2)]\boldsymbol{m}(\tau_1, \ldots, \tau_n)$, hence $\langle \boldsymbol{m}\cdot\boldsymbol{M}\rangle = [9\varepsilon_r/(2\varepsilon_r+1)(\varepsilon_r+2)]\,\langle m^2\rangle_{\varepsilon_r}$ where $\langle\ \rangle$ and $\langle\ \rangle_{\varepsilon_r}$ are averages in a vacuum and in a continuum of relative permittivity ε_r. For a macroscopic sphere $(n\to\infty,\ N/n\to\infty)\ \langle \boldsymbol{m}\cdot\boldsymbol{M}\rangle = (n/N)\langle M^2\rangle = \langle m^2\rangle$ and we should be forced to the relationship $\langle m^2\rangle = [9\varepsilon_r/(2\varepsilon_r+1)(\varepsilon_r+2)]\,\langle m^2\rangle_{\varepsilon_r}$ which is valid only if the molecules are not polarisable[20, 28] or if fluctuations in electronic and nuclear coordinates are included in m^2 and if these high-frequency internal motions are treated in accordance with the laws of classical statistical mechanics[29]. It is true that Mandel used rigid dipoles as a molecular model, but the Kirkwood expression for $\langle M_a(\tau)\rangle$ (equation (7.31)) is also valid for non-rigid molecules[23]; the above analysis shows that the model of a continuum surrounding the inner sphere is not appropriate to Kirkwood's theory, except as a means to relate $\langle M_a(\tau)\rangle$ to $\overline{m_a}$.

7.2.1.6 The permittivity of a fluid comprised of rigid dipoles

If the molecules are rigid dipoles of moment μ

$$\varepsilon_0(\varepsilon_r-1)\boldsymbol{E} = \boldsymbol{P} = NV^{-1}\overline{\boldsymbol{\mu}(\tau_1)\cdot\boldsymbol{e}} = NV^{-1}\mu\,\overline{\cos\theta}$$

Using equation (7.41), this yields

$$\varepsilon_r - 1 = \frac{3\varepsilon_r}{2\varepsilon_r + 1} \frac{N\boldsymbol{\mu} \cdot \overline{\boldsymbol{m}}}{3\varepsilon_0 VkT} \tag{7.42}$$

which is Kirkwood's equation for the permittivity of a polar fluid comprised of rigid dipoles. Two microscopic dipole moments, $\boldsymbol{\mu}$ and $\overline{\boldsymbol{m}}$, enter the expression. Their scalar product may be written in the form[22]

$$\boldsymbol{\mu} \cdot \overline{\boldsymbol{m}} = \mu^2 \left[1 + NV^{-1} \int^{V_0} \cos \gamma_{12} \exp\left[-V(\tau_1, \tau_2)/kT \right] \mathrm{d}\tau_2 \right] \tag{7.43}$$

where γ_{12} is the angle between the dipoles of molecules 1 and 2, $\cos \gamma_{12} = \boldsymbol{\mu}_1 \cdot \boldsymbol{\mu}_2/\mu^2$, and where $\int^{V_0} \mathrm{d}\tau_2$ is an integral over all positions and orientations of molecule 2 when it is inside the inner sphere centred on molecule 1, and $\int \mathrm{d}\tau_2$ has the dimensions of volume. The energy $V(\tau_1, \tau_2)$ is the 'potential of averaging force and torque' acting on an arbitrary pair of molecules,

$$V(\tau_1, \tau_2) = \frac{\int V(\tau_1, \tau_2, \tau_3 \ldots, \tau_N) \exp\left[-V(\tau_1, \tau_2, \tau_3, \ldots, \tau_N)/kT \right] \mathrm{d}\tau_3 \ldots \mathrm{d}\tau_N}{\int \exp\left[-V(\tau_1, \tau_2, \tau_3, \ldots, \tau_N)/kT \right] \mathrm{d}\tau_3 \ldots \mathrm{d}\tau_N} \tag{7.44}$$

and $V(\tau_1, \tau_2)$ is equal to the Helmholtz free energy when molecules 1 and 2 are in the configurations τ_1 and τ_2:

$$V(\tau_1, \tau_2) = -kT \ln \left[\int \exp\left[-V(\tau_1, \tau_2, \tau_3, \ldots, \tau_N)/kT \right] \mathrm{d}\tau_3 \ldots \mathrm{d}\tau_N / \int \mathrm{d}\tau_3 \ldots \mathrm{d}\tau_N \right] \tag{7.45}$$

Oster and Kirkwood[30] introduced the 'correlation parameter' g:

$$g = \frac{\boldsymbol{\mu} \cdot \overline{\boldsymbol{m}}}{\mu^2}$$

$$= 1 + NV^{-1} \int^{V_0} \cos \gamma_{12} \exp\left[-V(\tau_1, \tau_2)/kT \right] \mathrm{d}\tau_2 \tag{7.46}$$

so $g = 1$ if there is no angular correlation between neighbouring molecules, as in the Onsager theory. Elongated molecules in which the dipoles are along the major axis, for example acetonitrile or nitrobenzene, probably have g values smaller than unity while disc-shaped polar molecules, for example trimethylamine and paraldehyde, have $g > 1$ [23]. In water, hydrogen bonding leads to a strong correlation between the dipole axes of neighbouring molecules; in Pople's model of liquid water[31], g varies from 2.60 at 0 °C to 2.46 at 83 °C. Dannhauser and Cole[32] deduced approximate values of g for the liquid butyl alcohols over a wide temperature range, using a modified Onsager equation in which $g\mu_0^2$ replaces μ_0^2 in equation (7.23); the resulting g values of n-, iso- and s-butyl alcohol, varying from over 3 at −140 °C to *c.* 1.3 at the boiling points, were interpreted in terms of equilibria among chains of hydrogen-bonded polymers of different length.

Kirkwood[22] supposed that polarisability made only a small contribution to the relative permittivity of a polar liquid, and he treated it in a very approximate manner. Actually, the contribution of α can be very significant[17], as can be seen by comparing $(\varepsilon_r-1)/(\varepsilon_r+2)$ to $(\varepsilon_\infty-1)/(\varepsilon_\infty+2)$; if $\varepsilon_r=\infty$ and $\varepsilon_\infty=2$, the ratio is only 4 to 1.

7.2.2 The role of polarisability

Although Kirkwood's theory leads to the expression (7.41) for $\overline{\cos\theta}$, it has not proved possible to deduce a general microscopic theory of dielectric polarisation when, as in reality, the molecules are polarisable. Many attempts have been made to extend Kirkwood's approach but all involve approximations or simplified models.

Fröhlich[33] introduced an ingenious model in which the fluid is represented by rigid dipoles immersed in a continuum of relative permittivity ε_∞. Unfortunately the limitations inherent in such a model are difficult to assess and to lessen.

There are two distinct effects of polarisability – firstly there is a direct contribution to the polarisation arising from the distortion of the molecules by the applied field, and secondly the molecular dipole moments are affected by interaction with neighbours. The general statistical theory begins with Kirkwood's macroscopic sphere in the external field $\boldsymbol{E}_0$. Then, from equation (7.25),

$$\frac{\varepsilon_r-1}{\varepsilon_r+2}\boldsymbol{E}_0=\tfrac{1}{3}(\varepsilon_r-1)\boldsymbol{E}=\frac{1}{3\varepsilon_0}\boldsymbol{P}=\frac{1}{3\varepsilon_0 V}\overline{\boldsymbol{M}} \tag{7.47}$$

where $\overline{M}$ is the mean moment of the sphere in the external field $\boldsymbol{E}_0=E_0\boldsymbol{e}$. In an isotropic fluid $\overline{\boldsymbol{M}}$ has the direction of $\boldsymbol{E}_0$ and (see equation (7.27))

$$\overline{\boldsymbol{M}\cdot\boldsymbol{e}}=\frac{\int\boldsymbol{M}(\tau,\boldsymbol{E}_0)\cdot\boldsymbol{e}\exp\left[-V(\tau,\boldsymbol{E}_0)/kT\right]\mathrm{d}\tau}{\int\exp\left[-V(\tau,\boldsymbol{E}_0)/kT\right]\mathrm{d}\tau} \tag{7.48}$$

From equations (7.47) and (7.48) and the equation $\partial V(\tau,\boldsymbol{E}_0)/\partial E_0=-\boldsymbol{M}(\tau,\boldsymbol{E}_0)\cdot\boldsymbol{e}$,

$$\begin{aligned}\frac{\varepsilon_r-1}{\varepsilon_r+2}&=\frac{1}{3\varepsilon_0 V}\left[\frac{\partial}{\partial E_0}\overline{\boldsymbol{M}\cdot\boldsymbol{e}}\right]_{E_0=0}\\&=\frac{1}{3\varepsilon_0 V}\left[\left\langle\frac{\partial\boldsymbol{M}(\tau,\boldsymbol{E}_0)}{\partial E_0}\cdot\boldsymbol{e}\right\rangle+\frac{1}{3kT}\langle M(\tau)^2\rangle\right]\end{aligned} \tag{7.49}$$

The first term on the right-hand side of equation (7.49) is proportional to the mean extra moment induced by the external field when all the molecules are fixed in position and orientation; it represents the distortion polarisation. This term was equated to $(\varepsilon_\infty-1)/(\varepsilon_\infty+2)$ by Harris and Alder[17], and this is justified by the fact that ε_∞ is the relative permittivity at frequencies sufficiently high that orientation has no time to occur. $\boldsymbol{M}(\tau)$ is the moment of the sphere when $E_0=0$ and the molecules are in the particular positions and orientations τ; the electronic and vibrational motion adjusts to the configuration τ and fluctuations in these internal coordinates do not contribute to $\langle M(\tau)^2\rangle$. Hence equation (7.49) may be written

$$\frac{\varepsilon_r-1}{\varepsilon_r+2}-\frac{\varepsilon_\infty-1}{\varepsilon_\infty+2}=\frac{1}{9\varepsilon_0 VkT}\langle M(\tau)^2\rangle \tag{7.50}$$

If $\boldsymbol{M}(\tau)$ is written as the sum of N molecular moments, $\langle M(\tau)^2\rangle$ may be equated to $N\langle \boldsymbol{m}(\tau)\cdot\boldsymbol{M}(\tau)\rangle$ where $\boldsymbol{m}(\tau)$ is the moment of molecule 1 which, because of distortion, is dependent on the positions and orientations of all the other $N-1$ molecules. Thus it is not possible to use electrostatic arguments of the type employed in Section 7.2.1.4 to reduce $\langle \boldsymbol{m}(\tau)\cdot\boldsymbol{M}(\tau)\rangle$ to microscopic form. Various approximations have been suggested. Harris and Alder[17] and Harris[18] replaced $\boldsymbol{m}(\tau)$ by the mean moment $\boldsymbol{\mu}$ of a molecule in the fluid and then used equation (7.31)[17, 18]

$$\langle M(\tau)^2\rangle \to N\mu\langle M_\mu(\tau)\rangle = N[9\varepsilon_r/(2\varepsilon_r+1)(\varepsilon_r+2)]\boldsymbol{\mu}\cdot\overline{\boldsymbol{m}}$$

The reader should be cautioned that Harris's paper[18] contains errors resulting from his equations (13) and (23) which are not generally valid. Buckingham[34] put $\boldsymbol{m}_i(\tau) = \boldsymbol{\mu}_{0i}+\boldsymbol{m}_i'(\tau)$ where $\boldsymbol{\mu}_{0i}$ is the dipole moment of the free gaseous molecule and $\boldsymbol{m}_i'(\tau)$ the extra moment induced in the molecule by the others in the sample. On neglecting terms in $m_i'(\tau)^2$ (Ref. 34)

$$\langle M(\tau)^2\rangle \to N\frac{9\varepsilon_r}{(2\varepsilon_r+1)(\varepsilon_r+2)}\boldsymbol{\mu}_0\cdot\overline{\boldsymbol{m}}+N\left[\frac{3(2\varepsilon_r+\varepsilon_\infty)}{(2\varepsilon_r+1)(\varepsilon_\infty+2)}\boldsymbol{\mu}_0\cdot\overline{\boldsymbol{m}}-\boldsymbol{\mu}_0\cdot\overline{\boldsymbol{\mu}_0}\right]$$

where $\overline{\boldsymbol{\mu}_0}$ is the mean 'permanent' moment of the inner sphere, that is, $\boldsymbol{\mu}_0\cdot\overline{\boldsymbol{\mu}_0} = g\mu_0^2$; hence $\overline{\boldsymbol{m}}-\overline{\boldsymbol{\mu}_0}$ is the mean extra moment of a molecule and its neighbours in the inner sphere due to distortion of the electronic and nuclear motion. Both $\overline{\boldsymbol{m}}$ and $\overline{\boldsymbol{\mu}_0}$ could be calculated for simple models of a fluid. In this approximation, three microscopic moments, $\boldsymbol{\mu}_0$, $\overline{\boldsymbol{\mu}_0}$ and $\overline{\boldsymbol{m}}$, are needed to describe dielectric polarisation. For the Onsager model, the total extra moment of the spherical sample is zero[34], and as there are no fluctuations in $\boldsymbol{m}'(\tau)$ the term in $\langle m'(\tau)^2\rangle$ is also zero. Thus, for this model[9]

$$\langle M(\tau)^2\rangle \to N\frac{9\varepsilon_r}{(2\varepsilon_r+1)(\varepsilon_r+2)}\boldsymbol{\mu}_0\cdot\boldsymbol{\mu} = N\frac{3\varepsilon_r(\varepsilon_\infty+2)}{(2\varepsilon_r+\varepsilon_\infty)(\varepsilon_r+2)}\mu_0^2$$

where equation (7.19) for μ/μ_0 has been used.

Cole[35] used an isotropic classical harmonic oscillator model to represent the effects of polarisability. Neglecting fluctuations in internuclear distances, as for a lattice with a regular molecular spacing Cole deduced that $\langle\boldsymbol{\mu}_0\cdot\boldsymbol{M}'\rangle = 0$ where $\boldsymbol{M}'$ is the extra moment of the spherical sample. He obtained[35]

$$\langle M(\tau)^2\rangle \to N\frac{9\varepsilon_r}{(2\varepsilon_r+1)(\varepsilon_r+2)}\boldsymbol{\mu}_0\cdot\overline{\boldsymbol{m}} = N\frac{3\varepsilon_r(\varepsilon_\infty+2)}{(2\varepsilon_r+\varepsilon_\infty)(\varepsilon_r+2)}g\mu_0^2$$

where the final equality is dependent upon an additional assumption that $\overline{\boldsymbol{m}} = g\boldsymbol{\mu}$ where $\boldsymbol{\mu}$ is related to $\boldsymbol{\mu}_0$ by equation (7.19).

Mandel and Mazur[36] developed a general theory in which the dipole moment $m_{i\alpha}$ of molecule i is approximated by

$$\mu_{0i\alpha}+\alpha_i\left[E_{0\alpha}+\sum_{j\neq i}(3r_{ij\alpha}r_{ij\beta}-r_{ij}^2\delta_{\alpha\beta})r_{ij}^{-5}m_{j\beta}\right]$$

where $\boldsymbol{r}_{ij}$ is the vector from an origin in molecule i to an origin in j. This expression neglects the dipole induced (i) by the fields of the higher multipoles of j, (ii) by derivatives of the field at i, (iii) by non-linear distortion, that is through the hyperpolarisabilities of i [37], and (iv) by distortion due to interactions which are not of an electrostatic nature, for example, dispersion[38] and overlap[39] effects of the type responsible for the dipole moment of an interacting pair of different inert gas atoms (e.g. He–Ar)[40, 41]. The theory[36] gives a molecular interpretation of the permittivity in the sense that the polarisation is related to a statistical average over all the individual molecules of the sample, but it is not microscopic in that the sums are not rapidly converging as in $\overline{m}_a$ in equation (7.41).

Van Vleck[42] started from the partition function $Z = \Sigma_n \exp(-W_n/kT)$ of an array of polar molecules in an external field $\boldsymbol{E}_0$, where W_n is the energy of the nth quantum state of the sample in the field. The polarisation is (see equation (7.48))

$$\boldsymbol{P} = kTV^{-1}(\partial \ln Z/\partial \boldsymbol{E}_0)_V \tag{7.51}$$

The molecules are in fixed positions and interact with one another via their dipole moments. The Hamiltonian

$$H = -\boldsymbol{E}_0 \cdot \sum_i \boldsymbol{m}_i - \sum_{i<j} m_{i\alpha}(3r_{ij\alpha}r_{ij\beta} - r_{ij}^2\delta_{\alpha\beta})r_{ij}^{-5}m_{j\beta} \tag{7.52}$$

and the energy was evaluated for a molecular model comprising a dipole moment and an isotropic harmonic oscillator to represent the polarisability. The partition function is expanded as a power series in W_n/kT:

$$Z = \sum_n [1 - W_n/kT + \tfrac{1}{2}W_n^2/(kT)^2 - \ldots] \tag{7.53}$$

which may be helpful at high temperatures or low densities. Since the sum of the diagonal elements of a matrix is independent of the basis set, it is not necessary to determine the eigenvalues of H and

$$Z = \sum_n [1 - \langle n | H | n \rangle/kT + \tfrac{1}{2}\langle n | H^2 | n \rangle/(kT)^2 - \ldots] \tag{7.54}$$

The part of Z that is quadratic in $\boldsymbol{E}_0$ is needed, but unfortunately the calculation is difficult. The terms in T^{-3} and T^{-4} in Z give a dependence of the polarisation on r_{ij}^{-3} and r_{ij}^{-6}, respectively.

7.3 DILUTE SOLUTIONS IN NON-POLAR SOLVENTS

The dipole moment of a non-volatile compound is normally determined through measurement of the permittivity of a dilute solution of it in a non-polar solvent. The effective moment μ_s is computed via the Debye equation

$$\frac{\varepsilon_r - 1}{\varepsilon_r + 2} = \frac{\varepsilon_\infty - 1}{\varepsilon_\infty + 2} + \frac{N_s \mu_s^2}{9\varepsilon_0 VkT} \tag{7.55}$$

where ε_r and ε_∞ are the static and high-frequency permittivities of a dilute solution containing N_s/V solute molecules in unit volume. The moment μ_s may vary slightly with the solvent[43].

The general statistical equation (7.50) may be used to provide a microscopic interpretation of μ_s. If there is just one polar solute molecule in the entire spherical sample, and if fluctuations are neglected, then $\boldsymbol{M}(\tau) = [9\varepsilon_r/(2\varepsilon_r + 1)(\varepsilon_r + 2)]_0 \overline{\boldsymbol{m}}$ where the subscript 0 refers to the pure solvent and where $\overline{\boldsymbol{m}}$ is the mean dipole moment of the solute molecule and its solvent neighbours in a small macroscopic sphere centred on the solute and immersed in the solvent. Thus equation (7.50) yields the approximate expression

$$\frac{\varepsilon_r - 1}{\varepsilon_r + 2} = \frac{\varepsilon_\infty - 1}{\varepsilon_\infty + 2} + \frac{N_s}{9\varepsilon_0 VkT}\left(\frac{9\varepsilon_r}{(2\varepsilon_r + 1)(\varepsilon_r + 2)}\right)_0^2 \overline{m}^2 \tag{7.56}$$

implying that μ_s in equation (7.55) is actually $[9\varepsilon_r/(2\varepsilon_r+1)(\varepsilon_r+2)]_0\,\overline{m}$; the factor $[9\varepsilon_r/(2\varepsilon_r+1)(\varepsilon_r+2)]_0$ is 0.87 for a typical non-polar solvent having $\varepsilon_r = 2.25$.

The expression for $\langle M(\tau)^2\rangle$ in equation (7.56) can be appreciated by considering the mean square moment of a sphere containing a single dipolar molecule in a non-polar solvent. Write $\boldsymbol{M}(\tau) = \boldsymbol{m} + \boldsymbol{m}_{\text{solvent}}$ where $\boldsymbol{m}_{\text{solvent}}$ is the dipole moment of the part of the spherical sample outside the inner sphere of moment $\boldsymbol{m}$. Then if there are no fluctuations for a single solute

$$\boldsymbol{M}(\tau) = \left[\frac{9\varepsilon_r}{(2\varepsilon_r+1)(\varepsilon_r+2)}\right]_0 \overline{\boldsymbol{m}},$$

$$\boldsymbol{m}_{\text{solvent}} = \left[\frac{9\varepsilon_r}{(2\varepsilon_r+1)(\varepsilon_r+2)}\right]_0 \overline{\boldsymbol{m}},$$

$$\langle M(\tau)^2\rangle = \left[\frac{9\varepsilon_r}{(2\varepsilon_r+1)(\varepsilon_r+2)}\right]_0^2 \overline{m}^2$$

molecule and this must be multiplied by N_s for a dilute solution containing N_s independent polar molecules.

Van Vleck[44] has considered a model comprised of a system of harmonic oscillators mounted on a spherical sample of a simple cubic lattice containing a single dipole of moment μ_0. If this dipole were mounted on one of the sites *with a harmonic oscillator*, then the mean extra total moment $\langle \boldsymbol{\mu}_0 \cdot \Sigma_i\, \boldsymbol{m}_i'\rangle/\mu_0$ is zero. This can be shown through an electrostatic calculation similar to that of Cole[35] and is consistent with an analysis of the Onsager model[34] (see page 255). Thus for N_s independent dipoles $\langle M(\tau)^2\rangle = N_s\mu_0^2$. This is consistent with equation (7.56) if the Onsager relation (7.19) for μ/μ_0 is used with $\varepsilon_\infty = \varepsilon_r$ for the non-polar solvent.

Other approximate descriptions of dielectric polarisation of solutions in non-polar solvents have been reviewed by Brown[21].

7.4 THE IMPERFECT GAS

For a gas at a very low density the Debye equation (7.3) is valid and useful for determining gas-phase dipole moments μ_0. There are small corrections arising from the quantisation of angular momentum and from non-rigidity[45]. For a diatomic rigid-rotator the orientation polarisation is actually $(N\mu_0^2/9\varepsilon_0 VkT)\,[1-\frac{1}{3}(B_0/kT)+\frac{2}{45}(B_0/kT)^2+\ldots]$ where B_0 is the rotational constant[46] (so that the Jth rotational energy level is $(J^2+J)B_0$) and all the polarisation results from the molecules that are in the ground rotational state $J = 0$ [45, 46].

If the density is not so low, interactions between the gas molecules may be significant. Then it is convenient to expand the function $(\varepsilon_r - 1)V_m/(\varepsilon_r + 2)$ as a power series in V_m^{-1} as in equation (7.24). Rigorous classical statistical-mechanical equations may be written for the first and second dielectric virial coefficients[15]:

$$A_\varepsilon = \frac{N_A}{3\varepsilon_0}(\alpha_0 + \mu_0^2/3kT) \tag{7.57}$$

$$B_\varepsilon = \frac{N_A^2}{3\varepsilon_0 \Omega} \int \left\{ \tfrac{1}{2}\alpha_{12}(\tau_2) - \alpha_0 + \frac{1}{3kT}\left[\tfrac{1}{2}(m_{12}(\tau_2))^2 - \mu_0^2\right] \right\} \exp(-V_{12}(\tau_2)/kT)\, d\tau_2 \tag{7.58}$$

where N_A is Avogadro's number, α_0 is the polarisability of a free molecule, $\alpha_{12}(\tau_2)$, $m_{12}(\tau)$ and $V_{12}(\tau_2)$ are the polarisability, dipole moment and interaction energy of the pair of molecules 1, 2 in the relative position and orientation τ_2; the constant Ω is defined so that $\int d\tau_2 = \Omega V_m$. The integration in equation (7.58) is over an infinite sphere centred on molecule 1.

Measurements of second dielectric virial coefficients over a range of temperature yield information about molecular interactions; the topic has recently been reviewed by Sutter[16].

Polarisability has a very large influence on B_ε through its effect on $\boldsymbol{m}_{12}(\tau)$; the change from a rigid dipole model to a polarisable dipole model of CH_3F increases B_ε from 1700 to 7500 $cm^6\ mol^{-2}$ at 50 °C [15].

The long-range diffuse polarisation leading to the moment $-\{2(\varepsilon_r - 1)^2/[(2\varepsilon_r + 1)(\varepsilon_r + 2)]\}\,\overline{\boldsymbol{m}}$ in a spherical sample outside the inner sphere contributes to $(\varepsilon_r - 1)V_m/(\varepsilon_r + 2)$ a term proportional to V_m^{-2} at low densities. Care is therefore needed in evaluating integrals contributing to the third dielectric virial coefficient C_ε; the long-range part of the two-particle correlation function discussed by Nienhuis and Deutch[25] must be included, but because the Clausius–Mossotti function $(\varepsilon_r - 1)V_m/(\varepsilon_r + 2)$ was chosen for expansion in powers of the gas density only a spherical shape need be considered.

7.5 DIELECTRIC POLARISATION IN ANISOTROPIC MEDIA

In crystalline materials the molecules experience an anisotropic interaction with their environment – the potential energy of a polar molecule is dependent on its orientation relative to the crystal axes. The sample may exhibit special properties – it may be ferroelectric[47], an electret[48] or a thin film – and it may be an ionic, molecular, polymeric, or liquid crystal. In this section, the emphasis is placed on the effect of the anisotropic potential on the orientation of polar molecules by an electric field. The dielectric properties of some solids were reviewed by Brown[21], Cole[49] and Davies[50].

In the most general case the permittivity is a non-symmetric complex second-rank polar tensor relating the displacement $\boldsymbol{D}$ to the field $\boldsymbol{E}$, $D_\alpha = \varepsilon_{\alpha\beta}\varepsilon_0 E_\beta$; there are therefore 18 independent components of $\boldsymbol{\varepsilon}$ but in most cases of practical interest symmetry reduces this number. For a uniaxial optically inactive material

$$\boldsymbol{\varepsilon} = \begin{pmatrix} \varepsilon_\perp & 0 & 0 \\ 0 & \varepsilon_\perp & 0 \\ 0 & 0 & \varepsilon_{||} \end{pmatrix}$$

and the field inside a spherical sample is related to the external field $\boldsymbol{E}_0$ by (see equation (7.25))

$$\begin{aligned} E_{||} &= \frac{3}{\varepsilon_{||}+2} E_{0||} \\ E_\perp &= \frac{3}{\varepsilon_\perp+2} E_{0\perp} \end{aligned} \qquad (7.59)$$

The equations analogous to equation (7.50) are

$$\left.\begin{aligned} \frac{\varepsilon_{||}-1}{\varepsilon_{||}+2} &= \frac{\varepsilon_{\infty||}-1}{\varepsilon_{\infty||}+2} + \frac{1}{3\varepsilon_0 VkT}\langle M_{||}(\tau)^2\rangle \\ \frac{\varepsilon_\perp-1}{\varepsilon_\perp+2} &= \frac{\varepsilon_{\infty\perp}-1}{\varepsilon_{\infty\perp}+2} + \frac{1}{3\varepsilon_0 VkT}\langle M_\perp(\tau)^2\rangle \end{aligned}\right\} \qquad (7.60)$$

The transition from the macroscopic spherical averages $\langle M_\alpha(\tau)^2\rangle$ to microscopic form follows similarly to that in Section 7.2.1.4, with $\varepsilon_{||}$ and $\varepsilon_\perp$ replacing the isotropic ε_r. This yields in the rigid dipole approximation

$$\left.\begin{aligned} \langle M_{||}(\tau)^2\rangle &= N\frac{9\varepsilon_{||}}{(2\varepsilon_{||}+1)(\varepsilon_{||}+2)}\langle \mu_{||}\overline{\mu_{||}}\rangle \\ \langle M_\perp(\tau)^2\rangle &= N\frac{9\varepsilon_\perp}{(2\varepsilon_\perp+1)(\varepsilon_\perp+2)}\langle \mu_\perp\overline{\mu_\perp}\rangle \end{aligned}\right\} \qquad (7.61)$$

In equation (7.61) the quantities $\langle \mu_{||}\overline{\mu_{||}}\rangle$ and $\langle \mu_\perp\overline{\mu_\perp}\rangle$ involve an average over all orientations of the polar molecule in the crystalline field when E_0

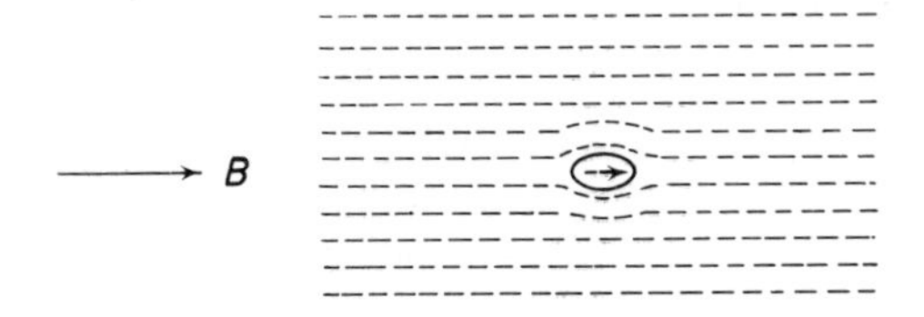

Figure 7.4 Possible structure of a prolate solute molecule in a nematic liquid crystal solvent in a magnetic field ***B***

is zero; in the isotropic case each of the three cartesian components is equal to one-third of the scalar product.

As an illustration of these generalities, consider a dipolar molecule dissolved in a nematic liquid crystalline solvent[51]. A magnetic field orients the liquid crystal which then provides a uniaxial environment for the polar solute (see Figure 7.4). The orientation distribution function of the partially orientated solute may conveniently be described by 'order parameters' $\boldsymbol{S}$ [52, 23]; the probability $\mathscr{P}(\theta)$ that the dipole axis in a solute molecule with a threefold or higher rotation axis is at an angle θ to the favoured direction of the solvent (that of the magnetic field in a nematic phase) is[23]

$$\mathscr{P}(\theta) = \tfrac{1}{2}+\tfrac{3}{2}S_a \cos\theta+\tfrac{5}{2}S_{aa}(\tfrac{3}{2}\cos^2\theta-\tfrac{1}{2})+\tfrac{7}{2}S_{aaa}(\tfrac{5}{2}\cos^3\theta-\tfrac{3}{2}\cos\theta)+\ldots \quad (7.62)$$

where

$$S_a = \langle\cos\theta\rangle = \int_0^\pi \cos\theta\mathscr{P}(\theta)\sin\theta\, d\theta$$

$$S_{aa} = \langle\tfrac{3}{2}\cos^2\theta-\tfrac{1}{2}\rangle = \int_0^\pi (\tfrac{3}{2}\cos^2\theta-\tfrac{1}{2})\mathscr{P}(\theta)\sin\theta\, d\theta$$

$$S_{aaa} = \langle\tfrac{5}{2}\cos^3\theta-\tfrac{3}{2}\cos\theta\rangle = \int_0^\pi (\tfrac{5}{2}\cos^3\theta-\tfrac{3}{2}\cos\theta)\mathscr{P}(\theta)\sin\theta\, d\theta$$

In an isotropic system $S_a = S_{aa} = S_{aaa} = 0$ and in a completed aligned situation $S_a = S_{aa} = S_{aaa} = 1$. In a weak electric field $E_{||}$ parallel to the axis of the system, the mean value of the cosine of the angle between the solute dipole $\boldsymbol{\mu}$ and the field is

$$\overline{\cos\theta_E} = \frac{\displaystyle\int_0^\pi \cos\theta_E\left(1+\frac{3\varepsilon_{||}}{2\varepsilon_{||}+1}\frac{\mu E_{||}}{kT}\cos\theta_E\right)\mathscr{P}(\theta)\sin\theta\, d\theta}{\displaystyle\int_0^\pi \left(1+\frac{3\varepsilon_{||}}{2\varepsilon_{||}+1}\frac{\mu E_{||}}{kT}\cos\theta_E\right)\mathscr{P}(\theta)\sin\theta\, d\theta} \quad (7.63)$$

where, in accord with the general equation (7.41), the Onsager cavity field equation (7.5) is used as the orientating field. If the field is static, $\cos\theta_E = \pm\cos\theta$ and from equations (7.62) and (7.63)

$$\overline{\cos\theta_E} = \pm S_3+\frac{3\varepsilon_{||}}{2\varepsilon_{||}+1}\frac{\mu E_{||}}{3kT}(1+2S_{33}-3S_3^2) \quad (7.64)$$

For an alternating field the terms in S_3 and S_3^2 in equation (7.64) vanish. If the field is perpendicular to the axis of the system, then in a static and an alternating field the mean value of the cosine of the angle α between $\boldsymbol{\mu}$ and the field is

$$\overline{\cos\alpha} = \frac{3\varepsilon_\perp}{2\varepsilon_\perp+1}\frac{\mu E_\perp}{3kT}(1-S_{33}) \quad (7.65)$$

Thus if S_{33} is positive, the permittivity $\varepsilon_{||}$ is enhanced while $\varepsilon_\perp$ is diminished by the coupling of the axis of the molecule to that of its environment; the situation is reversed if S_{33} is negative implying that the dipole favours the plane at right angles to the axis of the system. Thus the polarisation of a dipole is influenced by its coupling to an anisotropic environment, thus providing a means of studying this coupling. Using a combination of dielectric and high-resolution nuclear magnetic resonance techniques, Chapman, Long and McLauchlan[51] found that S_{aa} for acetonitrile in a nematic solvent at 60 °C is +0.090, and Chapman and McLauchlan[51a] investigated the hydration structure of collagen. The above theory should also be applicable to the polarisation of polar molecules in quinol and similar clathrates[53].

The interaction of dipoles on lattice sites has provided a classic problem in statistical mechanics. The work of Van Vleck[42] and Cole [35] has already been mentioned; that of Lax[54–56] is also relevant, particularly to phase transitions. The merit of the dipolar lattice is that it is a well-defined and tractable model system. Other simple models of molecules and condensed phases can help in the development of rigorous statistical mechanical theories. For example, Jepsen and Friedman[57] and Jepsen[58] summed re-arranged Mayer graphs for rigid dipoles in calculations of the permittivity of a polar fluid, using both Kirkwood's theory (which uses the laws of classical electrostatics) and an entirely microscopic approach through the evaluation of the interaction energy of two test charges or dipoles immersed in the fluid. Wertheim[59] employed the 'mean spherical model' for fluids of hard spheres with rigid dipole moments and deduced, using Kirkwood's theory, a simple closed form expression for ε_r that is independent of the hard-sphere-diameter.

The Monte-Carlo and molecular-dynamics methods of simulating with a computer the equilibrium and non-equilibrium properties of fluids[60] would appear to provide an important means of investigating dielectric phenomena.

The local electric and electro-optic fields in anisotropic molecular crystals and particularly in ice have been considered recently by Eisenberg and Kauzmann[61], Minton[62] and Dunmur[63].

7.6 THE ALIGNMENT OF MOLECULES IN FLUIDS

In previous sections the emphasis has been placed on the *polarisation* of molecules by an external field, that is, on $\overline{\cos\theta}$. This is because $\overline{\cos\theta}$ is intimately related to the permittivity. The *alignment* of molecules is defined as $\overline{\frac{3}{2}\cos^2\theta-\frac{1}{2}}$ and this is related to observables such as the electro-optic Kerr effect[64] and dipolar splittings in high-resolution nuclear magnetic resonance spectroscopy[65].

In a uniform electric field $\boldsymbol{E}$ in a fluid that is isotropic in the absence of the field $\overline{\frac{3}{2}\cos^2\theta-\frac{1}{2}}$ is an *even* function of $\boldsymbol{E}$, for unlike $\overline{\cos\theta}$ it must be unaffected by changing $\boldsymbol{E}$ into $-\boldsymbol{E}$. In the absence of saturation, the alignment is proportional to E^2. A general theory of the Kirkwood type has been developed for the alignment in a uniform field[66]. It appears that a general and rigorous expression like equation (7.41) for $\overline{\cos\theta}$ does not exist. A static field may cause alignment through interaction with the dipole moment or the anisotropic polarisability but a high-frequency alternating field, such as an optical field, acts only on the polarisability. The alignment is dependent upon the two- and three-particle distribution functions[23, 66, 67].

A statistical theory of alignment in an electric field has been developed by Hilbers, MacLean and Mandel[68]; it is subject to the same difficulties as the analogous theory of polarisation[36] (see Section 7.2.2). Ramshaw, Schaefer, Waugh and Deutch[69] have discussed polarisation and alignment in an electric field and their significance for the structure of polar fluids.

Accurate values for $(\overline{\frac{3}{2}\cos^2\theta-\frac{1}{2}})/E^2$ in polar fluids have recently been determined through observations by Hilbers and MacLean[70] of the effects of E on high-resolution nuclear magnetic resonance spectra, particularly of the quadrupolar nuclei ^{14}N and ^{2}H.

In an electric field-gradient $\boldsymbol{E}'$ it *is* possible to obtain a rigorous expression for the alignment[24]. The expression is analogous to equation (7.41) and is

$$\overline{(\tfrac{3}{2}\cos^2\theta - \tfrac{1}{2})}_{E'} = \frac{5\varepsilon_r}{3\varepsilon_r + 2}\,\frac{\overline{\Theta_{aa}}E'_{zz}}{10kT} \tag{7.66}$$

where $\overline{\Theta_{aa}}$ is the mean value of the component in the direction of the a axis of molecule 1 of the quadrupole moment of the inner sphere centred on molecule 1, when $\boldsymbol{E}' = 0$. Equation (7.66) can be extended to give the alignment of any molecular axis in a field gradient of any symmetry[24]. For an Onsager-type model[9], the quadrupole moment Θ in the medium is related to the gas-phase value Θ_0 by

$$\Theta = \left[1 - \frac{6(\varepsilon_r - 1)C}{(3\varepsilon_r + 2)4\pi\varepsilon_0 a^5}\right]^{-1} \Theta_0 \tag{7.67}$$

where C is the quadrupole polarisability of the spherical molecule (that is, its induced quadrupole in a field $\boldsymbol{E}'_0$ is $C\boldsymbol{E}'_0$). If we put (see equation (7.18)) $C/4\pi\varepsilon_0 a^5 = (\varepsilon_\infty - 1)/(2\varepsilon_\infty + 3)$, equation (7.67) reduces to

$$\Theta = \frac{(3\varepsilon_r + 2)(2\varepsilon_\infty + 3)}{5(3\varepsilon_r + 2\varepsilon_\infty)}\,\Theta_0 \tag{7.68}$$

which is analogous to equation (7.19). For $\varepsilon_r = \varepsilon_\infty = 2.2$, the reaction field-gradient increases the quadrupole moment by 16%.

Nienhuis and Deutch[72] have studied the three-particle correlation function of a polar fluid comprised of rigid dipoles; they deduced an expression for the long-range part of the correlation function and used it to show that $\overline{(\tfrac{3}{2}\cos^2\theta - \tfrac{1}{2})}/E^2$ is independent of the shape of the sample. This result follows from electrostatics by a similar argument to that used in Section 7.2.1.1 to show that $\overline{(\cos\theta)}/E$ is an intrinsic property of the fluid. Nienhuis and Deutch[25, 72] used a simplified model of rigid dipoles, but the laws of electrostatics require that $\overline{(\cos\theta)}/E$ and $\overline{(\tfrac{3}{2}\cos^2\theta - \tfrac{1}{2})}/E^2$ are true properties of the fluid regardless of the rigidity of the molecules comprising it. The significance of these two papers by Nienhuis and Deutch would seem to lie in their treatment of the long-range behaviour of the two- and three-particle distribution functions leading to consistency between the statistical mechanical results and those deduced from electrostatics.

References

1. Mossotti, O. F. (1850). *Mem. Mat. Fis. Soc. Ital. Sci.*, Modena, **24** (Part II), 49
2. Clausius, R. (1879). *Die mechanische Wärmetheorie,* Band 2. (Braunschweig: F. Vieweg und Sohn)
3. Lorentz, H. A. (1909). *The Theory of Electrons,* 303. (Leipzig: Teubner)
4. Stratton, J. A. (1941). *Electromagnetic Theory,* (New York: McGraw-Hill)
5. Debye, P. (1912). *Phys. Z.,* **13,** 97
6. Debye, P. (1929). *Polar Molecules,* (New York: Chemical Catalog Company)
7. Debye, P. (1935). *Phys. Z.,* **36,** 100
8. Fowler, R. H. (1935). *Proc. Roy. Soc. A.,* **149,** 1
9. Onsager, L. (1936). *J. Amer. Chem. Soc.,* **58,** 1486

10. Hill, N. E., Vaughan, W. E., Price, A. H. and Davies, M. (1969). *Dielectric Properties and Molecular Behaviour,* (London: Van Nostrand Reinhold Co. Ltd.)
11. Wilson, J. N. (1939). *Chem. Rev.,* **25,** 377
12. Scholte, Th. G. (1949). *Physica,* **15,** 437, 450
13. Böttcher, C. J. F. (1952). *Theory of Electric Polarisation,* (Amsterdam: Elsevier)
13a. For a recent example see Black, H. and Hayes, E. F. *Trans. Faraday Soc.,* **66,** 2512
14. Buckingham, A. D. (1955). *J. Chem. Phys.,* **23,** 2370
15. Buckingham, A. D. and Pople, J. A. (1955). *Trans. Faraday Soc.,* **51,** 1029, 1179
16. Sutter, H. (1972). *Specialist Periodical Reports,* 31, Vol. 1, Chap. 3 (Chemical Society: London)
17. Harris, F. E. and Alder, B. J. (1953). *J. Chem. Phys.,* **21,** 1031; (1954). **22,** 1806
18. Harris, F. E. (1955). *J. Chem. Phys.,* **23,** 1663
19. Fröhlich, H. (1954). *J. Chem. Phys.,* **22,** 1804
20. O'Dwyer, J. J. (1957). *J. Chem. Phys.,* **26,** 878
21. Brown, W. F. (1956). *Handbüch der Physik,* **17,** 1 (S. Flügge, editor)
22. Kirkwood, J. G. (1939). *J. Chem. Phys.,* **7,** 911
23. Buckingham, A. D. (1967). *Discuss. Faraday Soc.,* **43,** 205
24. Buckingham, A. D. and Graham, C. (1971). *Molec. Phys.,* **22,** 335
25. Nienhuis, G. and Deutch, J. M. (1971). *J. Chem. Phys.,* **55,** 4213
26. Fröhlich, H. (1958). *Theory of Dielectrics* (Second Edn.), Section 7. (Oxford: Clarendon Press)
27. Mandel, M. (1972). *Physica,* **57,** 141
28. Fröhlich, H. (1958). *Theory of Dielectrics* (Second Edn.), 181 (Oxford:Clarendon Press)
29. Fröhlich, H. (1958). *Theory of Dielectrics* (Second Edn.), eqn. (7.44) (Oxford: Clarendon Press)
30. Oster, G. and Kirkwood, J. G. (1943). *J. Chem. Phys.,* **11,** 175
31. Pople, J. A. (1951). *Proc. Roy. Soc. A.,* **205,** 163
32. Dannhauser, W. and Cole, R. H. (1955). *J. Chem. Phys.,* **23,** 1762
33. Fröhlich, H. (1958). *Theory of Dielectrics* (Second Edn.), Sections 7 and 8 (Oxford: Clarendon Press)
34. Buckingham, A. D. (1956). *Proc. Roy. Soc. A.,* **238,** 235
35. Cole, R. H. (1957). *J. Chem. Phys.,* **27,** 33
36. Mandel, M. and Mazur, P. (1958). *Physica,* **24,** 116
37. Buckingham, A. D. and Orr, B. J. (1967). *Quart. Rev. Chem. Soc.,* **21,** 195
38. Byers Brown, W. and Whisnant, D. M. (1970). *Chem. Phys. Letters,* **7,** 329
39. Matcha, R. L. and Nesbet, R. K. (1967). *Phys. Rev.,* **160,** 72
40. Kiss, Z. J. and Welsh, H. L. (1959). *Phys. Rev. Letters,* **2,** 166
41. Bosomworth, D. R. and Gush, H. P. (1965). *Canad. J. Phys.,* **43,** 751
42. Van Vleck, J. H. (1937). *J. Chem. Phys.,* **5,** 556, 991
43. Le Fèvre, R. J. W. (1953). *Dipole Moments* (Third Edn.)(London:Methuen)
44. Van Vleck, J. H. (1970). Personal communication
45. Van Vleck, J. H. (1932). *The Theory of Electric and Magnetic Susceptibilities,* Chap. 7 (Oxford University Press)
46. Buckingham, A. D. (1970). *Physical Chemistry—An Advanced Treatise,* Vol. 4, Chap. 8, (D. Henderson, editor). (New York: Academic Press)
47. Zheludev, I. S. (1971). *Solid State Phys.* **26,** 429
48. Gross, B. (1971). *Endeavour,* **30,** 115
49. Cole, R. H. (1960). *Ann. Rev. Phys. Chem.,* **11,** 149
50. Davies, M. (1969). *Dielectric Properties and Molecular Behaviour,* Chap. 5 (London: Van Nostrand Reinhold Co. Ltd.)
51. Chapman, G. E., Long, E. M. and McLauchlan, K. A. (1969). *Molec. Phys.,* **17,** 189
51a Chapman, G. E. and McLauchlan, K. A. (1969). *Proc. Roy. Soc. B,* **173,** 223
52. Saupe, A. (1964). *Z. Naturforsch,* **19a,** 161
53. van der Waals, J. H. and Platteeuw, J. C. (1959). *Advan. Chem. Phys.,* **2,** 1
54. Lax, M. (1952). *J. Chem. Phys.,* **20,** 1351
55. Lax, M. (1955). *Phys. Rev.,* **97,** 629
56. Toupin, R. A. and Lax, M. (1957). *J. Chem. Phys.,* **27,** 458
57. Jepsen, D. W. and Friedman, H. L. (1963). *J. Chem. Phys.,* **38,** 846
58. Jepsen, D. W. (1966). *J. Chem. Phys.,* **44,** 774: **45,** 709
59. Wertheim, M. S. (1971). *J. Chem. Phys.,* **55,** 4291

60. McDonald, I. R. and Singer, K. (1970). *Quart. Rev. Chem. Soc.,* **24,** 238
61. Eisenberg, D. and Kauzmann, W. (1969). *The Structure and Properties of Water,* Section 3.4(a) (Oxford University Press)
62. Minton, A. P. (1972). *J. Phys. Chem.,* **76,** 886
63. Dunmur, D. A. (1972). *Molec. Phys.,* **23,** 109
64. Le-Fèvre, C. G. and Le Fèvre, R. J. W. (1972). *Physical Methods of Chemistry, Part 3C,* p. 399. A. Weissberger and B. W. Rossiter, eds) (New York: Wiley–Interscience)
65. Buckingham, A. D. and McLauchlan, K. A. (1967). *Progr. im N.M.R. Spectroscopy,* **2,** 63
66. Buckingham, A. D. and Raab, R. E. (1957). *J. Chem. Soc.,* 2341
67. Deutch, J. M. and Waugh, J. S. (1965). *J. Chem. Phys.,* **43,** 2568
68. Hilbers, C. W., MacLean, C. and Mandel, M. (1971). *Physica,* **51,** 246
69. Ramshaw, J. D., Schaefer, D. W., Waugh, J. S. and Deutch, J. M. (1971). *J. Chem. Phys.,* **54,** 1239
70. Hilbers, C. W. and MacLean, C. (1969). *Molec. Phys.,* **16,** 275: **17,** 433, 517
71. Buckingham, A. D. (1967). *Advan. Chem. Phys.,* **12,** 107
72. Nienhuis, G. and Deutch, J. M. (1972). *J. Chem. Phys.,* **56,** 235